IET TRANSPORTATION SERIES 15

Cybersecurity in Transport Systems

Other related titles:

Volume 1 **Clean Mobility and Intelligent Transport Systems** M. Fiorini and J-C. Lin (Editors)
Volume 2 **Energy Systems for Electric and Hybrid Vehicles** K.T. Chau (Editor)
Volume 5 **Sliding Mode Control of Vehicle Dynamics** A. Ferrara (Editor)
Volume 6 **Low Carbon Mobility for Future Cities: Principles and Applications** H. Dia (Editor)
Volume 7 **Evaluation of Intelligent Road Transportation Systems: Methods and Results** M. Lu (Editor)
Volume 8 **Road Pricing: Technologies, economics and acceptability** J. Walker (Editor)
Volume 9 **Autonomous Decentralized Systems and their Applications in Transport and Infrastructure** K. Mori (Editor)
Volume 11 **Navigation and Control of Autonomous Marine Vehicles** S. Sharma and B. Subudhi (Editors)
Volume 12 **EMC and Functional Safety of Automotive Electronics** K. Borgeest
Volume 15 **Cybersecurity in Transport Systems** M. Hawley
Volume 16 **ICT for Electric Vehicle Integration with the Smart Grid N. Kishor and J. Fraile-Ardanuy** (Editors)
Volume 17 **Smart Sensing for Traffic Monitoring** Nobuyuki Ozaki (Editor)
Volume 18 **Collection and Delivery of Traffic and Travel Information** P. Burton and A. Stevens (Editors)
Volume 20 **Shared Mobility and Automated Vehicles: Responding to socio-technical changes and pandemics** Ata Khan and Susan Shaheen
Volume 23 **Behavioural Modelling and Simulation of Bicycle Traffic** L. Huang
Volume 24 **Driving Simulators for the Evaluation of Human-Machine Interfaces in Assisted and Automated Vehicles** T. Ito and T. Hirose (Editors)
Volume 25 **Cooperative Intelligent Transport Systems: Towards high-level automated driving** M. Lu (Editor)
Volume 26 **Traffic Information and Control** Ruimin Li and Zhengbing He (Editors)
Volume 30 **ICT Solutions and Digitalisation in Ports and Shipping** M. Fiorini and N. Gupta
Volume 32 **Cable Based and Wireless Charging Systems for Electric Vehicles: Technology and control, management and grid integration** R. Singh, S. Padmanaban, S.Dwivedi, M. Molinas and F. Blaabjerg (Editors)
Volume 34 **ITS for Freight Logistics** H. Kawashima (Editor)
Volume 36 **Vehicular ad hoc Networks and Emerging Technologies for Road Vehicle Automation** A. K. Tyagi and S Malik
Volume 38 **The Electric Car** M.H. Westbrook
Volume 45 **Propulsion Systems for Hybrid Vehicles** J. Miller
Volume 79 **Vehicle-to-Grid: Linking Electric Vehicles to the Smart Grid** J. Lu and J. Hossain (Editors)

Cybersecurity in Transport Systems

Edited by
Martin Hawley

The Institution of Engineering and Technology

Published by The Institution of Engineering and Technology, London, United Kingdom

The Institution of Engineering and Technology is registered as a Charity in England & Wales (no. 211014) and Scotland (no. SC038698).

First published 2022

The Institution of Engineering and Technology
Futures Place
Kings Way, Stevenage
Herts, SG1 2UA, United Kingdom

www.theiet.org

British Library Cataloguing in Publication Data
A catalogue record for this product is available from the British Library

ISBN 978-1-78561-668-6 (hardback)
ISBN 978-1-78561-669-3 (PDF)

Typeset in India by Exeter Premedia Services Private Limited

Cover Image: EschCollection/Stone via Getty Images

Contents

About the Editor

Dr Martin Hawley is the founder of Winsland, an international aviation consulting company. He has been developing and applying security methodology in the air transport industry since 2010, focusing on air traffic control and airports. His work includes developing security management systems, conducting security risk assessments and developing methods for partners in transport networks to collaborate and share security information. Martin maintains an interest in furthering the academic side of security and has co-authored a number of papers. Aside from his security interests, Martin's work focuses on collaborative digital innovation projects, and he has been involved in several start-ups including a new focus on mitigating climate change through a digital transformation of airspace management. His consulting work also includes strategy, policy, economic and regulatory aspects in air traffic management. Martin has a BSc and PhD in Physics from the University of Surrey. He is a member of the IET and a Liveryman of the Worshipful Company of Information Technologists.

Introduction

1.1 Why this book?

This book is for professionals in the transport sector, but much of the content is applicable to other sectors, particularly where operations are safety critical and there is widespread use of 'operational technology'. Our specific goal is to assist readers in developing a cyber security context for their work. We are not trying to make the reader into an overnight cyber expert, nor are we trying to replace existing study programmes such as those of $(ISC)^2$ [i]. Neither is this book a technical guide to implementing cyber technologies. Instead, we have drawn on our collective experiences to give professionals within the transport sector context and direction for avoiding cyber security problems. This includes coverage of what sort of management actions need to be taken and what types of technology may be deployed to defend against cyberattacks. There are myriad cyber technologies, and we have endeavoured to address a few that may seem familiar to you. We hope that the book will equip you to ask good questions, better engage with cyber experts, and perhaps even study this fascinating subject more deeply yourself.

The idea for this book arose from a small group working in the cyber security of air traffic management; it has since been enriched by a wide group of authors who are thought leaders in cyber security and transport. Through a long road of developing methodology and guidance within aviation, we became increasingly convinced that cyber security must become a part of the business and operations of transport. The issues we found were echoed by others working in different transport modes, which gave us the confidence that the book should address the wider transport sector, sharing our collective knowledge and experience. However, the issues the book addresses are not limited to transport and are common with other sectors, particularly where there is a mix of operational and information technologies deployed.

An underlying concern is that the adoption of effective cyber security into transport systems appears to be slow. A 2020 report [ii] raises concerns that *far too many agencies have not implemented adequate cybersecurity measures and are ill prepared to respond to a cyber incident* and that *cybersecurity is not yet widely seen as a critical issue among public transit leadership.* We interpret this concern as not only relevant to board-level leadership but to leadership at all levels across all disciplines within an organisation. It is therefore critical that every member of an organisation's staff play a role in effective cyber security. This is particularly the case for professionals, whether managers or subject matter experts, who can make the difference in cyber security across the whole system lifecycle, from R&D to operations to decommissioning.

To address a relatively fast changing subject matter, the book has focused more on technology principles and concepts than details, and by doing so, we also hope this book will appeal to a wide range of readers, including those without a formal technical background. Since cyber security is becoming increasingly addressed through formal frameworks and methods, such as risk analysis and security cases, we also address these.

1.2 Following in the path of safety knowledge

While the book is focused on cyber security, we draw many parallels with safety. Since the 1990s, there has been significant development and engagement with formal safety methods, and the parallels with cyber security are all too clear. Who now would not consider safety as an integral part of a transport business? Beyond the introduction of formalised safety methods, which have continuously driven down incidents, there is a cultural acceptance that safety is part of everyone's job. Such formal methods are not conducted solely by an isolated team of experts, but bring together all parts of an organisation to identify hazards and ensure and maintain safety.

The same can become true for cyber security, but at the time of writing, cyber security expertise is concentrated in too few experts, each using their own disparate terminology, struggling to keep up with relentless new threats, prioritise their expenditure, and gain appropriate levels of investment. The cultural change that happened in organisations for safety must also be possible for security.

Hence, for the future safety and security of our transport systems, something has to change. Transport professionals should take note: as industry modernises and becomes more connected, it will be increasingly difficult to maintain the safety and resilience of operations without robust cyber defences. All transport professionals have a role to play, not just the occasional cyber expert.

1.3 How bad could it get?

Cyberattacks can be devastating. On 27th June 2017, the entire global network of Danish shipping giant Maersk was crippled by one of the most devastating pieces of malware ever seen 'in the wild' [iii]. However, Maersk was not the intended target of this virus. The virus began in a server stack in Ukraine and spread across Europe and the Americas to cause devastation to networks in all sectors.

While NotPetya does not appear to have had any safety implications, it is not difficult to imagine the consequences of such an attack on more 'operational technology'. The incident highlights how viruses, even those not directly targeted, can propagate and have a significant impact on connected networks. Further details are in Box 1.

Box 1 The 'NotPetya' cyber attack

To gain a foothold in computer networks, NotPetya used two 'exploits' together, both previously seen in isolation:

- The first was developed by the US National Security Agency (NSA) known as 'EternalBlue' and stolen from the NSA in a security breach. EternalBlue took advantage of a Microsoft Windows vulnerability to gain access to any unpatched Windows machines, of which there were many at the time [iv].
- The second, Mimikatz, is a utility that finds passwords stored in plain text in a computer's random access memory (RAM). This enabled NotPetya to gain access to computers that had been patched, as it was able to extract user credentials and passwords from memory.

These two components enabled it to spread rapidly. NotPetya simply required a user on an unpatched terminal to interact with an infected email; it would then pull the user's credentials from memory. Using these credentials, the virus could then hop to any other terminal on which those credentials were valid, even if the terminal was already patched against EternalBlue. NotPetya then ran its destructive payload, known as a 'thrasher', which irreversibly encrypted the computer's master boot records.

The virus is believed to have been Russian in origin and developed to attack the Ukrainian 'M.E.Doc' software, widely used within Ukraine. At least 300 companies were hit, including hospitals, power companies, airports and banks. The White House estimated the total global damage of NotPetya to be around $10 billion [v].

However, the impacts of NotPetya were not limited to Ukraine; they propagated through the global networks of multinationals and associated companies. For Maersk, it led to its container ships being stranded and all 76 of the company's ports being stalled. When operating at close to a fifth of the World's shipping capacity, the knock-on economic impacts of this standstill were significant. To recover, manual reversion was initially used, then systems were slowly rebuilt and brought back online.

Eight days after the incident, the company resumed its online bookings after replacing an estimated 4,000 servers, 45,000 PCs and 2,500 applications [vi]. Many others were indirectly impacted, as the process of goods in and out of port terminals worldwide was slowed or halted, hampering industries that are now increasingly adhering to the 'just-in-time' principle.

Maersk were not the only casualties of the virus. The list of those directly impacted is extensive, including law firm DLA Piper, FedEx, pharmaceutical giant Merck (reported costs $310M) and snack company Mondelez.

Furthermore, it is critical that cyber security be designed and built into projects from the start. In 2018, San Francisco's Bay Area Rapid Transit (BART) reviewed the cyber security of its capital projects. One of the projects, Silicon Valley Berryessa Extension (SVBX), led to a shocking revelation; of the network's ~1000 Cisco devices, 86% had been previously installed in hostile nations. Investigations revealed [1] *hidden backdoors on the devices, as well as a persistent 'ping' where data are sent to a foreign nation hostile to American interests. In short, BART discovered intentionally planted spyware.*

Contents

Each chapter has been written to be somewhat standalone, but there are clear links between them. The early chapters set the scene for what is driving cyber security (Chapter 1), and the regulatory frameworks and guidance that support its adoption (Chapter 2). We then turn to risk management (Chapter 3), which is the major component of cyber security management (Chapter 4). In chapter 5, we look at the interplay between safety and security and how to address them together. Chapters 6 and 7 look at some security 'controls', the mitigations against cyber security attacks. As these are not always effective, Chapter 8 addresses the processes around 'response and remediation' to recover from cyberattacks. Chapter 9 covers autonomous systems, with a particular look at autonomous cars and maritime vessels. Chapter 10 broadens the cyber security focus to consider data use in future systems.

Authors

Paul Thomas BSc, MSc. Paul spent his earlier career supporting the RAF, initially through operations research and then as a programme manager for IDS and ADV Tornado security-sensitive upgrades. For UK NATS, he created the ATM development centre and delivered the infrastructure upgrade design and build programme. He spent 4 years as manager for the ATM software programme, which focused his understanding and knowledge related to cyber security. As a consultant, he now specialises in the operational context of cyber security.

Dr John Hird. John has a BSc in Electrical and Electronic Engineering and a PhD in Signal Processing. He has worked extensively in R&D, and since 2006, he has focused on Air Traffic Management Security. John has participated in numerous aviation research programmes, including SESAR, the Single European Sky ATM R&D programme. He has developed and delivered cyber security training material on methods, tools and regulations to ANSPs, Regulators, State Authorities and Industry. John is a member of the Advisory Council for Aeronautical Research in Europe (ACARE).

Dr Rory Hopcroft. Rory is a researcher at the University of Plymouth. He is currently working on the EU Horizon 2020 CyberMAR Project, helping develop cyber security awareness training. Prior to this, he researched for his PhD within the Centre of Doctoral Training in Cyber Security at Royal Holloway University.

His research primarily focuses on the regulatory aspects of maritime cyber security, and he has a keen interest in how the international community uses regulation and governance to increase security.

Alžbeta (Betty) Helienek. Betty is a certified engineer and cyber security expert, specialising in national critical infrastructure security. She served on the UK's Cyber Security Council formation project and is part of ISA's technical committee for cyber security in the IoT. She founded a cyber security start-up after delivering for 25+ years many technology firsts in the railway industry, including the initial developments in cyber security. She has now moved on to build a wider intelligent transport systems cyber security capability at WSP.

Christina Rose. Christina is a State Aviation Security Manager who, in 2020, was named 'Australia's most outstanding woman in protective security/ resilience. Christina's responsibilities cover aviation and precinct security operations at airports in Canberra and Albury, including the management of contracts, staff, service, and support. She has also jointly headed the Australian women in security network's Canberra chapter, contributing to a strong pipeline into the security sector. Previously, Christina held senior positions within the Australian Government, leading on aviation security policy and regulatory reform. Christina holds an MA in Terrorism & Security Studies, a Graduate Diploma in Media and a B.Ed.

Dr Amy Ertan. Amy is a researcher at the NATO Cooperative Cyber Defence Centre of Excellence and a cyber security fellow at Harvard's Belfer Center for Science and International Affairs. Amy holds a PhD in Information Security from Royal Holloway University of London, where she researched the security implications of artificial intelligence in military contexts. Her research interests are in cyber strategy and emerging security challenges. These have led to publications on topics including national cyber capabilities, military exercises, offensive cyber and cyber risk management.

Mauricio Duarte. Mauricio is a Security Operations Centre Analyst. He is a graduate of Tallinn University of Technology, where he earned an MSE in Cyber Security. Inspired by Yanis Varoufakis, he considers cyber security is too important to be left only to the experts. He believes this is a call to action to ensure cyber security reaches a wider audience.

Dr Rainer Koelle. Rainer holds an MSc in Electrical Engineering and a PhD in Communication System. In 2005/2006, Rainer conducted research and development for a real-time pan-European security incident management system and set up and established ATM Security within air traffic management. He headed the security-related work packages during the SESAR Development Phase. His current work revolves around open air traffic data for incident management or operational performance monitoring.

Chris Machin. Chris has been involved in risk management for over 20 years. He began examining ATC subsystems such as surveillance and communications equipment and moved on to large parts of the ATM network, such as the Network Management Function at EUROCONTROL. He evolved his activities into the SESAR framework, defining and applying the security management system now applied throughout the SESAR Programme. He has worked with many large organisations on ATM matters, including EUROCONTROL, the FAA, NASA and national providers of air navigation service providers.

James Hunt. James is a director at PwC, leading their transport cyber security competency. He has over 14 years of experience managing and leading large teams of highly skilled cyber security consultants, engineers and operators located in the UK and globally. He has supported numerous transport companies to prioritise their investment in cyber security. He has also been seconded as the CISO of an FTSE 100 manufacturing organisation.

Professor Chris Johnson, FRSE, FRAeS, FBCS. Chris is Pro Vice Chancellor, Engineering and Physical Science at Queen's University Belfast. Chris has more than 20 years' international experience at the interface between safety and security, and he has held fellowships with NASA and the US Air Force. He has also helped to establish cyber security laboratories for UK civil nuclear license holders. In 2021, he was retained as an expert witness to the Grenfell Tower fire inquiry, where he was responsible for analysing the digital and analogue communications failures that exacerbated the loss of life. In 2022, he was responsible for the development of pragmatic means to ensure the ethical use of machine learning in autonomous weapons systems, working with Thales for DSTL/MoD. He is a fellow of the Royal Society of Edinburgh, the Royal Aeronautical Society and the British Computer Society.

Lucy Caldwell. Lucy is a cyber security professional with experience in both the public and private sectors in information security management and cyber threat intelligence. Lucy has a BSc in Criminology from Cardiff University and has studied information security management and open-source intelligence techniques. She was involved in the initial conception and planning phases of this book, as well as being a co-author on several chapters.

Andrew Watson. Andrew is an ethical hacker and security researcher with over 25 years of IT/InfoSec experience and is currently working as a security architect. Realising his inherent aptitude for circumventing computer security, Andrew advanced in exploit development, penetration testing, vulnerability management, and cryptography. Andrew earned an MSc in Information Security from Royal Holloway University of London and regularly returns as a guest lecturer. Andrew has discovered and privately disclosed numerous security vulnerabilities, while publicly releasing two proof-of-concept zero day exploits.

Pearl Noble-Mallock. Pearl is currently Head of Product Security for BAE Systems and has a wealth of knowledge in technology, security and engineering. She has won numerous awards in cyber security, achieving the Royal Academy of Engineering's rising star award and the Top 10 Cyber Security Leaders globally in 2021, amongst many others. Pearl speaks regularly on security, management, autonomy, and engineering. She is passionate about diversity and often encourages hidden talent into the industry by publicly speaking about her 'superpower', dyslexia.

Stuart Southern. Stuart is a security and performance engineer specialising in Microsoft technologies with over 20 years' experience in software development. He is currently CTO for LAM Jersey, an aviation maintenance organisation, and a director of LAM Digital, which provides technology solutions to the aviation sector in the Channel Islands.

Patrick Mana. Patrick is the EUROCONTROL Cyber Security Program Manager and EATM-CERT Manager (European Air Traffic Management Computer Emergency Response Team). He has spent 35 years working in air traffic management (ATM). He started working with Thales on aviation software development and project/product management. In 1999, he joined EUROCONTROL, where he led the safety assessment activities. In 2008, he moved to the SESAR JU, where he was the Head of the Development Framework and SJU Programme Manager for all transverse activities, including security, for six years.

Andreas F. Wehowsky. Andreas is the co-founder and CEO of Muninn (formerly named wehowsky.com), a Danish cyber-tech company that builds AI-based cyber threat detection and response systems. Andreas has a background as an engineer in computer science and aerospace engineering from the Massachusetts Institute of Technology and has 20 years of experience in IT, AI and cyber. He is a member of the Danish Council for Digital Security at the Danish Business Authority, and he is a regular speaker and thought leader on cyber security.

David Damato. David is a security leader with two decades of experience and holds a master's degree in Information Assurance from the University of Maryland Global Campus. He is currently the Chief Security Officer (CSO) at Gemini, a leading cryptocurrency exchange and custodian. David developed a passion for incident response and remediation as an early leader at Mandiant, where he helped hundreds of organisations - including those in the transport industry - recover from serious breaches.

Jadee Hanson. Jadee is the chief information security officer and chief information officer at Code42 and leads global risk and compliance, security operations, incident response, and insider threat monitoring and investigations. She previously worked at Target Corporation and Deloitte. She is a co-author of 'Inside Jobs: Why Insider Risk is the Biggest Cyber Threat You Can't Ignore' and is the founder and CEO of the non-profit organisation Building Without Borders, which serves those in poverty-stricken areas throughout the world through housing services.

Andy Jones. Andy is a recently retired cyber security executive. He has held Chief Information Security Officer and leadership positions for a number of large organisations and worked as a researcher for one of the world's leading authorities on cyber security. He is an experienced speaker and published contributory author, specialising in the transport and supply-chain sectors. Andy holds a BSc in Physics and an MBA and continues to help a small number of organisations on an ad hoc basis.

Dr Hongmei (Mary) He FHEA, SIEEE. Mary is an associate professor of cyber security specialising in AI-driven IoT systems at the School of Computer Science and Informatics, De Montfort University. Previously, she was a lecturer in AI and cyber security in manufacturing at Cranfield University. She received her PhD in Computer Science from Loughborough University, UK, in 2006. Her current research focuses on trustworthy robots and autonomous systems in relation to safety, cyber security, system health, human-machine interaction and ethics.

Dr Raja Naeem Akram. Raja is an Associate Professor (SL) in Cyber Security and Artificial Intelligence at the University of Aberdeen and the CEO and founder of Seclea, a technology-for-good company that is building a novel explainable AI platform. Raja was awarded his PhD from Royal Holloway, University of London for his work in Smart cards, including common criteria evaluation, smart card operating system security and security of GlobalPlatform communication channel and execution environment. As a postdoctorate, he was associated with three major projects: SHAWN, Payment Security and DICE (Data to Improve Customer Experience).

Miriam Howe. Miriam has over 20 years' experience as a cyber security professional, having spent the majority of her career helping government, defence, and law enforcement organisations to protect themselves against cyberattack. Her current role involves leading cyber programmes for the UK and international governments, particularly relating to national cyber defence capabilities. At the same time, she directs the development of BAE System's national cyber solutions to support its global client base in these areas. She is an active thought leader and public speaker on the topics of national cyber security and cyber power.

Dr Richard J. Thomas. Richard is an Industrial Fellow in Data Integration and Cyber Security and leads the Rail Cyber Security research theme within the University of Birmingham Centre for Railway Research and Education. His PhD thesis reviewed the cyber security of some UK and European railway standards. His research interests include the security of the 'industrial internet-of-things' and how we can leverage the wealth of data available from the rail sector. Richard is a member of the Research Institute in Trustworthy Interconnected Cyberphysical Systems, addressing the challenges the rail sector faces when moving towards compliance with the EU Network and Information Systems Directive.

Dr Tom Chothia. Tom is a Senior Lecturer in Cyber Security for the School of Computer Science at the University of Birmingham. His research involves the development of new mathematical analysis and their application to cyber security problems.

Dr Hafizah Binti Mansor. Hafizah is a Senior Lecturer of Security Systems at the International Islamic University, Malaysia. She completed her PhD in Cybersecurity at Royal Holloway, University of London, focusing on Security and Privacy Aspects of Automotive Systems. She has also worked as an electronics engineer at Malaysia Microelectronic Solutions, an IC design company that designs smart card chips. Her current research focuses on automotive system security, machine learning and IoT systems.

Acknowledgements

I not only owe a huge debt of gratitude to the authors but also the numerous people who helped me along the way. I would never have started the book if it had not been for Lucy Caldwell. As well as authoring, she provided great editorial support on content, coordination, suggesting edits and proofreading. Another big thank you is due to Dr Karol Götz for his support in research, drawings and the critical work of checking references.

Numerous other people influenced the scope and contents of this book and assisted in its development. In particular, I must thank Dr Howard Chivers who I and the 'ATM gang' enjoyed working with and learning from over the years. Dr Sally Ernst has been an inspiration and leader in demystifying cyber security. I also need to mention Jack Jones, Chris Starnes, Robert Orr, Emyr Thomas, Lucy Shenton, Sarah Armstrong-Smith, Jessica Williams and Cédric Levy-Bencheton. Thank you for your contributions. I had the privilege of joining the Worshipful Company of Information Technologists in 2020 and managed to rope in several members in reviewing – thank you to Dr Suzanne Harkins, Jonathan Sinclair, Professor Roy Isbell, Martin Hogg, Professor James Davenport and Robert Bielen.

A special thank you is due to Emeritus Professor Fred Piper for his encouragement that the book is needed to broaden the knowledge of people working with, but not necessarily on, cyber security. Fred told me that he would be amazed and impressed, if we were to complete on time. He was right to be doubtful!

A large thank you is also due to the wonderful and patient staff at the IET! Olivia Wilkins, John Walker, Natalie Harper, Ranga and Lekha.

The final thanks must go to my wife Caroline and my family. They supported me all the way, but in 2021, I was diagnosed with cancer, and their support went to another level. I'm incredibly lucky in so many ways.

Dr Martin Hawley
December 2022

References

[i] https://www.isc2.org/.

[ii] Belcher S, Belcher T, Greenwald E, Thomas B. Is the Transit Industry Prepared for the Cyber Revolution? Policy Recommendations to Enhance Surface Transit Cyber Preparedness. Mineta Transportation Institute. 2020.

[iii] This expression is described by Courteney J. O'Connor as referring to 'general cyberspace': "To release code into the wild is to employ it, or publish it. Once a code is in the wild it cannot be taken back, and can be further exploited or modified by those with the skill to do so. Control of that particular code is lost to the original user." Source: O'Connor CJ. 5. Cyber counterintelligence: Concept, actors and implications for security.

[iv] https://en.wikipedia.org/wiki/EternalBlue. Accessed 23 February 2020.

[v] Greenberg A. The Untold Story of NotPetya, the Most Devastating Cyberattack in History. August 2018. Wired. Available at https://www.wired.com/story/notpetya-cyberattack-ukraine-russia-code-crashed-the-world/.

[vi] Chirgwin R. IT 'heroes' saved Maersk from NotPetya with ten-day reinstallation blitz, January 2018, The Register. https://www.theregister.co.uk/2018/01/25/after_notpetya_maersk_replaced_everything/. Accessed 23 February 2020.

Chapter 1
Modernisation in transport
John Hird, Rory Hopcraft, Alžbeta Helienek, and Paul Thomas

1.1 Introduction

In this chapter, we first look at the need for modernisation in all transport modes and the issues created as they modernise, considering the integration of operational and information technologies, the 'Internet of Things' (IoT) and the adoption of artificial intelligence (AI). We then take a look at some of the new technologies specific to aviation, maritime, road and rail.

The transport sector has been relatively slow to embrace technological change, and we can take the example of aviation to illustrate some of the reasons for this. In aviation, safety is of prime importance, and there has been much investment in the development of a safety culture across the stakeholder community. This has been driven by regulation and the application of rigorous standards and best practices albeit with a 16% increase in accidents between 2018 and 2019 [1].

However, a key consequence of this rigour is that the time required for a new concept to be agreed upon by the aviation community, to go through the research and development life cycle, and to be deployed in an operational environment is many years. Such long timescales result from several factors:

- The need to obtain international agreement prior to work commencing
- The requirement for relevant technical and operational standards and best practices to be developed and proven
- The creation or adaptation of regulations to accommodate the new concept
- The need to meet certification and interoperability requirements in a global context.

In addition, time is also required to install and test the resulting system in its target environment, while training operational and technical staff prior to the commencement of operations. The investment required to retro-fit aircraft and the supporting infrastructure is also significant, particularly where global interoperability is a requirement.

The potential benefits of embracing the digital revolution in transport systems are evident from developments in next-generation systems, currently undergoing research and development or at various stages of deployment. In aviation, new approaches to air traffic management are being developed by the SESAR programme [2] in Europe and the NextGen programme [3] in the United States. For example, the System Wide Information Management (SWIM) [4] is being addressed jointly by both programmes, and it will revolutionise information sharing between aviation stakeholders, delivering more connectivity and data sharing, while incorporating new developments in communications, navigation and surveillance.

In rail, the European Shift2Rail programme [5] is seeking to accelerate the integration of new and advanced technologies into innovative rail solutions.

Such digital transformations will impact customer relations, operational processes and the business models of service providers. It is achieved by optimising the use of digital data, increasing the connectivity of systems, introducing automation where feasible, and enhancing the variety of digital services available to customers.

The increased openness of transport systems brought about by this transformation will, however, make them potentially more vulnerable to exploitation by increasing the attack surface* due to the use of commercial-off-the-shelf (COTS) products, open standards and open protocols.

The integration of new systems with legacy systems is also a cause for concern. Many legacy systems were designed to be protected primarily by physical means, with connections to the outside world not being envisaged in their original requirements and design. Examples are computers running unsupported or unpatched operating systems with known vulnerabilities, or control system components with security controls either lacking or not enabled. Once connected to a network, perhaps to enable remote monitoring and maintenance, such legacy components may possess exploitable vulnerabilities within the system.

1.2 Drivers of change in the transport sector

1.2.1 Introduction

In this section, we explore the main drivers of change that are shaping the transport sector:

- Meeting the growth in demand while decreasing unit costs
- Improving performance
- Managing the challenge of network effects
- Adapting to new regulatory approaches and requirements.

These trends have led to widespread modernisation programmes across the transport sector, leading to the adoption of new technologies, which often interface with legacy systems. This complex intermingling of systems creates significant

*Attack Surface – a metaphor for the complete set of vulnerabilities that exist within a system.

challenges for cybersecurity, at the same time that cyberattacks are growing in sophistication and frequency.

1.2.2 Growth as a driver for change

Growth in transport has been more or less continuous for decades, with occasional downturns linked to recessions, as shown in Figure 1.1.

This growth has not necessarily been accompanied by continuous investment, so over time legacy systems have been under strain.

Growth is not only linked to gross domestic product (GDP) but also to energy costs. For instance, growth in the total number of passenger kilometres travelled in 2015 (2.6%) is linked to both GDP growth and lower fuel costs [6]. Although the COVID-19 pandemic may reduce transport growth for a number of years, previous trends towards urbanisation may continue to drive transport growth within and between urban areas, within which 55% of populations were living in 2018 [7]. Additionally, the gradual lowering of transport costs for freight and passenger travel has driven increased demand.

1.2.3 Performance drivers

As transport demand grows, we need capacity to increase in anticipation to avoid congestion or delays in the system. As demand approaches capacity, delays can grow exponentially. This was particularly the case for air traffic in the mid-1990s, when delays began to soar. In the first six months of 1999, a 5.6% increase in traffic led to a 145% increase in delays [8]. This increase in delays was becoming untenable and led to the creation of the 'Single European Sky' programme

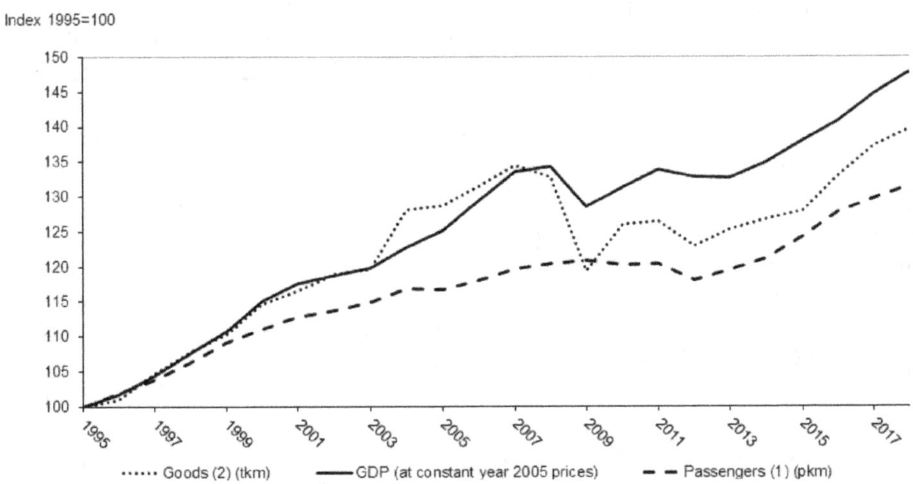

Figure 1.1 Transport growth in the EU 1995–2018 comparing GDP, tonne km and passenger km. Source Eurostat EU 27 Growth Tables 2.22 and 2.32. 2020.

of reforms by the European Commission. The combined performance need can be described as the 'cost-effective addition of new capacity', in order to reduce unit costs, while simultaneously improving in safety and reducing environmental impact.

Safety is a significant driver of performance. New technologies promise improvements in safety through the elimination of human error, as autonomous vehicles become common in all transport modes and reduce the frequency of accidents. The level of tolerable risk differs between transport modes, with aviation perceived as having the smallest risk appetite.

The International Civil Aviation Organisation (ICAO) has defined 11 key performance areas [9], which include safety, security, capacity, cost-effectiveness, flexibility, predictability, and the environment. However, a security incident may result in a number of consequences, which include the following potential impacts: a disruption of service provision, resulting in congestion, delays or service termination; a reduction in safety margins leading to stress, injuries or fatalities; financial consequences to transport operators or service providers; an unwanted impact on the environment. Clearly, inadequate performance in the area of security may result in detrimental impacts on several other performance areas, including safety.

1.2.4 Network effects

Each transport mode has its own network in some form or another: road and rail networks, shipping lanes and air routes. These are generally linked to timetables that allow connection within each network, and these networks also interact with each other. Door to door, a trip may comprise air transport with road and rail at both ends, and each mode is part of a network of its own.

As these networks grow with respect to the volume of traffic handled, a disruption may result in a substantial impact. This has led to procedures and technologies to manage the flow and control of traffic. Examples include variable speed limits on motorways [10, 11], air traffic flow management and control [12], and rail traffic management and signalling systems [13]. Since these flow and traffic management systems rely on information technology (IT) and COTS products to connect and manage operational technology (OT), new vulnerabilities may be unknowingly introduced into critical operational systems. The increasing use of network connected sensors attached to operational technologies, commonly known as IoT devices, further progresses the convergence of IT and OT [14, 15]. Such convergence has the potential to increase the likelihood of transport systems being vulnerable to attack if not adequately protected, separated, and monitored.

A growing concern is the resilience of transport networks. Should a disruption occur, it is important that an incident is appropriately managed, minimising the degradation of operations and supporting the return to normal service provision quickly. Solutions to resilience generally require the implementation of a security management system, the performance of security risk assessments, the implementation of risk mitigations, and the implementation of an incident management process. For the purpose of staff training, exercises may be carried out using prepared threat scenarios, which may be based on information received on recent applicable cybersecurity incidents, or derived from threat catalogues. Consequently, when system protections fail and a cyberattack is successful, it is necessary to be able to respond and recover quickly in a controlled manner.

1.2.5 Regulatory drivers

The majority of the transport sector is subject to regulation covering a variety of areas, including safety and security. Many operators must obtain operating licences, which require them to demonstrate compliance, usually via an independent audit, with regulatory requirements which may include examining organisational structure (e.g. roles and responsibilities), management systems (e.g. quality, safety, security) and processes (e.g. recruitment, change management).

Regulators in transport have a variety of high-level goals:

- Preventing loss of life or serious injury (to the traveller, service operators, and the general public)
- Preventing the disruption of the service resulting from acts of unlawful interference
- Reducing environmental impact
- For monopoly service providers, controlling costs to users and ensuring an adequate quality of service

As described earlier, a security incident may result in an unwanted impact in a variety of areas, including safety. In order to address this, regulators require assurances that appropriate action is being taken on cybersecurity, and this is supported by a number of regulations which apply at national, regional, and global level. These regulations are described in Chapter 2 for different transport modes.

1.2.6 Trends in regulation

Regulation can be a blunt instrument. Regulatory rules are typically created from lessons learned and are adapted over time with the goal of creating optimal outcomes. However, occasionally they result in unintended consequences. The nature of regulatory oversight can make the application of regulations less blunt, but there are still challenges for regulators in creating rules that deliver the intended results, avoiding unwanted side-effects and costs. Hence, policymakers are increasingly focused on delivering 'better' regulations. Two developments in recent years are 'performance-based regulation' and 'risk-based oversight':

- In safety, risk-based oversight requires a focus on safety management systems and the sharing of specialist resources between regulators. Risk-based approaches were initially developed in Canada and the United Kingdom, where regulatory oversight is proportional to the perceived risk of the organisations being regulated. For example, larger organisations with mature safety management systems may be subject to less frequent audits than smaller ones.

Performance-based regulation and risk-based oversight may be viewed as developing together. They are also attractive to transport organisations as there is potential for regulatory requirements to be achieved more efficiently, with room for innovation. This is particularly important since the modification of a regulation may take a number of years.

1.2.6.1 Performance-based regulation

In recent years, the European Commission has been working on introducing performance-based rules and oversight through the European Aviation Safety Agency (EASA) [16]. This move to performance-based regulation is being addressed stepwise as regulations are updated. For example, in the 'Flight Time Limitations' regulation, there is flexibility on crew flight hours when managed with a Fatigue Management System [17]. EASA has also had its competence extended to the safety-related aspects of security, including cybersecurity, so in future there is the potential for performance-based cybersecurity rules to be developed.

Regulations are discussed in detail in Chapter 2, including the EU's Network and Information Systems (NIS) Directive, which will no doubt shape the future approach by transport regulators. An update of this directive, NIS 2.0, has recently been proposed to address, among other things, increasing resilience, greater consistency and joint situational awareness and response across states. In Chapter 4, we discuss how organisations can widen their security management practices for greater information sharing and collaboration.

It is likely that the approach and experiences around safety performance-based regulation will apply to cybersecurity, so it is worthwhile looking more deeply at the emerging features:

- *Focus on outcomes.* While prescriptive rules may initially be retained, new performance-based rulemaking seeks to define outcomes, with supporting standards and guidance to show how an outcome may be achieved. An example is the flight-time limitation regulation. While this sets limits on pilot hours, it also addresses a more fundamental point that pilot fatigue needs to be risk-managed, since pilots can still be fatigued even if they are below a maximum working-hours limit.
- *Performance monitoring.* The desired outcome may itself be difficult to measure – consider that security is an absence of an outcome. Performance indicators can be difficult to define and can be manipulated mercilessly, such as how queuing/waiting times are often 'managed' in any sector (e.g. health or transport delays). Moreover, it is difficult to link cause and consequence in loosely coupled and complex systems. To address some of these difficulties, organisations are now pooling resources to define metrics through 'networks

of analysts'. This bottom-up approach can deliver essential 'buy-in', making it more likely that a performance indicator is properly adopted.

- *Effective risk management systems.* Regulators want to see that organisations properly own the risks, and performance-based rules require the industry to be more thoughtful. Security management systems will therefore need to be effective and embody the spirit of the Deming cycle (plan, do, check and act) as well as the paperwork. Regulatory oversight can then become less about audit and more about judgement, and risk management can become a genuine exercise in uncovering and mitigating risks.
- *Reporting culture.* Perhaps the simplest to understand but the hardest to deliver is reporting culture. Without a culture of open reporting, problems are buried rather than learned from, and performance measurement becomes a measure of reporting culture rather than performance. Where safety of life is at stake, the concept of 'Just Culture' is critical – people need to report their own errors without an unjust threat of recrimination.
- *Collaboration.* Performance-based regulation heralds a new phase of collaboration, within and between the industry and regulators. The central need for successful collaboration is a major shared goal that requires a new style of leadership behaviour that supports collaboration across organisational boundaries. Information and ideas need to be shared for lessons to be learned. The free-flow of information between organisations and regulators may seem improbable in the area of security, but once a shared goal is established the question becomes how, not why.

1.2.6.2 Regulatory resources

A challenge for many states is how to properly resource their regulators, which are often funded from national budgets. This can lead to shortages of appropriate expertise, and this is particularly the case in cyber security. Some states, such as the UK, fund their regulators from licences or other fees, and as their industry changes and grows, they are able to keep pace with changing regulatory demands. To provide more cost-effective specialist services, regulators can pool resources, either at an institutional level, for instance with EASA, or within institutional frameworks.

Early adopters of performance-based regulation are convinced that this is the way to go, and they recognise the challenges. Both industry and regulators see the opportunity for more targeted resources and better outcomes. But they also recognise in each other the substantial challenges in developing the an appropriate culture and behaviours.

1.2.6.3 Privacy

In common with other organisations, those involved in transport must comply with regulations associated with the protection of personal data, including that of employees, passengers and third parties.

The General Data Protection Regulation (GDPR) [18] applies to all EU Member States and non-EU establishments, where personal data about data subjects in the EU is processed in connection with 'offering goods or services' to European data subjects or 'monitoring' their behaviour. Non-EU entities that are subject to the GDPR are required to designate a representative in an EU Member State.

Personal data is any information that relates to an identified or identifiable living individual. Different pieces of information, which collected together can lead to the identification of a particular person, also constitute personal data. Personal data that has been de-identified, encrypted or pseudonymised but can be used to re-identify a person remains personal data and falls within the scope of the GDPR. Personal data that has been rendered anonymous in such a way that the individual is not, or no longer, identifiable is not considered to be personal data. For data to be truly anonymised, the anonymisation must be irreversible. Examples of personal data include name, surname, Internet Protocol (IP) address, location data or an ID card number.

In contrast, the United States has several sector-specific and medium-specific national privacy or data security laws, including laws and regulations that apply to financial institutions, telecommunications companies, personal health information, credit report information, children's information, telemarketing and direct marketing. In addition, the individual states have their own privacy and data security laws.

1.3 Convergence of OT and IT

As already mentioned, there is a growing convergence between regular business IT and OT. Drivers for this convergence are described in the following sections, and include the Internet of Things (IoT) and 'Mobility as a Service' (MaaS).

1.3.1 Operational technology

Operational systems are often referred to as belonging to the class of OT. OT consists of the hardware and software systems that monitor and control physical equipment and processes.

OT is found in critical infrastructures, such as energy and utilities, and can be found in vehicles and control systems in all transport modes. OT systems have had long life cycles compared to enterprise systems, with typical operational lifetimes of 10 to 20 years or more. For example, a flight data processing system for an air traffic control (ATC) centre will be expected to operate for at least 15 years, and the expectation for a ship or rail rolling stock may be several decades. In contrast, enterprise computing systems may be considered to be obsolete after five years.

OT systems traditionally apply a variety of technologies, including centralised operational server-based systems, networks and local control systems, generally known as industrial control systems. Multiple standards and recommended practices are available to ensure that they can be securely developed and maintained.

OT systems are designed to perform a limited number of functions, with reliability and integrity being the primary considerations. Driven by stability and interoperability requirements, OT is the classic late adopter, only taking on new technology once it has matured. OT may use and maintain systems which are a generation or more behind those being applied in IT. Windows XP, for example, which ended mainstream support in 2009 and extended support in 2014, can still be found in many OT systems, as integral components in mission-critical systems.

For an isolated system, the use of obsolete software may not pose a threat, but connection to an external network could result in the creation of new vulnerabilities, which could be exploited by adversaries. This risk was highlighted by the worldwide impact of the Wannacry virus, against which Microsoft issued a patch in May 2017 [19].

Due to the long life cycles and age of some OT systems, expertise to maintain them tends to be scarce and the practitioners tend to be older, on average, than in the IT domain. Legacy transport systems also suffered from a number of practical limitations. For example, rail systems were originally designed to operate within restricted geographical limits, as local or national systems. They varied in architecture, were designed to different standards and required cross-border interoperability with neighbouring entities, without which there would be restrictions on vehicle movements, delays and capacity limitations.

1.3.2 Integration of IT into operations

In contrast to OT, IT is the application of computers to the processing, transmission and storage of data, usually in business or enterprise environments. The establishment of international computing and networking standards in IT has created an ecosystem of fast, interoperable information processing technologies, yielding continuous improvements in storage capacity and data processing speed, while exploiting the opportunities afforded by new developments such as cloud computing and data sciences.

Traditionally, IT and OT have not overlapped, but there are benefits to be gained from migrating towards more modern technologies in transport system development. These include minimising the cost of replacing legacy systems, reducing time-to-deployment, minimising project delays and exploiting the enormous improvements in performance which continue to be delivered in each new generation of IT systems.

The convergence of OT and IT facilitates the rapid development and deployment of new concepts, simplifies interconnection to other systems and eases interoperability with other systems with similar capabilities. Consequently, it has become standard practice to develop new transport systems and sub-systems using COTS software and hardware components, which use widely available hardware, operating systems and application software, and use open protocols for networking and communications.

Use of COTS in new transport systems is intended to leverage economies of scale and scope. COTS systems are relatively cheap, feature a broad variety of common interfaces as standard and are plug-compatible with a vast range of peripherals, communications and networking equipment. COTS systems thus open up new possibilities in system design and enable ambitious new concepts to be engineered. In addition, to support this ecosystem, there is a large population of IT professionals.

The vision is that, once in operation, new transport systems will deliver more vehicle movements and enhance coordination between traffic management systems, both intra-mode and inter-mode, across national borders. At the same time, new technology is intended to maintain or improve safety levels, thus increasing capacity while simultaneously reducing delays.

Enhanced data sharing across a greater geographical area will also improve coordination between traffic management systems and contribute to substantial performance enhancements. However, there are some potential disadvantages to be addressed while developing new and upgraded transport systems, which are addressed in subsequent chapters.

1.3.3 Mobility as a service

A new approach to transport is emerging, which seeks to enhance data sharing between transport modes to optimise travel times, enhance the passenger experience and optimise the transit of goods. This integration of various forms of transport services into a single mobility service accessible on demand is termed MaaS.

A MaaS provider would be able to offer a diverse menu of transport options drawn from various transport modes and add value to users by providing the best value proposition, via a single application and a single payment channel. Additional expected benefits of MaaS and the associated proliferation of self-driving vehicles are reductions in transport fatalities, cost per kilometre, passenger kilometre emissions and increased highway throughput rates [20].

1.3.4 IoT devices

The IoT involves the extension of connectivity via the Internet to devices and everyday objects containing embedded electronics and networking capabilities. Such devices can communicate and interact with others over the Internet, and they can be remotely monitored and controlled. Such capabilities are now routinely integrated into domestic appliances, such as washing machines, lighting solutions, door entry systems and central heating controllers. Such 'smart' devices can be monitored and controlled remotely via the Internet but could potentially be recruited as 'zombie' devices for use in Distributed Denial of Service attacks.

There are many applications for IoT in all transport modes. The monitoring of the location of a requested Uber or Lyft vehicle on a smartphone is just one example. Similar applications allow a transport operator to monitor the location of their entire fleet of mobile assets. Examples of the use of IoT in aviation include passenger flow management in airport hubs, aircraft engine monitoring and baggage tracking [21]. In general, IoT will facilitate the development of smart airports, seaports and smart roads, and streamline the management of traffic in all transport modes.

Maintaining the health of mobile assets can be simplified and made more efficient if diagnostic information can be retrieved and assessed continuously while the assets are in service. This enables predictive maintenance to be accomplished by scheduling maintenance slots depending on need, thus minimising downtime. The ability to monitor operating parameters, such as fuel consumption, lubricant levels and temperature, speed and other environmental factors, provides data which may be analysed and potentially used for performance optimisation.

1.3.5 AI – attack and defence

AI is expected to have a profound influence on cybersecurity in the future. Developments in machine and deep learning are expected to increase the resilience of systems and services, for example, in malware detection, network analysis, message filtering, and providing support to human operators [22].

However, AI can also be used as a tool to attack systems and will change the nature of the attack surface accessible to malicious actors. As AI systems extend further into domains commonly believed to be uniquely human (e.g. social interaction), more sophisticated social engineering attacks can be expected to result in an epidemic of network penetrations, intelligent computer viruses and the theft of personal data [23]. To counteract this, the best means of defence may be the automation of cyber defences using AI. Further thoughts on AI are explored in Chapter 10.

1.3.6 Growing hazards

Whereas new OT systems are based on COTS building blocks, many of these systems possess known vulnerabilities, with new ones being uncovered regularly. In addition, since the available means of exploiting system vulnerabilities are increasing constantly, so the likelihood of an attack being attempted and being successful is also increasing.

Potential attackers of transport systems may have a variety of motivations. For example, the goal of a cyberterrorist may be to intimidate, force political change or cause fear or physical harm. A hacktivist may be driven by political, economic or social goals, such as highlighting human rights issues or disseminating sensitive information. State-sponsored actors are motivated to advance a State's interests, perhaps by stealing intellectual property or disrupting computer systems. Cybercriminals are motivated by financial gain and often hide behind anonymous and peer-to-peer networks, such as The Onion Router (Tor) and OpenBazaar, using encryption technologies and digital currencies, for example, Bitcoin, to hide their communications and transactions. Some systems are attacked because they are easy to breach, some because they are deemed to be attractive targets, while others, which may not be specifically targeted, may suffer as collateral damage from undirected attacks.

The increased interconnectivity of transport systems may lead to an attack being capable of propagating through interconnected systems between different transport modes. This could lead to larger scale impacts over a wide geographical area.

Given these new hazards, there are concerns about how transport is evolving. For example, the application of similar technologies across modes could result in common vulnerabilities and more frequent attacks. Some traditionally distinct transport technologies, including communications, navigation and surveillance, are converging in terms of the services provided and the technology and infrastructure used, so if any of these services are compromised, the impact could have substantial knock-on effects.

1.4 Cross sector examples of modernisation

1.4.1 Introduction

In this section, we look at the introduction of new technologies that have impact across the transport sector. Our aim is to explore briefly what the technologies are and the cybersecurity concerns they raise, either on their own or when integrated with legacy systems. The following sections take a more focused look at transport sectors: aviation, maritime, rail and road.

1.4.2 Global navigation satellite systems

A key driver of innovation in transport has been the development of satellite navigation systems, known generically as Global Navigation Satellite Systems (GNSS). A GNSS comprises a constellation of satellites orbiting the earth and a network of ground stations. GNSS regularly broadcasts radio signals containing information which describes their orbit, status and time information. At the users' end, GNSS receivers determine the time for signals from multiple satellites to reach it, which is then used to determine the three-dimensional position of the receiver anywhere on the globe. Combined with a map, they can be used for navigation and are now a common application for smartphones. Currently operational GNSS systems include the USA's GPS, Russia's GLONASS, the EU's Galileo and China's BeiDou system.

GNSS-based navigation is ubiquitous in aviation, maritime and road transport, and is being introduced into rail. The precision of GNSS time information is used for synchronisation purposes in a variety of applications, from mobile communications networks, electrical power grids, stock trading systems, financial networks and military systems.

Increase utility from GNSS is obtained from 'augmentation systems', which monitor and improve the performance of the GNSS satellites in terms of accuracy, integrity, continuity and availability. For example, the European Geostationary Navigation Overlay Service (EGNOS) is a regional satellite-based augmentation service (SBAS), which improves accuracy and provides integrity by complementing the existing service provided by GNSS [24]. Differential corrections and integrity messages are broadcast by geostationary satellites as an 'augmentation' or 'overlay' of the original GNSS messages. EGNOS provides a wide area augmentation service over Europe, while other regional systems are the United States' 'Wide Area Augmentation System' (WAAS) [25] and Japan's MSAS [26]. OmniSTAR and StarFire are global SBAS systems.

In contrast to SBAS, the ground-based augmentation system (GBAS) is a safety-critical system specific to civil aviation. It supports local augmentation of the GNSS constellation to support precision services such as airport approach and landing operations. The main goal of GBAS is to provide integrity assurance, but it also increases accuracy.

Of particular concern, however, is the vulnerability of GNSS signals, which arises from their very low power level as they reach the receiver. This makes them susceptible to interference from faulty radio equipment or a Global Positioning System (GPS) jammer. GPS jammers broadcast noise-like signals at frequencies used by GNSS systems, which prevent receivers tracking authentic signals [27].

Spoofing of a GNSS receiver is possible by introducing a false signal that is either created by a signal generator or is a re-broadcast of a pre-recorded GNSS signal. Spoofing may be used to feed false positioning or timing information to a target receiver to covertly misdirect the receiver and its platform to some desired location [28]. While spoofing GNSS takes more sophisticated equipment than jamming, the equipment is readily available. Commercial GPS emulators exist, which can be used for GPS spoofing, but these are relatively expensive to purchase. However, using easily procured COTS products, it is possible to produce inexpensive spoofing devices with minimal effort [29]. Readymade 'GPS signal generators' are now available on the Internet and are very popular with Pokemon-Go enthusiasts [30]. These are used to trick cell phones into collecting rewards in the game from remote locations. Hence, GPS spoofing is now truly mainstream.

There are various means available to mitigate the effects of such vulnerabilities, although such security controls are not yet widespread in transport systems:

- Anti-jam antennae configurations, with digital beamforming and anti-spoofing algorithms
- Deploying multiple navigation sensors, such as inertial navigation system, odometers and altimeters
- Using multiple GNSS constellations.

1.4.3 Passenger information systems

Information systems, such as Passenger Information Display Systems, facilitate the flow of passengers through transport hubs, such as airports, seaports, railway stations and bus stations. The information provided allows passengers to make their way to the correct location in order to board for departure to a specific destination at a particular time. If this information is incorrect or unavailable, perhaps due to a cyberattack or a maintenance error, it could result in confusion, stress, congestion, delays and passengers missing their connections. Operators could be faced with performance and capacity reductions and associated economic impacts. In September 2018, such an attack impacted Bristol Airport [31].

1.4.4 On-board infotainment systems

On-board infotainment systems provide passengers with a wide variety of services: access to media including music, news, movies, television programmes and games, journey information and access to Wi-Fi for passengers' own devices. Such distractions can also reduce perceived journey times and improve passenger satisfaction, while providing transport providers with information on passenger preferences as they interact with the systems. Transport operators can also reap the rewards of partnerships with online retailers by receiving commissions on purchases completed during the journey.

While infotainment systems deliver clear benefits to transport operators, they are more open to the outside world via connections with passenger devices. They may also possess vulnerabilities, which could be exploited by attackers using wired or wireless connections. Consequently, such systems must be isolated from other systems that are critical to the operation of the transport mode in question.

1.4.5 Retail systems

In the rail, water and aviation sectors, it is common for food, drink and other items to be sold on board. Point-of-sale terminals and card readers, wired and wireless, may have exploitable vulnerabilities, so these systems should also be isolated from operationally critical systems.

1.5 Aviation modernisation

1.5.1 Overview

For the aviation system to operate, there is a need to share information between a wide variety of operators, including airport operations, ATC, airline operations, aircraft maintenance, ground handlers, operators of unmanned aircraft systems and so on. A key problem is that the systems developed and communications between them have been developed over time and often independently in industry silos. This has led to a plethora of legacy systems and different communications protocols supporting a complex system of systems, as shown in Figure 1.2.

Many legacy systems operating in aviation were developed long before cybersecurity became an imperative. The challenge of modernisation is therefore to secure legacy systems and at the same time integrate new technologies. A further complication is the rapid addition of types of aircraft, such as Remotely Piloted Aircraft Systems (RPAS) and Unmanned Aerial Vehicles (UAV), and vehicles supporting the concept of urban mobility.

A further aspect of modernisation is the increase in connectivity across the industry. New connected aircraft, such as the Airbus A350 and Boeing 787, are improving connectivity in both the cockpit and the cabin. The introduction of new networks and computer systems and the arrival of digital services, such as Wi-Fi, on aircraft for passengers, the use of electronic flight bags for pilots and the use of handheld devices by crew (e.g. to process credit card payments and communicate with airline operations centres), is making it possible to benefit from almost the same data and services in the air as on the ground, but it has also increased the number of possible ways of attacking air transport, and therefore increased the associated risks.

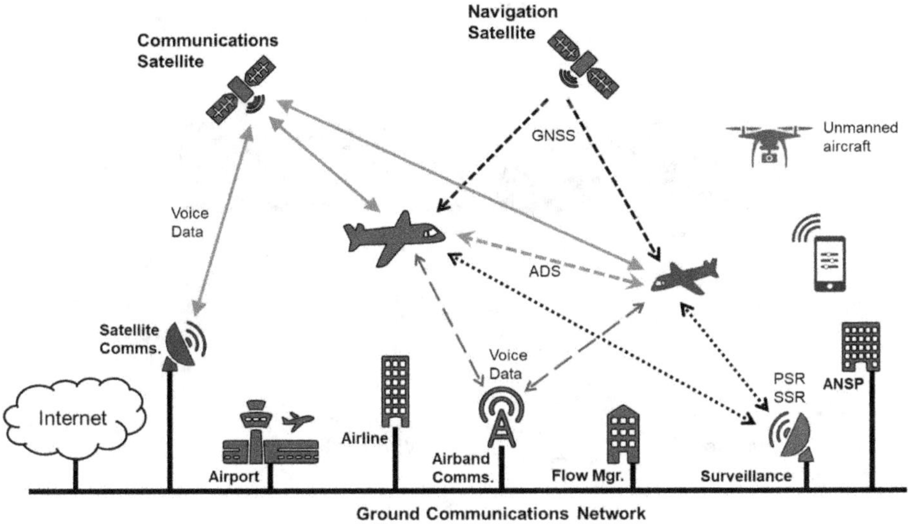

Figure 1.2 Aviation as a complex system of systems

1.5.2 The connected aircraft

We can view aircraft systems as being separated into three areas: responsible for control of the aircraft, for passenger cabin entertainment and services and for communication with the airline, as shown in Figure 1.3.

1.5.2.1 Control of the aircraft

Voice communications between controllers and pilots have historically been via analogue radio communications (VHF), on both ground and ground-to-air networks, and so are vulnerable to eavesdropping, jamming and spoofing [32]. Since there is no means of authentication, it is difficult to verify that messages are being transmitted by those authorised to do so, resulting in the possibility of false instructions being provided to aircraft on approach by 'pretend' controllers. This is not a regular occurrence [33], and since there are humans in the loop, such situations are usually identified and addressed by the pilot and legitimate controller.

Whereas voice communications from ATC to the aircraft are via analogue VHF, the ground-to-ground link between the ATC centre and VHF transmitter is increasingly digital, using 'Voice over Internet Protocol' (VoIP), prior to encoding for transmission as an analogue signal to the aircraft. In the other direction, analogue voice signals transmitted by the aircraft from the pilot are demodulated at the VHF receiver, then digitised, encoded and transmitted as VoIP to ATC ground receivers and networks, before being provided to the controller in analogue form at the relevant ATC centre. Communications between ATC centres are also increasingly achieved using VoIP.

The main security concern is that potential vulnerabilities could result from VoIP functions and need to be identified and addressed through risk management.

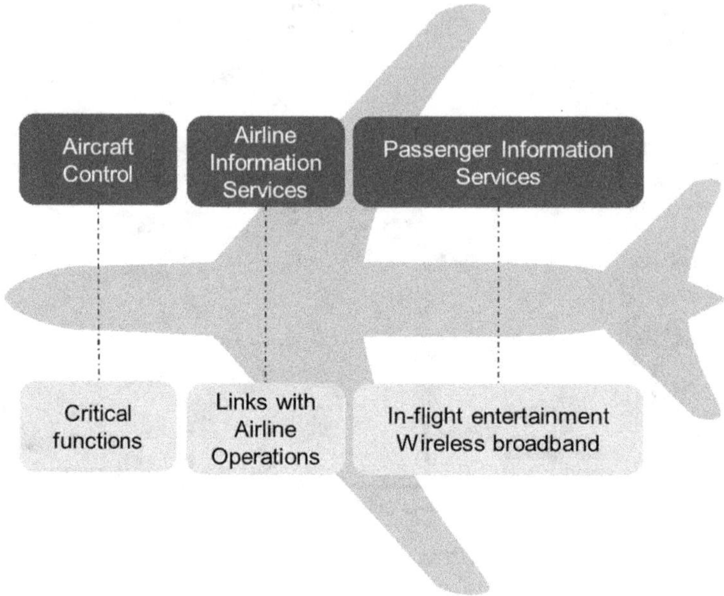

Figure 1.3 Connected aircraft systems

1.5.2.2 Airline information services

Electronic Flight Bags

Electronic flight bags are electronic devices that are used by pilots and store a variety of information previously carried by pilots in paper form. Some are installed on the aircraft, whereas others are portable consumer devices, typically tablet computers, which can be updated by Internet connection, typically via airport gate Wi-Fi. Electronic flight bags are also a source of vulnerability; a faulty aerodrome database on electronic flight bags resulted in the entire American Airlines Boeing 737 fleet being grounded for several hours while the error was being identified and fixed [34]. Since the new database had to be updated via Wi-Fi, the aircraft had to remain at their gates while awaiting the fix. This was an unintentional failure, possibly caused by inadequate quality processes in place for rolling out database updates for operational use. However, it illustrates dependencies on the supply chain, in this case, third party data providers.

Airline communications

There are a variety of communications between aircraft and ground networks, using ground-based networks, Wi-Fi, Wimax or GSM (Global System for Mobile communications) telecommunications at the airport and air–ground datalink or satellite-based networks in flight. The Aircraft Communications Addressing and Reporting System (ACARS) is a digital datalink system used to transmit short messages between aircraft and ground stations via radio or satellite. The protocol was designed by Aeronautical Radio Incorporated (ARINC), and deployed in 1978. It is used by airlines for operational communications with the aircraft, such as providing flight plans and weather information to the Flight Management System (FMS), and to send status messages on the aircraft's systems and sensors in real time.

1.5.2.3 Passenger cabin entertainment

Passenger in-flight entertainment systems are more open to the outside world than other aircraft sub-systems, so these systems must be isolated to prevent penetration into aircraft critical systems. In connected aircraft, passengers with their own devices are able to use on-board wireless broadband to connect to the Internet via satellite link. Security controls must be in place to ensure that such devices are not capable of interfering with critical aircraft systems.

1.5.3 Modernisation of communications networks

The need for new systems to be backwards compatible with older operational systems plays a significant role in the development of new technologies. A good example of the issues is found from the aeronautical fixed services network.

The fixed services comprise voice and data networks and circuits, enabling functions such as voice communications, the exchange of flight plans, meteorological data and flow management information between air traffic service providers. The major part of message interchange in the aeronautical fixed service is performed by the Aeronautical Fixed Telecommunications Network (AFTN). This is a message handling network [35] and was the world's first large-scale message handling system, dating back to the 1950s.

The technology upon which AFTN is based is completely outdated, so an alternative, the Common ICAO Data Interchange Network (CIDIN), was created in the 1980s and deployed in the 1990s to replace its core. AFTN services and procedures remain in place as a user application of CIDIN, which provides significantly higher bandwidth and quality of service than its predecessor. CIDIN technology is itself now nearing obsolescence, and both AFTN and CIDIN need to be replaced by more modern technology. To meet this requirement, ICAO specified the ATS Message Handling System [36], which has been deployed in many parts of the world.

More generally, there are numerous aeronautical communications links in operation today, as shown in Figure 1.4. Example links are as follows:

- ATC communicates with pilots by voice using analogue VHF radio communications;
- Surveillance information from dedicated surveillance networks provides location, velocity and altitude of aircraft, which are displayed on the 'Controller Working Positions';

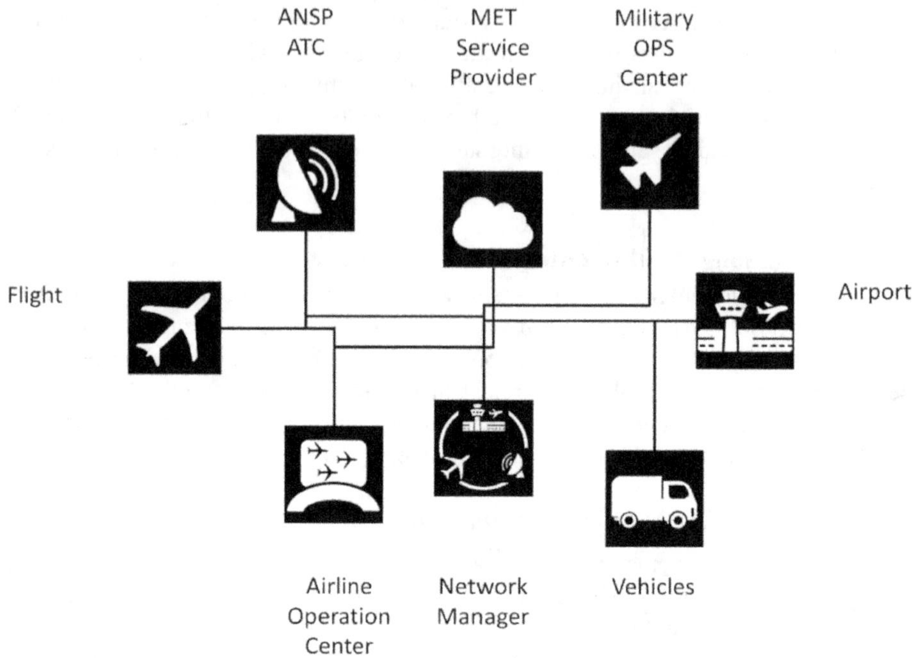

Figure 1.4 Legacy communications in aviation

- The 'Network Manager', which is a European-wide service, receives flight plans from airline operations centres and manages the traffic flow across Europe, communicating flight plans to ATC services and, where necessary, allocating departure slots to airlines to prevent overloading of the air traffic system.

The operation and maintenance of these disparate systems are costly and inefficient, so a new concept in sharing information is starting to be implemented – 'System Wide Information Management'.

SWIM [37, 38] provides for the efficient exchange of information between stakeholders, comprising aeronautical, flight, meteorological, air traffic flow and surveillance information. This is achieved by replacing the plethora of separate systems and connections by one SWIM system, which is based on standard specifications, a single governance framework and a common set of infrastructure components.

From an information viewpoint (Figure 1.5), SWIM will provide interoperable services using seamless data exchange between data providers and data users, by applying open standards on an IP-based network. The availability of SWIM is facilitating the introduction of new concepts in aviation, such as the 'Remotely Operated Tower', now known as a 'Digital Tower'.

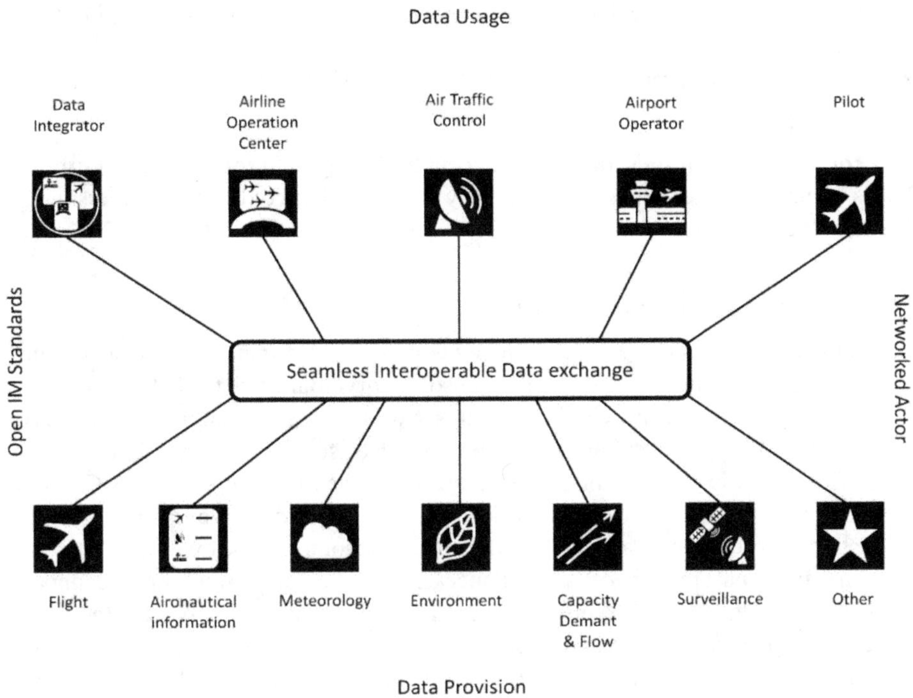

Figure 1.5 Communications in a SWIM environment

1.5.4 Digital towers

The control tower at an airport provides a variety of air traffic services, such as final approach and departure control, departure sequencing and airport surface management. However, operating a tower at a small aerodrome with a handful of aircraft movements per day has some efficiency challenges, particularly for airports with low hourly traffic.

A 'digital tower', also known as a 'Remotely Operated Tower', allows air traffic services at small and medium capacity airports to be provided remotely by personnel in a 'remote tower centre', which could be hundreds of kilometres away. Remote control is possible by using live video feeds to provide a virtual visual view of the remote aerodrome and surrounding airspace. Surveillance, communications, meteorological and flight plan information is fed to the remote tower centre so that it can perform all functions normally carried out locally. The virtual view may also be augmented by aircraft and object detection and tracking, tagging of aircraft with surveillance information or fusing infrared and visual information to improve situational awareness in bad weather. Benefits include the following:

- Reduced aerodrome infrastructure, substantially reducing building and maintenance costs;

- Human resources, particularly air traffic controllers and engineers can be used more efficiently;
- Certain systems previously required locally at an aerodrome can be centralised.

For larger aerodromes, the same concept has left benefit for efficiency purposes but is being implemented as a contingency tower, which could be required such as in the event of a fire or loss of power.

1.5.5 Surveillance in aviation

Surveillance systems are essential elements of air traffic management operations. Such a system includes surveillance sensors, surveillance data distribution systems, surveillance data processing systems and analysis and support tools. Over the last few years, surveillance has been modernised through the implementation of 'Automatic Dependent Surveillance – Broadcast' (ADS-B), ADS-Contract (ADS-C) and 'Wide-Area Multilateration' systems. Primary Surveillance Radar and Secondary Surveillance Radar will still be used for some time, however.

The principal means of surveillance in civil aviation remains Secondary Surveillance Radar, which requires the maintenance of an extensive ground infrastructure of sensors, communications links and radar data processing systems. Secondary Surveillance Radar is a 'dependent' surveillance technology, which requires the cooperation of the aircraft to provide its identity, position information and other parameters useful for ATC such as 'selected altitude'. By comparison, Primary Surveillance Radar is 'independent', requiring no cooperation from airborne systems.

ADS-B relies on an automated broadcast of an aircraft's position and is also 'dependent' as the surveillance information (e.g. GNSS position) is provided from the aircraft to the ground. The technology is also described as 'cooperative', as aircraft must be transmitting ADS-B signals to be detected. If the aircraft is not transmitting the ADS-B signals, it is invisible to an ADS-B receiver. ADS-C uses a different datalink, either ground-based VHF or via satellite links to transmit aircraft position, heading, identity and other information, and is widely used in oceanic regions, particularly the North Atlantic.

When first introduced, ADS-B was deemed to be the successor of radar and a new paradigm for ATC. With ADS-B, each aircraft determines its altitude, latitude, longitude, bearing and speed using an on-board GNSS receiver. This is periodically broadcast, along with aircraft identification information, by an 'ADS-B-Out' transmitter to ground receivers and to other aircraft in the vicinity which are equipped with 'ADS-B-In' receivers.

ADS-B has been developed to support air-to-air surveillance, improving pilots' situational awareness, which is an important safety benefit. Future applications include the potential for aircraft to 'self-separate'.

Two related services, which are part of the US Federal Aviation Administration's NextGen programme, are given as follows:

1. Traffic Information Services-Broadcast (TIS-B), which was developed to transmit aircraft position data to aircraft cockpit displays. In essence, TIS-B allows pilots to see what the air traffic controller sees – other aircraft, along with those aircraft's altitudes, direction and speed vectors on their aircraft's display screen. Potential users of TIS-B include airborne aircraft, aircraft operating in airports and airport surface vehicles.
2. Flight Information System Broadcast (FIS-B), which allows pilots to receive aeronautical information including weather and airspace restrictions through a data link to the cockpit.

ADS-B-equipped aircraft also support ATC where there may be gaps in secondary radar coverage, for instance, in remote and oceanic areas. The ground stations for ADS-B are small and easily maintained and can be placed in areas where radar was never possible. For example, the Gulf of Mexico now has complete coverage with ADS-B installations on oil platforms and along the coast. Since 2019, there is global ADS-B coverage through the Aireon [39] satellite-based ADS-B system.

1.5.5.1 Flight tracking applications

ADS-B is also behind the various flight-tracking applications, such as FlightRadar24 [40], FlightAware [41], Planefinder [42] and RadarBox [43]. These companies gather crowdsourced ADS-B and other meta data. ADS-B position information and aircraft ID are correlated with flight plans and aircraft registration databases, enabling information about a flight to be displayed with its track. The ADS-B information is obtained partly from sensors operated by commercial and government entities, and partly from rooftop and window receivers run by thousands of flight tracking enthusiasts across the globe. Aircraft owners and operators have the option of blocking their tail numbers from release through public databases, and state and military flights are typically concealed. In Europe, this service may be requested through EUROCONTROL.

These flight tracking services do not support the high integrity and availability needed for ATC services but can support operations by:

- Providing data for the operational analysis of flight routes to optimise their fuel efficiency;
- Providing take-off times of aircraft to improve traffic flow management, as once an aircraft is airborne its arrival time is fairly predictable, whereas substantial delays can occur prior to take-off.

1.5.5.2 ADS-B vulnerabilities

ADS-B operates by receiving and processing GNSS signals to obtain position information which is then used by ADS-B transceivers. Consequently, ADS-B may be impacted by events which affect the confidentiality, integrity or availability of GNSS information, resulting from the exploitation of GNSS vulnerabilities.

Table 1.1 ADS-B vulnerabilities

Vulnerability	Description	Potential impact (C, I, A)
Eavesdropping	The act of listening to unsecured broadcast transmissions, as made available to flight tracking service providers. Military and state flights are generally excluded from these services, and it is becoming possible to request that information on specified aircraft be blocked	• Loss of confidentiality of ADS-B message contents (C)
Jamming	Denial of service. Broadcasting with sufficiently high power at the frequencies used by ADS-B (1,090 MHz with Mode-S), preventing the transmission or reception of messages by one or more ADS-B participants	• Ground station flooding (denial of service) (A) • Aircraft flooding (A)
Message injection	The injection of correctly modulated and formatted, but false ADS-B messages into the communications systems. No authentication measures are implemented in ADS-B	• Ground station target ghost injection/flooding (I/A) • Aircraft target ghost injection/flooding (IA)
Message deletion	Legitimate messages can be 'deleted' from the wireless communication medium by using constructive or destructive interference	• Aircraft disappearance (I)
Message modification	Injection of arbitrary data into messages	• Virtual aircraft hijacking. (I) Virtual trajectory modification (I)

ADS-B messages are broadcast openly without encryption and are susceptible to eavesdropping, so there is essentially no confidentiality. ADS-B messages can be obtained by anyone with equipment which is both inexpensive and widely available. The vulnerabilities are listed in Table 1.1 in the order of increasing level of difficulty to exploit [44]. The potential impact is expressed in terms of impact on Confidentiality (C), Integrity (I) and Availability (A).

Other aviation communication protocols that are known to possess vulnerabilities similar to those of ADS-B [45] include Controller Pilot Data Link Communications (CPDLC), the Aircraft Communications Addressing and Reporting System (ACARS) and the Traffic Collision Avoidance System (TCAS).

1.6 Maritime modernisation

1.6.1 Overview

As shown in Figure 1.6, maritime transport can also be seen as a complex system of systems, through its port operations, vessel traffic services, shipping line operations,

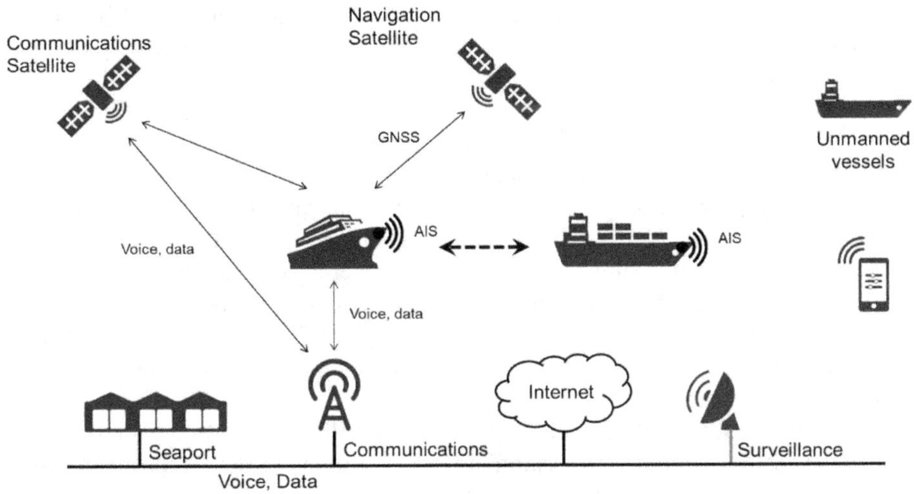

Figure 1.6 Maritime system of systems

vessel operations and unmanned maritime systems. Maritime supplies both cargo and passenger transport, and hands off passengers and cargo to other modes of transportation. Maritime users include commercial, military, individual, corporate and public sector vessels.

There are numerous maritime-specific communications systems used for navigation, ship-to-ship and ship-to-shore information exchange, vessel management and control, cargo scheduling and management, and safety and passenger entertainment. As in aviation, most of these systems were created without cybersecurity in mind and well before the widespread cyberattacks that now commonly occur.

Modern ships are obliged to carry an extensive array of navigation and control systems and equipment on the bridge, most of which have evolved at different periods over the past 60–70 years. This array of systems leads to increasing dependency on technology, so much so that the use of GNSS became so prevalent in the early 1990s that the US navy stopped teaching celestial navigation in 1996. However, this was re-introduced 20 years later, probably because GNSS-based tools were becoming increasingly vulnerable to cyberthreats.

The most recent system to have been mandated under SOLAS (the International Convention for the Safety of Life at Sea) [46] is an Electronic Chart Display and Information System (ECDIS).

ECDIS is a geographic information system for nautical navigation that complies with International Maritime Organization (IMO) regulations and is an alternative to paper nautical charts. It will still be some time before all vessels are required to be equipped and there will also be a significant number of 'smaller' vessels below 3,000 gross tonnage that are not required to install one. However, due to the functionality provided by such systems and ever-decreasing prices, many vessels are equipping voluntarily.

As a consequence of the continual addition of new equipment, many ships have a bridge comprising a number of separate, stand-alone systems, a variety of subsystems and networks for communications, navigation, control and passenger services. On newer vessels, it is possible to integrate systems so that two or more can share data or sensor inputs, with most of the very latest vessels having Integrated Navigation Systems or Integrated Bridge Systems.

The bridge navigation systems include GNSS receivers, the ECDI and other navigation systems, including the automatic identification system (AIS) and the long-range identification and tracking system. External communications systems include satellite communications and VHF radio transceivers, including the Global Maritime Distress and Safety System (GMDSS) [47, 48]. The GMDSS is a combination of systems, safety procedures and communications protocols that helps support distressed vessels at sea.

The expectation has been that the widespread roll-out of 'e-navigation' would reduce the cost of maintaining existing aids to navigation. However, problems with the availability or integrity of GNSS data, or a failure in on-board receivers or the ECDIS, can effectively leave the crew of a ship blind to hazards, unless they were able to fall back on more traditional methods of navigation. These include inertial navigation systems, ready-reckoning, the analysis of coastal features, the use of radio beacons or celestial navigation. Consequently, as in aviation, legacy systems remain in coexistence with the new systems.

On passenger vessels, passenger services include entertainment systems and the provision of Wi-Fi for passenger-owned devices, each with their associated wired and wireless networks, so it must be ensured that these services are isolated from the vessel's critical systems. Point-of-sale terminals and supporting networks on passenger vessels must similarly be secured.

1.6.2 Automatic identification system

The primary method of collision avoidance in water transport continues to be the marine radar; however, the AIS is intended to assist a vessel's watchstanding officers to maintain situational awareness and to allow maritime authorities to track and monitor vessel movements.

AIS was introduced in the 2002 IMO SOLAS convention, where Chapter 5, 'Safety of Navigation', mandates that ships meeting certain characteristics must carry AIS transceivers as an additional safety measure. The IMO SOLAS convention requires AIS to be fitted aboard ships voyaging internationally with a gross tonnage of 300 tonnes or more and all passenger ships regardless of size.

AIS allows ships to view marine traffic in their vicinity and to be seen by that traffic. Each AIS integrates a VHF transceiver with a positioning system (GNSS receiver) and other navigation sensors, for instance, a gyrocompass or a rate-of-turn sensor, and regularly transmits information including a unique identifier, position, course and speed. AIS transceivers in the vicinity are able to receive the information and display it on a screen or ECDIS. A key component of AIS is the use of precise

GNSS positioning technology. However, the range of AIS is typically only slightly greater than line-of-sight and is only designed for short-range operation as a collision avoidance and navigational aid.

Vessels fitted with AIS transceivers can be tracked by AIS base stations located along coastlines or, alternatively, through satellites (S-AIS) equipped with AIS receivers.

1.6.2.1 Ship tracking applications

Paralleling the evolution of ADS-B in aviation, an emerging use of AIS data is to make it available for viewing publicly over the Internet, without the need for an AIS receiver. For this purpose, global AIS transceiver data is collected from satellite and interconnected shore-based stations, aggregated, and made available on the Internet by a number of service providers, providing global, real-time information on ships, which generally includes the vessel name, location, speed and heading on a map.

Shore-based AIS receivers contributing to the Internet are generally run by volunteers, echoing the crowdsourcing model of ADS-B data in aviation. Similarly, AIS smartphone apps are available, and examples include MarineTraffic [49], SEAiq [50], iNavx [51] and AIS Radar [52].

1.6.2.2 AIS vulnerabilities

There are many similarities between AIS used in maritime and ADS-B used in aviation. For example, AIS messages are unencrypted, unauthenticated, have no mechanism for integrity checks and deliver vessel identification, position and speed information. As for ADS-B, encrypted AIS is used for military and law enforcement purposes, but such products are not generally available to civilian classes of vessels.

As a consequence of the similarities in concept and implementation, the vulnerabilities of AIS can be considered to be identical to those for ADS-B shown in Table 1.1 [53]. An additional concern is the extent that AIS can be used by pirates to navigate towards their targets. In high-risk areas, for example, the Gulf of Aden, a solution is to deactivate AIS [54].

1.7 Rail modernisation

1.7.1 Overview

Rail is a central part of the transport system and part of the critical infrastructure in many countries. Rail systems have traditionally relied upon bespoke, stand-alone systems, in the form of a variety of legacy 'switching' technologies, which manage signals and points in order to regulate rail traffic through the network.

In recent years, there has been a move towards standardised equipment with COTS components and networked control and automation systems that can be accessed via public and private networks. This means that without appropriate action, rail transport will become increasingly vulnerable to cyberattacks. For example, safety could be impacted if systems responsible for vital signalling, automatic train protection or interlocking were compromised by a cyberattack. Operations could also be affected by attacking passenger information displays, either by blanking the screens or by displaying incorrect information on them.

Within the EU, there are still more than 20 train control systems in use. Each train used by a national rail company has to be equipped with at least one system, simply to run safely within that country. As each system is stand-alone and non-interoperable, for cross-border operations extensive integration and engineering effort are required. When compared to road transport, this hampers the competitiveness of the rail sector by creating technical barriers and costs for international journeys. For example, it was reported that Thalys trains running between Paris–Brussels–Cologne and Amsterdam were equipped with seven different types of train control systems [55].

Because of these diverse systems, the rail industry is examining new approaches to streamline processes, to improve energy efficiency, reliability and punctuality and to save costs, all while maintaining safety and security levels. This is being achieved through digitalisation of the rail system, as shown in Figure 1.7, which is aiming to:

- Provide real-time information on train movements
- Enable predictive maintenance on rolling stock and fixed assets
- Implement automatic train operation, which is already well established in certain metro systems around the world, and would be possible to implement in main-line railways.

In Europe, the European Rail Traffic Management System (ERTMS) is currently being deployed to assist in attaining this vision.

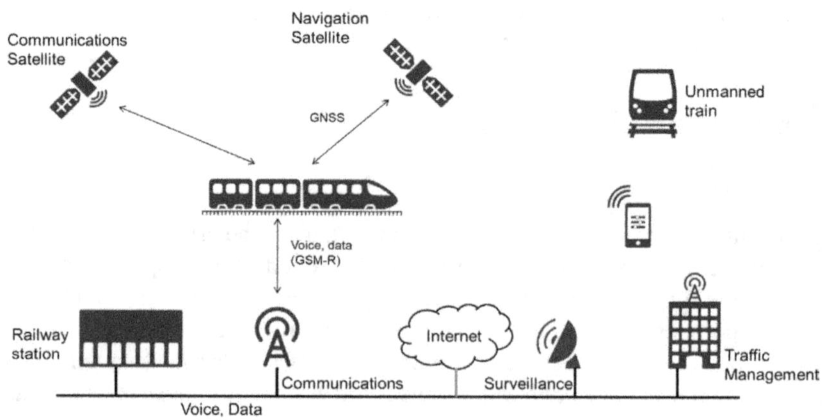

Figure 1.7 Rail Transport.

1.7.2 The European rail traffic management system

ERTMS was first outlined in a paper in 2008 [56]. It is a major industrial project developed by eight rail industry stakeholders in close cooperation with the EU, other railway stakeholders and the GSM-R industry [57, 58]. ERTMS is designed to gradually replace the many different national train command, control and signalling systems in Europe with a standardised system. This is expected to bring advantages in terms of reduced maintenance costs and improved safety, reliability, punctuality and traffic capacity. ERTMS is also being implemented in China, India, Taiwan, South Korea and Saudi Arabia.

ERTMS has four main components:

1. ETCS – the European Train Control System. This is an automatic train protection system to provide a single replacement for the existing national automatic train protection systems. ETCS is an intermediary between the vehicle and the track which conveys driving instructions from the track to the vehicle via devices called 'balises', which process this information, along with precise positional data.
2. GSM-R – Global System for Mobile Communications – Railway. This is a radio system providing a common means of voice and data communication between the track and the train across the rail network, based on standard GSM, and using frequencies specifically reserved for rail applications with certain specific and advanced functions.
3. ETML – the European Traffic Management Layer. The ETML aims to standardise the interfaces between railway companies so that the exchange of operational data is possible, especially for cross-border train connections. The system should allow data to be shared between participating railway undertakings, including position, expected schedule, delay information and estimated arrival times.
4. INESS – the Integrated European Signalling System. This comprises functions and technical devices for setting and protecting routes and sequencing and controlling train movements.

In common with other digitalisation programmes, potential cybersecurity issues have been discovered in ERTMS [57]. These include vulnerabilities in GSM-R, in which network subscribers are protected by two security mechanisms, namely network authentication and message encryption. Communications with a particular handset and Subscriber Identity Module (SIM) are encrypted by 64-bit keys, which can nowadays be broken in minutes, resulting in the possibility of eavesdropping on messages, breaching their confidentiality. It is also possible to 'hijack' an attacked handset, in other words redirect the data traffic destined for it. Since authentication only takes place one way on the network, GSM-R base stations could also be imitated in an unauthorised manner.

In common with other wireless communications protocols, GSM-R is also vulnerable to intentional jamming and to radio frequency interference, specifically from 3G and 4G public networks, which are replacing GSM in the 900 MHz band.

This situation occurs where the 3G and 4G transmitters are close to the railway infrastructure [59]. The modernisation of GSM-R to address such vulnerabilities is being addressed by industry stakeholders.

Potential security vulnerabilities have also been identified in other ERTMS subsystems and appropriate security controls are now being implemented [60]. These vulnerabilities reflect the importance of designing in security from the start and assuming an increasing sophistication of attackers' capabilities. Hence we provide a detailed case study of rail system security in the Appendix.

1.7.3 GNSS in rail

As an example of the potential direction in rail situational awareness, research is ongoing in ERTMS into the application of GNSS localisation to the control of train spacing, in railway networks equipped with the ERTMS ETCS [61]. This technology could play a pivotal role in reducing rail infrastructure equipment requirements by enabling trains to determine their positions autonomously using GNSS, instead of relying on equipment installed all along the tracks, for example, in the form of track circuits and axle counters. There are several advantages to such an approach:

- By moving equipment from the track side to the train, infrastructure requirements are reduced, resulting in reduced maintenance costs;
- Securing infrastructure on a train is potentially easier than securing it alongside the track;
- Safety and capacity can potentially be enhanced.

There are, however, some issues which have the potential to adversely impact such systems if they were to be exploited by adversaries, including the known vulnerabilities of GNSS and GSM-R.

1.8 Road modernisation

1.8.1 Overview

The road transport sector has an enormous impact on society, the environment and the economy, and is the focus of substantial research and development aimed at improving safety, increasing capacity and reducing congestion and environmental impact.

In England, for example, the average driver spends 235 hours driving every year, the equivalent of 6 working weeks, during which the driver must concentrate on driving 100% of the time. In future, people will be able to hand the task of driving to the vehicle itself [62] and apply that time to other tasks. As human error is a factor in over 90% of collisions, significant reductions in accidents and associated injuries and fatalities are expected as highly automated vehicles (HAV) are introduced. This should reduce insurance claims and the impacts on emergency and health services. By communicating with the environment and other vehicles, HAV are expected to use the road more effectively, identifying optimum routes and spreading demand to minimise energy consumption and emissions.

Vehicles today contain a multitude of electronic systems used to control the vehicle and monitor its state, controlled by between 50 and 120 electronic control units (ECUs), each of which comprises a central processing unit, memory and input/output devices. The embedded ECUs run both safety-critical and non-safety-critical software applications and control electronic actuators to deliver a wide variety of functions. ECUs, sensors and actuators are generally clustered together in interconnected sub-domains, for example, as shown in Figure 1.8.

The most widely used communication bus in vehicle systems is the Controller Area Network (CAN) bus, which was developed by Bosch in collaboration with industrial and academic partners. This was presented to the SAE Congress in Detroit in 1986 as the 'Automotive Serial Controller Area Network' [63]. The first CAN controller chip was manufactured in 1987, with the first production Mercedes-Benz vehicle using the technology being produced in 1992. The CAN bus was also applied in other domains, such as elevator control (Kone) and internal networking in X-ray machines (Philips).

In common with many other legacy systems in transport, the CAN bus was designed and developed at a time when few could have predicted potential security issues in the future. The protocol is inherently insecure, since it was designed without authentication, encryption or plausibility checks being included. Today, there is still no implicit support for secure communications, and there are limitations in doing this given the low bandwidth, short message lengths and timing constraints.

Vehicle control systems rely on components from a diverse range of manufacturers and rely on real-time communications between ECUs. Different manufacturers may implement security features with proprietary variations, relying on security through obscurity, which means that CANs are vulnerable to reverse engineering and spoofing.

To improve data rates for new applications, and enhance reliability and robustness, a number of alternative bus systems have arisen over the years. CAN FD [64], for example, provides increased bandwidth (1 Mb/s) and reduced latency relative to

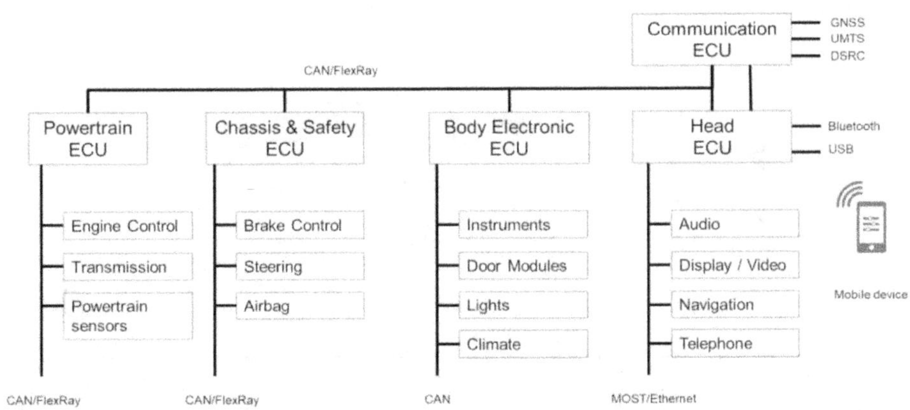

Figure 1.8 In-vehicle network

CAN. However, to reliably support new safety-critical applications, including adaptive cruise control, drive/steer/brake-by-wire, requirements include high bandwidth, fault tolerance and deterministic performance. One solution which meets these needs is the FlexRay [65] bus which has a 10 Mb/s bandwidth.

In modern vehicles, the transmission of audio and video between vehicle systems and passenger devices is increasingly a basic requirement. For multimedia and infotainment networking, the MOST bus [66] is a popular standard. Given the multitude of different applications in the automative space and the availability of a variety of networking solutions, it is not unusual for more than one bus system to be simultaneously in operation in a vehicle, as indicated in Figure 1.8.

1.8.2 Highly automated vehicles

Enhanced levels of vehicle automation are possible by using embedded intelligence to fuse sensor information to improve situational awareness, by enhancing communications between vehicles (vehicle to vehicle or V2V) and other systems (vehicle to everything or V2X), including traffic management systems (Figure 1.9). Currently, several car manufacturers are developing highly automated vehicle prototypes and corporations including Google, Waymo and Uber have testing programmes in place.

V2V and V2X aim to improve safety through direct communications between vehicles and sensors. For example, collision warnings are likely to become standard components in a vehicle's specification through economies of scale and market forces.

Connected vehicles communicate externally using a wide range of communications technologies, and over-the-air software updates are typically used for non-critical software and firmware from original equipment manufacturers.

The Society of Automotive Engineers (SAE) classification levels for autonomous driving are shown in Table 1.2. Vehicles rely on a broad range of sensors and, as the level of vehicle autonomy increases, so does the need for sensors, processor

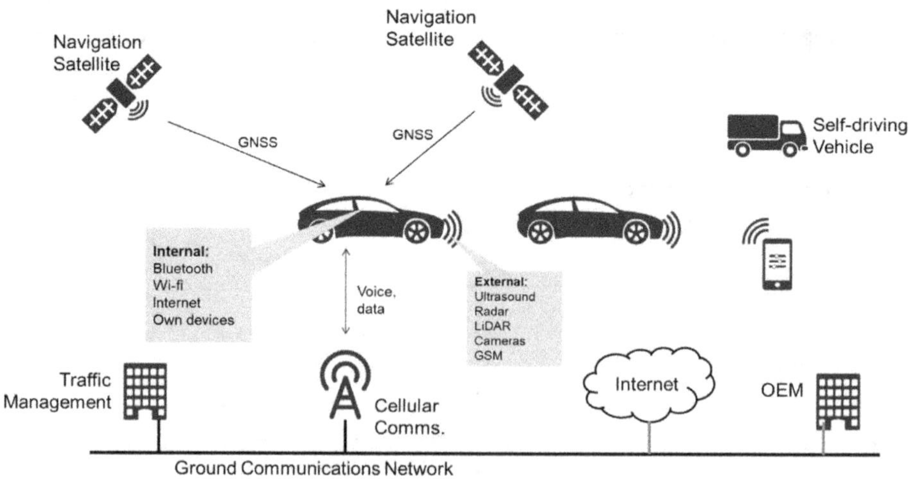

Figure 1.9 Road transport

Table 1.2 SAE classification levels for autonomous driving

SAE level		Description
0	Driver only	A human driver controls all functions independently
1	Assisted driving	Systems assist a human driver during certain vehicle operations (e.g. cruise control, anti-lock braking, speed limiting)
2	Partial automation	At least one system is fully automated, but a human driver must monitor the system at all times (e.g. cruise control and lane centring)
3	Conditional automation	A human operator monitors the system and intervenes when required. Certain safety-critical functions are delegated to the vehicle in certain circumstances
4	High automation	The vehicle operates all safety-critical functions and monitors road conditions for an entire journey without human intervention. A human may be required to take over if a driving scenario is not within the operational scope of the vehicle
5	Complete automation	The vehicle controls the car without any human intervention

performance and embedded intelligence. In the majority of cases, the human driver monitors the driving environment, and most aspects of driving are performed by the driver (SAE Levels 0–2). In future, however, the monitoring of the driving environment is expected to be performed by the vehicle (SAE Levels 3–5).

SAE Levels 1 and 2 vehicles typically use cameras and radar (radio detection and ranging), which, for example, provide information to the vehicle on distances to other vehicles and obstacles. Currently, available production vehicles comprise a variety of sensors and systems delivering functions to enhance the driving experience and improve safety. For example, ultrasound sensors and cameras are used to assist with parking, short-range radar sensors and surround view cameras detect vehicles in the blind spots of rear-view mirrors, long-range radar is used in adaptive cruise control, while cameras provide the information for lane-departure warning systems. Light Detection And Ranging (LiDAR) is currently used in the development of most vehicles for SAE Level 3 and above. However, in contrast to the approaches applied by the majority of aspiring HAV developers, Tesla committed to a solely camera-based approach in the development of its vehicles in 2021.

GNSS systems have been deployed in production of road vehicles for some time and have primarily been used to provide information to the driver on their current position, which supports navigation on the optimum route to a destination. Although GNSS signals are vulnerable to a variety of exploits, at SAE Levels 0–2, the driver in the loop would be able to detect discrepancies between actual position and 'compromised GNSS' position. At SAE Levels 3 and above, it is likely that information from other sensors, such as cameras, radar and LiDAR, could be used to detect if GNSS information was erroneous.

1.8.3 Threats and vulnerabilities

Safety-critical vehicle functions, which include throttle, braking and steering, are increasingly being controlled by electronic control systems (drive-by-wire) as traditional mechanical and hydraulic systems are replaced.

Vehicle entertainment systems are increasingly connected wirelessly to the outside world, and these often operate using embedded COTS software products. Integrated SIM cards provide on-demand access to cellular communications networks, Internet connectivity and passenger Wi-Fi access. Owners have access to vehicle information and vehicle control facilities via their personal devices. Cars are now routinely physically accessed using remote key fobs and personal devices. The proliferation of such sub-systems increases the attack surface, providing opportunities for malicious attacks which could impact security, safety and privacy.

Connected vehicles are often updated wirelessly ‘over the air’, with manufacturers sending software and firmware updates to vehicles to address system bugs and upgrade features and to deliver security updates. Such updates may be vulnerable to interception, jamming (Denial of Service), eavesdropping and data manipulation.

Current production vehicles, even at SAE Levels 0–2, have been shown to be vulnerable to attack, with vehicle sub-systems such as those controlling entertainment systems, windscreen wipers, drive-train, braking and steering being able to be manipulated remotely [67].

Given the variety of sensors and sub-systems typically integrated into a vehicle and the number of original equipment manufacturers and suppliers involved, the complexity of the supply chain is also of concern. To minimise the potential attack surface, appropriate controls should be applied to address areas such as hardware and software development, system integration, vehicle maintenance and decommissioning.

User confidence in passenger and non-passenger safety in automated vehicles may need to be developed, particularly given recent publicised incidents with semi-autonomous vehicles in production, such as Tesla cars equipped with the Tesla Autopilot (SAE Level 2) [68] and the Uber incident of March 2018, which was the first recorded case of a pedestrian fatality involving a self-driving car in March 2018 [69].

1.8.4 Data protection and privacy

Connected vehicles provide many opportunities for the transmission of information to and from the vehicle. However, any processing of such data where an individual could be identified should comply with applicable data protection rules, one example being GDPR in Europe [18].

Various devices within vehicles are capable of recording data which could be linked to an individual. For example, navigation systems may store historical information on programmed routes and historical vehicle locations.

Vehicle ECUs may be able to record and store data, while event data recorders (EDRs) are also used to record selected information. EDRs are on-board event recorders which retain data on speed, acceleration, brake use and so on, before, during and after an incident. The data can subsequently be downloaded from the EDR for analysis in compliance with data protection rules. This data may be useful for various purposes, including the technical analysis of vehicle system performance during an event, event reconstruction, and the collection of evidence for use in a legal process.

1.8.5 The Vienna convention

The 'Vienna Convention on Road Traffic' [70] requires that 'every moving vehicle or combination of vehicles shall have a driver' and that 'every driver shall at all times be able to control his vehicle'. These statements are considered to be barriers to the introduction of automated vehicles; however, several signatories have not ratified the convention, including the United Kingdom, the United States, Malaysia and China. Recent amendments to the convention have begun to address these issues [71].

The United States was the first country to introduce legislation permitting the testing of automated vehicles. In 2013, the US National Highway Traffic Safety Administration issued recommendations for the testing of Levels 3 and 4 automated vehicles. Several states have authorised testing, and each has their own set of requirements; California began issuing licences in 2014.

1.9 Conclusions

This chapter has considered the cross-sector issues with modernisation. The key themes have been the integration of operational and information technologies, which work towards improved efficiency but introduce new and common vulnerabilities. The IoT and the adoption of AI are newer challenges, and it may be some time to come before AI is approved for safety significant applications. Indeed, safety has generally slowed down the adoption of new technologies, as increasingly complex systems may have emergent properties that do not immediately manifest as safety issues. Hence, where passenger lives are at stake, the transport sector may continue to be slow to adopt the latest technology.

This chapter has also explored technologies that are common across the transport sector, most noticeably GNSS (e.g. GPS and Galileo). The reliance on satellite positioning and timing in so many applications means that protection of the system and its 'signals in space' is vital. This is why older technology such as inertial navigation has been retained in aircraft, and ship navigators must retain their skills in celestial navigation. Systems such as Passenger Information Display Systems are also common across the transport sector, in airports, seaports, and railway and bus stations.

The examples of modernisation specific to different transport modes also show a high degree of commonality, two examples being ADS-B in aviation and AIS in maritime. Given that many of the technology principles are similar, we should be mindful of the propensity for attackers to copy and adapt attack methods from one transport mode to another.

References

[1] ICAO safety report 2020 [online]. Montreal, Canada: International Civil Aviation Organisation; 2020. Available from www.icao.int [Accessed April 2022].

[2] SESAR Joint Undertaking (SESARJU). *Single European Sky Air Traffic Management Research (SESAR) homepage* [online]. Available from https://www.sesarju.eu/ [Accessed April 2022].

[3] Federal Aviation Administration (FAA). *Modernization of US Airspace, NextGen* [online]. Available from https://www.faa.gov/nextgen/ [Accessed April 2022].

[4] EUROCONTROL. *EUROCONTROL specifications for system-wide information management* [online]. Available from https://www.eurocontrol.int/concept/system-wide-information-management [Accessed April 2022].

[5] *Shift2Rail homepage* [online]. Available from https://shift2rail.org/ [Accessed April 2022].

[6] Thomas S., Rosenow J. 'Drivers of increasing energy consumption in Europe and policy implications'. *Energy Policy*. 137; 2020.

[7] United Nations Department of Economic and Social Affairs. *World urbanization prospects 2018: highlights*. New York: United Nations; 2019 Aug 29. Available from https://www.un-ilibrary.org/content/books/9789210043137

[8] 'Philip Hogge'. *ATC in Europe: an IATA Action Plan for the New Millennium*. 1999.

[9] ICAO Doc 9883. *Manual on global performance of the air navigation system*. ICAO, the International Civil Aviation Organisation; 2009.

[10] Lopez A., Jin W., Al Faruque M.A. 'Security analysis for fixed-time traffic control systems'. *Transportation Research Part B: Methodological*. 2020;139(2):473–95.

[11] Ermagun A., Kelarestaghi K.B., Finney M., Heaslip K. 'Speed up to hit the worker: impact of hacked road signs on work zone safety'. *International Journal of Transportation Science and Technology*. 2020;10(1):49–59.

[12] International Airport Review. *Half of flights in Europe's air traffic network delayed due to system failure* [online]. Available from https://www.internationalairportreview.com/news/68202/half-europes-system-failure/ [Accessed April 2022].

[13] Soderi S., Hämäläinen M., Iinatti J. *Cybersecurity Considerations for Communication Based Train Control*. Florence, Italy: Alstom Signalling Solutions; 2016.

[14] Fujitsu. *Always connected whitepaper* [online]. 2016. Available from https://www.fujitsu.com/uk/Images/always-connected-whitepaper.pdf [Accessed April 2022].

[15] Zeadally S., Tsikerdekis M. 'Securing Internet of things (IoT) with machine learning'. *International Journal of Communication Systems*. 2020;33(1):e4169.

[16] 'Regulation (EU) 2018/1139 of the European parliament and of the council on common rules in the field of civil aviation and establishing a European union aviation safety agency'. *European Union*. 2018.

[17] CAA U.K. 'CAP1826: EASA FTL regulations combined document and CAA guidance to developing an FTL scheme'. *Civil Aviation Authority*. 2019.

[18] General data protection regulation. *Regulation (EU) 2016/679 of the European parliament and of the council on the protection of natural persons with regard to the processing of personal data and on the free movement of such data*. European Union; 2016.

[19] Deka L., Chowdhury M. *Transportation Cyber-Physical Systems*. 1. Elsevier; 2018.

[20] Van Audenhove F.J., Rominger G., Korn A., *et al.* 'The future of mobility 3.0—Reinventing mobility in the era of disruption and creativity'. *Arthur D. Little*. 2018 Mar.

[21] Drinkwater D. *Ten stellar real-life examples of IoT taking flight in aviation, Internet of Business* [online]. Available from https://internetofbusiness.com/10-real-life-examples-iot-aviation/ [Accessed April 2022].

[22] Artificial Intelligence. *A European perspective, European Commission, joint research centre*. European Union; 2018.

[23] Brundage. *The malicious Use of Artificial Intelligence: Forecasting, Prevention and Mitigation*. Future of Humanity Institute: University of Oxford; 2018.

[24] European Space Agency (ESA). *What is EGNOS?* [online] Available from http://www.esa.int/Our_Activities/Navigation/EGNOS/What_is_EGNOS [Accessed Apr 2022].

[25] Federal Aviation Administration (FAA). *Satellite Navigation – Wide Area Augmentation System (WAAS)* [online]. Available from https://www.faa.gov/about/office_org/headquarters_offices/ato/service_units/techops/navservices/gnss/waas/ [Accessed Apr 2022].

[26] European Space Agency (ESA). *MSAS general introduction* [online]. Available from https://gssc.esa.int/navipedia/index.php/MSAS_General_Introduction [Accessed Apr 2022].

[27] FAA Navigation Team AJP-652 Results. *GPS Privacy Jammers and RFI at Newark* [online]. 2011. Available from http://laas.tc.faa.gov/documents/Misc/GBAS%20RFI%202011%20Public%20Version%20Final.pdf [Accessed Apr 2022].

[28] Psiaki M., Humphreys T. 'Protecting GPs from Spoofers is critical to the future of navigation'. *IEEE Spectrum*. 2016.

[29] Lin H., Qing Y. 'GPS Spoofing - Low Cost GPS Simulator'. *Def Con 23*.

[30] Gartenberg C. *This Pokémon Go GPS hack is the most impressive yet, The Verge* [online]. 2016. Available from https://www.theverge.com/circuitbreaker/2016/7/28/12311290/pokemon-go-cheat-gps-signal-spoofing [Accessed Apr 2022].

[31] *Cyber attack led to Bristol Airport blank screens, BBC News* [online]. 2018. Available from https://www.bbc.com/news/uk-england-bristol-45539841 [Accessed April 2022].

[32] Stelkens-Kobsch T.H., Hasselberg A., Mühlhausen T., Carstengerdes N., Finke M., Neeteson C. 'Towards a more secure ATC voice communications system'. *2015 IEEE/AIAA 34th Digital Avionics Systems Conference*; IEEE, 2015. pp. 4C1–1.

[33] Iasiello E. 'Getting ahead of the threat'. *Aerospace America*. Jul 2013.

[34] O'Connor F. *iPad app glitch causes travel troubles for American Airlines, Computer World* [online]. Available from https://www.computerworld.com/article/2916636/ipad-app-glitch-causes-travel-troubles-for-american-airlines.html [Accessed Apr 2015].

[35] ICAO. *The Postal History of ICAO: Annex 10 – Aeronautical Telecommunications* [online]. Available from https://www.icao.int/secretariat/PostalHistory/annex_10_aeronautical_telecommunications.htmAnnex 10 to the ICAO Convention.

[36] *EUROCONTROL Specification on the Air Traffic Services Message Handling System (AMHS)* [online]. 2018. Available from https://www.eurocontrol.int/publication/eurocontrol-specification-air-traffic-services-message-handling-system-amhs [Accessed Apr 2022].

[37] *EUROCONTROL Specifications for System-Wide Information Management* [online]. Available from https://www.eurocontrol.int/publication/eurocontrol-specifications-system-wide-information-management-swim [Accessed Dec 2017].

[38] FAA. *System Wide Information Management (SWIM)* [online]. Available from https://www.faa.gov/air_traffic/technology/swim/ [Accessed Apr 2022].

[39] *Aerion. Global ATS Surveillance* [online]. Available from https://aireon.com/products/global-ats-surveillance/ [Accessed Apr 2022].

[40] *Flightradar24* [online]. Available from https://www.flightradar24.com [Accessed Apr 2022].

[41] *FlightAware* [online]. Available from https://flightaware.com/ [Accessed Apr 2022].

[42] *planefinder* [online]. Available from https://planefinder.net/ [Accessed April 2022].

[43] *AirNav RadarBox* [online]. Available from https://www.radarbox.com/ [Accessed Apr 2022].

[44] Strohmeier M., Lenders V., Martinovic I. *On the Security of the Automatic Dependent, Surveillance-Broadcast Protocol* [online]. 2014. Available from https://arxiv.org/pdf/1307.3664.pdf [Accessed Apr 2022].

[45] Strohmeier M. Security in Next Generation Air Traffic Communication Networks [DPhil Thesis]. University of Oxford, Trinity College, 2016.

[46] 'International maritime organisation'. *International Convention for the Safety of Life at Sea (SOLAS)*; International Maritime Organisation, 1974.

[47] *Canadian Coast Guard, Global Maritime Distress and Safety System (GMDSS)* [online]. Available from https://www.ccg-gcc.gc.ca/search-rescue-recherche-sauvetage/distress-sys-detresse-eng.html [Accessed Apr 2022].

[48] Bhattacharjee S. *Introduction to the Global Maritime Distress and Safety System (GMDSS) – What You Must Know, Marine Insight* [online]. 2022. Available from https://www.marineinsight.com/marine-navigation/introduction-gmdss-global-maritime-distress-safety-system/ [Accessed Apr 2022].

[49] *Marine Traffic homepage* [online]. Available from https://www.marinetraffic.com/ [Accessed Apr 2022].

[50] *SEAiq Pilot homepage* [online]. Available from http://seaiq.com/ [Accessed Apr 2022].

[51] *iNavX homepage* [online]. Available from https://inavx.com/ [Accessed Apr 2022].

[52] *AIS Radar homepage* [online]. Available from https://obiltschnig.com/aisradar/ [Accessed Apr 2022].

[53] Kessler G.C., Craiger P., Haass J.C. 'A taxonomy framework for maritime Cybersecurity: a demonstration using the automatic identification system'. *TransNav, the International Journal on Marine Navigation and Safety of Sea Transportation*. 2018;12(3):429–37.

[54] Kerbiriou R., Lévêque L., Rajabi A., Serry A. The Automatic Identification System (AIS) as a data source for studying maritime traffic. *In International Maritime Science Conference*; Apr 2017.

[55] A unique signalling system for Europe "The long way towards an interoperable railway system". Available from https://www.railwaypro.com/wp/a-unique-signalling-system-for-europe-%E2%80%9Cthe-long-way-towards-an-interoperable-railway-system%E2%80%9D/ [Accessed 22 Apr 2022].

[56] Kaiser W., Nielsen S. 'The core of ATP – data engineering'. *IRSE Technical Meeting 'All about ATP'*. March 2008.

[57] Gabriel A., Brauner F., Lotter A., Fiedrich F., Mudimu O.A. Cyber security flaws and deficiencies in the European Rail Traffic Management System towards cyber-attacks. *Cybersecurity Issue – Rochester and Innovations for Crisis Response Proceedings of the 15th ISCRAM Conference*; NY, USA; May 2018.

[58] Rockman S. *Pwned trains on an Insecure Platform, Forbes* [online]. October 2018. Available from https://www.forbes.com/sites/simonrockman1/2018/10/02/pwned-trains-on-an-insecure-platform/#60c8c1667cfa [Accessed Apr 2022].

[59] Giles L. *The Future of GSM-R?, Rail Engineer* [online]. November 2015. Available from https://www.railengineer.co.uk/the-future-of-gsm-r/ [Accessed Apr 2022].

[60] Bloomfield R.E., Bendele M., Bishop P.G., Stroud R., Tonks S. The Risk Assessment of ERTMS-based Railway Systems from a Cyber Security Perspective: Methodology and Lessons Learned. *First International Conference, RSSRail*; Paris, France; 2016. pp. 28–30.

[61] Beugin J., Legrand C., Marais J., Berbineau M., El-Koursi E.-M. 'Safety appraisal of GNSS-based localization systems used in train spacing control'. *IEEE Access: Practical Innovations, Open Solutions*. 2018.

[62] SAE J3101:2012. *Requirements for hardware-protected security for ground vehicle applications*. SAE International; 2012.

[63] *History of CAN Technology, CAN in Automation* [online]. Available from https://www.can-cia.org/can-knowledge/can/can-history/ [Accessed Apr 2022].

[64] *CAN FD – The Basic Idea, CAN in Automation* [online]. Available from https://www.can-cia.org/can-knowledge/can/can-fd/ [Accessed Apr 2022].

[65] *Basics of the FlexRay Bus, Test & Measurement Tips* [online]. Available from https://www.testandmeasurementtips.com/flexray-basics/ [Accessed Apr 2022].

[66] *MOST Cooperation - Specifications. Media Oriented Systems Transport* [online]. Available from https://www.mostcooperation.com/specifications/ [Accessed Apr 2022].

[67] Bradbury D. *How to Hack a Jeep Cherokee – But Don't Try This at Home, Kids.* Naked Security [online]. Available from https://nakedsecurity.sophos.com/2017/05/10/how-to-hack-a-jeep-cherokee-but-dont-try-this-at-home-kids/ [Accessed Apr 2022].

[68] *Tesla Autopilot Crashes and Causes. Autopilot Review* [online]. September 2021. Available from https://www.autopilotreview.com/tesla-autopilot-accidents-causes/ [Accessed Apr 2022].

[69] *Self-driving Uber Car Kills Pedestrian in Arizona, Where Robots Roam. New York Times* [online]. March 2018. Available from https://www.nytimes.com/2018/03/19/technology/uber-driverless-fatality.html Waymo incident – cyclist? [Accessed Apr 2022].

[70] The Vienna Convention on Road Traffic. United Nations; November 1968.

[71] UNECE. Report of the global forum for road traffic safety on its seventy-seventh session. ECE/TRANS/WP.1/165; October 2018.

Chapter 2

Navigating the transport system security landscape: threats, responses and governance

John Hird, Rory Hopcraft, and Christina Rose

2.1 Introduction

This chapter attempts to navigate the transport security landscape by examining its evolution in terms of assets and services, the threats to which transport systems are exposed, and the means of responding to these challenges.

Transport systems are evolving rapidly as a result of digitalisation, with new subsystems being incorporated into legacy systems to improve performance, reduce costs, and deliver new services. One component of these developments is the potential introduction of new vulnerabilities into the transport system, which may be exploited by malicious actors, and may result in undesirable impacts on systems and the services that they provide.

In order to illustrate the importance of addressing cyber security in several transport modes, a variety of historical cyber-security incidents are presented. These show the exploitation of a broad variety of system vulnerabilities by different types of attackers and explain the main issues associated with the incidents.

The chapter then goes on to address how to respond to the challenge of ensuring that transport systems are cyber-resilient, describing the need for a systematic, system-wide, holistic approach, which promotes security-by-design throughout the system life cycle. Transport stakeholders should converge towards a common level of security, ensure that information is shared securely between them, and seek to develop a high level of security culture.

The role of national, regional, and global regulations, standards and guidance material is then addressed. The majority of regulations are mode-specific, but some apply across all transport modes and encompass other industry sectors, adding complexity to the regulatory environment. The standards and guidance material applied also tend to be mode-specific, and tailored standards are being developed which meet the individual requirements of each mode. This section provides information on how the material has evolved and highlights the main developments for each mode. As such, it may be a useful reference for those seeking insights in this area.

The chapter concludes with a summary of the issues addressed.

2.2 Context

As discussed in Chapter 1, cyber security is a sensitive issue for all transport providers and users. It addresses a vast swathe of activities associated with unlawful acts of interference in aviation, maritime, rail, road and inland waterways. The issues to be considered include protecting against terrorist attacks, ensuring the safety of passengers and employees, maintaining the confidentiality of sensitive information, securing operational systems from cyber attacks, and ensuring the provenance and integrity of the information used.

The wide range of issues and actors means that a systematic approach to managing security is essential. Many organisations have therefore adapted formal 'security management systems', framed from standards such as the ISO 27000 series. The wider, external framework for transport is set primarily by national governments and agencies, who provide advice and intelligence to organisations to help them prepare for cyber attacks, take measures to prevent them from being successful, and to respond effectively to successful attacks to ensure rapid recovery.

Support also comes from Computer Emergency Response Teams (CERTs), also known as Computer Security Incident Response Teams (CSIRTs). These have been set up as national or industry-level organisations to be the first point of contact in handling a security incident. They also coordinate nationally and provide advice, typically to organisations delivering critical national infrastructure. The European Union Agency for Cybersecurity (ENISA) provides a comprehensive list of CERTs/CSIRTs on its website [1].

Most transport systems are regarded as critical national infrastructure and have for years been striving to support uninterrupted operations, guaranteeing levels of safety, security, capacity, environment and cost-effectiveness, and addressing service disruptions as quickly as possible. As will be seen in this chapter, there is a mixture of national and international regulatory frameworks in place. These aim to ensure that in the event of security-related crises, the impact on the transport network and the time to recover to full operations are minimised. There is therefore a need for the regular involvement of security experts from all transport modes in a range of national and international security teams.

2.3 Transport system evolution

Securing transport systems has to be achieved in an increasingly complex and changing environment. As discussed in Chapter 1, a key issue is the prevalence of legacy systems. To a certain extent, legacy systems have benefited by being 'bespoke' and therefore achieving 'security through obscurity'. Legacy systems were often custom-developments based on expensive platforms, such as mainframe computers or mini-computers, which were interconnected point-to-point, often using proprietary protocols. They were therefore difficult to access unless you were an insider, and a pre-requisite for interacting with them was knowledge of the proprietary operating systems, obscure communications protocols and programming languages, which were known only to a small community of experts. The systems were physically separated from corporate information systems and did not have connections to the outside world either, so that the protection of the systems was focused on physical security. Nevertheless, security through obscurity may be a comforting delusion in the face of an intelligent adversary.

2.4 What are we trying to protect?

2.4.1 Self-protection and collaborative support

A transport mode should consider both the resilience of its operations and the potential provision of support to other actors should there be a security incident. In the air traffic management (ATM) sector, security has therefore been defined in two parts:

1. *Self-protection* is concerned with limiting the effects of unlawful interference (i.e. deliberate acts) on the system, such as attacks on critical assets. The aim is to ensure resilience (rapid response and recovery) in the event of a successful attack. Self-protection is therefore concerned with protecting the provision of the transport service itself, and ensuring the safety of staff, passengers and third parties who may be impacted by an act of unlawful interference. This can be achieved by protecting the infrastructure and information exchanges which support the provision of the service.
2. *Collaborative support* is concerned with providing outside agencies with support during a security incident, such as the provision of services or information to law enforcement, military agencies, search and rescue, or incident investigation agencies relating to an act of unlawful interference. Collaborative support is accomplished by having appropriate processes, procedures, and communications means in place which have been agreed and rehearsed with external partners in advance with trials and exercises.

The concepts of self-protection, resilience and collaborative support are considered to be generally applicable to all transport modes, although collaborative support may be of limited concern in other modes.

2.4.2 Assets

An asset is something which we care about [2] and which needs to be protected. Such assets include human, physical, information and organisational assets, as Figure 2.1 illustrates.

The variety and nature of these different types of assets are described in some detail in the following paragraphs.

2.4.2.1 Physical assets

For all transport modes, these include buildings and their associated services, such as power, water, heating, ventilation, and air conditioning (HVAC) and other

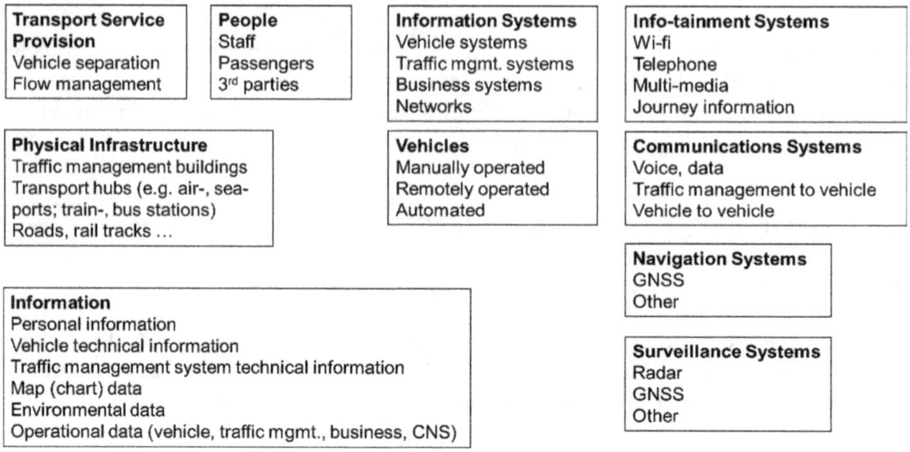

Figure 2.1 Assets to be protected

supporting infrastructures, such as computer systems, networks, access control systems, intrusion detection systems, authentication systems and operational technology (OT) components such as industrial control systems.

Examples of physical assets in specific transport modes are the following:

- Aviation – aircraft, airports, air traffic flow management systems, meteorological systems, air traffic control (ATC) centres, communications systems, surveillance systems, navigation systems, and the computer systems and utilities which support them.
- Road transport – vehicles, roads and motorways, road traffic management systems, road traffic signalling systems, and traffic information display systems.
- Rail transport – trains, rail network, signalling systems, train stations and rail traffic control centres.
- Maritime transport is, in many respects, similar to aviation, comprising ships, ports, traffic flow management systems, meteorological systems, maritime traffic control centres, communications systems, surveillance systems, navigation systems, and the computer systems and utilities which support them.

2.4.2.2 Human assets

These include the operational, engineering and other staff employed in each transport mode in addition to passengers. Maintaining the safety of human assets is paramount in any transport service; consequently, security incidents which could have an impact on safety must be anticipated and measures put in place to reduce their likelihood of occurrence and their impact.

2.4.2.3 Information assets

This refers to information used during the life cycle of the transport mode, from design through to deployment, operations and decommissioning. This could include information related to the architecture and design of the transport infrastructure and supporting systems, the technologies applied, the exchange of information during normal operations (in aviation, this could include flight plans and surveillance data), details on operational processes and procedures, and the personally identifiable information (PII) of staff and passengers. In a heightened threat environment, perhaps requiring the exchange of sensitive information on security incidents, the importance of protecting information assets is enhanced.

2.4.2.4 Organisational assets

These include finances and reputation, which encompasses brand management and future operations.

2.4.2.5 Service provision

This refers to the services that are provided within transport modes. For example, a vehicle separation service and a flow management service are typically provided in aviation and rail transport. In a system which has been adequately protected, a security incident may nevertheless result in a degradation of service provision, perhaps causing congestion, delays or termination of services, resulting in capacity reductions and inconvenience to passengers. The level of degradation should be limited, and the time taken to identify the attack, to respond, and to recover to normal operations should be similarly constrained.

2.5 Threats and vulnerabilities

2.5.1 Threats

Paraphrasing from ISO 22399:2007 [3], a threat can be defined as 'A potential cause of an unwanted incident which could result in harm to a system or function' and is the potential root cause of an unwanted impact. An attacker, also referred to as a 'threat agent', may attempt to exploit a weakness in a system, potentially resulting in a number of undesirable impacts. Threats can be categorised as different types of malicious acts:

- *Denial of service* (DoS) attacks impact system availability. Examples include the jamming of radio frequencies, which could result in the unavailability of services relying on wi-fi, mobile telephony or Global Navigation Satellite System (GNSS) signals. DoS attacks on systems with wired connections are also possible by flooding a system with data, preventing legitimate users from accessing information.

- *Exploitation of software vulnerabilities:* Vulnerabilities may exist due perhaps to security patches not being applied to address known issues, or due to the exploitation of a vulnerability for which a patch does not yet exist. The nature of the impact depends on the skills of the attacker, the type of exploit and the motivation of the attacker, but the impact could range from *minor* to *catastrophic.*
- *Network attacks* may include wiretaps and physical tampering, e.g., using key loggers or modifying internal assets, which could provide attackers with sufficient information to compromise systems or communications.
- *Social engineering:* This is a non-technical approach used by potential attackers to persuade or coerce people inside an organisation to break security procedures by, e.g., revealing sensitive information such as usernames and passwords. This could be via a telephone call, or interaction on social media. The technical security controls on a well-protected system can be entirely subverted by such means.
- *Misuse of authorisation attacks:* These result when an attacker is in possession of credentials and/or authorisation and misuses their privilege. This level of access could, e.g., be obtained through social engineering, a network attack or result from an insider misusing privileges.
- *Malware:* Malicious software exists in a number of variant forms and is designed to access or damage equipment, including passenger and staff devices, information and communications technology (IT) or OT computer systems, and other smart components, such as those which are part of industrial control systems (ICS). Malware may exploit vulnerabilities in operating systems or application software to acquire a presence on a system prior to executing its underlying code. Malware can result in a negative impact on infrastructure and is often capable of propagating autonomously, or via 'phishing' emails sent to potential victims on other systems. There are many different types of malware depending on their characteristics, e.g., spyware, viruses, worms, trojans, adware and ransomware.
- *Physical attacks* can result in assets being damaged or stolen, impacting availability and resulting in denial of service.
- *Human errors* are another major category of threat. For example, computer system configuration errors could lead to system downtime, performance reductions, or the introduction of new exploitable vulnerabilities, e.g., by installing a device with a default password. Lost hardware containing sensitive information, such as passwords or VPN certificates, could result in subsequent attacks if the information gets into the wrong hands. Staff may be directly targeted, such as in social engineering attacks and phishing email campaigns. Consequently, training all staff in security is fundamental to raise awareness, supported by policies and procedures that seek to minimise human error.
- *System failures* are normally addressed within the scope of safety, since they result not from intentional acts but from the passive failure of system components. Failures may occur in system hardware or software components, power supplies, or communications links, and they can also have an impact on security, potentially creating unanticipated vulnerabilities. In the world of connected

transport, third-party service providers play an increasingly critical role in operations, and interruption of those services may have an impact on capacity and performance.

- *Natural phenomena*, such as extreme weather, earthquakes and volcanic eruptions, may also impact the security posture of transport, as can social disruptions, pandemics and political instability.

2.5.2 Threat agents

The source of a threat is often referred to as an attacker, a threat actor or a threat agent. We use threat agent in this chapter in order to encompass a person, entity, or incident, with or without malicious intent, responsible for enabling the threats described above. It is important for organisations to understand who or what are their threat agents.

Threat agents will generally belong to one of the following categories:

- *Insiders* are staff with malicious intent who have legitimate access to security-critical assets. Insiders may be computer system administrators, system users with legitimate roles, or employees of external organisations who interact with the system, perhaps as users or maintainers. Disgruntled employees or activists may be motivated to disrupt operations or to inflict reputational damage on their organisation by, e.g., publishing sensitive information and attracting media attention [4].
- *Criminals* may be motivated by profit to sell stolen, sensitive, commercial or industrial information. They may demand ransoms for disabled systems to be returned to an operable state, e.g., in return for decryption keys to recover information encrypted by ransomware. Criminals have also breached systems in order to facilitate the transport of prohibited items, such as drugs, an example of which is highlighted by the cyber attack on the port of Antwerp, discovered in 2013 [5].
- *States and state-sponsored organisations* may be motivated to gain knowledge about transport systems and other critical infrastructures for the purpose of espionage, or to be in a position to cause disruption or economic impact for political gain.
- *Terrorists* may also use cyber attacks to reduce public confidence in the safety, security and resilience of the transport infrastructure, thus causing economic and political impact.
- *Travelling passengers* who are physically present within the transport system are also potential threat agents. For example, activation by a passenger of a radio-frequency jamming device on an aircraft or a ship could impact communication or navigation systems, while passenger entertainment systems or access to wireless or wired services could be used to attempt to breach critical systems.

- *Lone wolf* threat agents tend to operate as anonymous individuals on the 'Dark Web', often gaining financially from the popularisation of cyber crime-as-a-service, developing and selling malware and supporting tools to cyber criminals.

An attack may be specifically targeted at a selected organisation or transport mode, typically involving intelligent planning using specific tools and techniques. An opportunistic attack, however, takes advantage of a vulnerable target that was previously unidentified to the attacker, and may result from the exploitation of tools and techniques which are generally available on the Internet, and are capable of proliferating and locating systems with known vulnerabilities which can be exploited.

2.5.3 Vulnerabilities

A vulnerability can be defined as 'an intrinsic property of something resulting in susceptibility to a risk source that can lead to an event with a consequence' (ISO Guide 73:2009) [2]. This could be a weakness in a system, physical controls, security procedures or implementation that could be exploited by a threat agent. An attacker will attempt to exploit a vulnerability to achieve their goal. For example:

- Inadequate malware protection software could result in an employee unintentionally injecting malware into a system via a compromised USB key, perhaps found in a public place, provided as a gift, or distributed free at a conference.
- Inadequate vetting of personnel could result in a 'sleeper' being employed whose goal is to acquire confidential information and pass it on to competitors for the purpose of industrial espionage.

An attack may employ a sequence of different types of threats in order to achieve its goals, and may also take a substantial amount of time while an attacker consolidates a presence on a targeted system. Vulnerability to attacks is often identified by articulating 'Attack Surfaces' [6]. These are where system boundaries are exposed to an attacker, often correlating with trust boundaries [7], where the system exchanges information with other systems and users. The many potential attack surfaces in transport include those requiring physical interaction with assets, wired or wireless communication with assets, and those employing social engineering by interaction with transport staff (including those providing third-party services) and passengers.

2.6 Impacts

The result of a threat agent exploiting a vulnerability to harm an asset could be an impact on the confidentiality, integrity or availability of the asset. As defined in ISO 22399:2007 [3], impact is 'the extent to which a loss of confidentiality, availability or integrity of an asset affects the achievement of business objectives' and as 'an evaluated consequence of a particular event'. Definitions for confidentiality, integrity and availability are provided by ISO 27001 [8] as follows:

- *Confidentiality* – the property that information is not made available or disclosed to unauthorised individuals, entities or processes. For example, information should not be capable of being read or copied by an unauthorised person;
- *Integrity* – the property of safeguarding the accuracy and completeness of assets. For example, it should not be possible for unauthorised changes to be made to information;
- *Availability* – the property of being accessible and usable upon demand by an authorised entity. For example, it should be avoided that information is erased or becomes inaccessible, or that services cannot be accessed.

There are many different types of impact imaginable, as shown in Figure 2.2. As described earlier, service provision could be disrupted, causing congestion, delays or termination of services, resulting in inconvenience to the travelling public. In certain scenarios, safety may be compromised, and the impact on passengers or personnel could include stress, injury, or, in the worst case, fatalities. A security incident could also result in financial consequences and, in severe cases, bankruptcy. Reputation could suffer from bad publicity, leading to fewer passengers and impacting future revenues.

Disruptions to transport systems could also have a wider societal impact, resulting in disruption to trade and the wider economy. In a highly interconnected system of systems, a security incident could potentially impact multiple transport modes and extend across a large geographical area.

Given the plethora of regulations in force which require the transport industry to apply due diligence in risk management and incident management, it is also possible that a security incident could reveal non-compliance with regulations, potentially resulting in legal action and related financial, reputational and operational consequences. For example, the standards specified in International Civil Aviation Organisation (ICAO) Annex 17 [9] are applicable to aviation stakeholders

Figure 2.2 Potential impacts of a security incident

globally, while regional regulations may additionally be applicable, such as Reg (EU) 2017/373 [10] and Reg (EC) 300/2008 [11] in Europe.

To consider the effects of network-wide disruption, we take the example of aviation and, more specifically, Air Traffic Management (ATM). Future plans envisage the connection of multiple, disparate systems into a 'system-of-systems'. The aim is to share data for the provision of ever more complex functions and services. The transition to the new system will require the piecewise integration of new subsystems, and their associated functions and services with existing legacy systems. This in itself may result in the regular appearance of new vulnerabilities. Considering that in July 2019, there were a record number of daily movements* (35,282) in Europe, the impact of a security breach could have a wide geographical impact, affecting a large number of airlines, passengers and staff through delays inconvenience, and financial losses. It is also conceivable that safety margins could be reduced in such a situation.

In the remainder of this chapter, we look at:

- historical cyber-security incidents in transport;
- how to respond to the challenges of cyber security;
- applicable regulations, standards and guidance material.

2.7 Cyber-security incidents in transport

2.7.1 Introduction

Cyber-security incidents are often viewed as relatively new phenomena in transport, however, they have been occurring for many years. This section presents selected examples of incidents that have occurred in transport over the past 25 years and highlights lessons that may be learned in each area.

From the transport cyber-security literature, there are a number of key principles which recur with respect to promoting and ensuring the resilience of systems, including the following:

- establishment of a security policy;
- implementation of a security management system;
- ensuring effective coordination with the quality management system;
- applying a whole life-cycle approach, ensuring that security is designed-in and maintained;
- applying a holistic approach and implementing defence-in-depth;
- performing security risk assessments and implementing security controls to reduce risk to an acceptable level;

*https://www.eurocontrol.int/publication/network-operations-report-july-2019.

- monitoring systems to detect security breaches, contain them, respond, and recover in a timely fashion;
- establishing processes for continuous review and improvement;
- ensuring the provision of appropriate training and awareness for staff and developing a security culture;
- ensuring supply-chain security;
- ensuring that information is shared securely between stakeholders.

In order to support the above requirements, senior management must support the security policy and its objectives, and provide the necessary personnel and financial resources for them to be achieved.

2.7.2 Malware

The examples that follow reveal a variety of failings resulting in security incidents. These include system maintenance issues, such as delays in implementing patches to address known vulnerabilities, and failure to update antivirus databases. Inadequate system monitoring, anomaly detection, response and recovery are also cited.

Inadequate security awareness of personnel is highlighted with respect to phishing emails and the insecure connection of critical systems to the Internet. A failure caused by buggy antivirus software reveals the importance of supply-chain security.

2.7.2.1 Rail signalling systems immobilised

In 2003, CSX Corporation operated a control system for train signalling systems in the United States and became a victim of the Sobig.F virus and the Blaster worm [12, 13]. The Sobig.F virus was a mass-mailer that quickly spread on networks in 2002 and 2003 through malicious emails, while the Blaster worm took advantage of a Windows security flaw to infect systems. On 20 August, both spread simultaneously through the control system, causing train signalling systems to slow down and eventually stop, resulting in disruption to passenger and freight services in 23 states in the Eastern US. The incident could have had a serious impact, but no serious consequences resulted. The attack was not targeted. Post-incident analysis revealed that various controls could have been put in place to prevent the incident or to limit the impact, including

- building awareness among users of cyber hygiene and dealing with malicious emails (phishing);
- updating antivirus databases on a regular basis;
- adequate monitoring of industrial control systems;
- correcting flaws in procedures once the breach was detected.

Consequently, improvements in system maintenance, monitoring, and staff training and awareness could have prevented the incident from occurring. This is also an example of the potential issues which may result from the ill-considered connection of IT and OT systems.

2.7.2.2 Flight-planning computer immobilised

Also in 2003, a virus attacked an Air Canada Jazz flight-planning computer at an operations centre in Halifax that provided essential information on fuelling, weather and other variables [14]. Without this information, aircraft were unable to depart, affecting regional operations and resulting in delays to approximately 200 flights.

2.7.2.3 Air traffic control system loss of integrity

In 2006, a viral attack originating from the Internet spread from administrative networks to ATC systems, forcing the Federal Aviation Administration (FAA) to shut down a portion of its ATC system in Alaska [15]. Here the virus impacted the system's integrity and the subsequent shutdown impacted availability. Improvements in system maintenance and monitoring of the administrative systems, combined with measures to ensure the separation of IT and OT systems, could have prevented the incident from occurring.

2.7.2.4 Airport-targeted phishing scam

Up to 75 US airports were found to have been targeted by a phishing scam which occurred in 2013 [16]. Advanced Persistent Threats (APT) were found to be present in at least four airports. An APT occurs when an unauthorised person or group gains access to an organisation's network and remains hidden inside that network while pilfering sensitive information for an extended period of time. The Centre for Internet Security organisation attributed the attack to an undisclosed Nation State seeking to breach aviation networks.

An APT uses continuous, clandestine, and sophisticated hacking techniques to gain access to a system and remains inside for a prolonged period of time, with potentially destructive consequences. Due to the level of effort needed to carry out such an attack, APTs are usually levelled at high-value targets, such as nation states and large corporations, allowing them to quietly enhance their level of access to systems, to harvest information, or to use a system as a stepping stone to gain access to others. When an APT is discovered and the immediate threat addressed, hackers may have left multiple backdoors open, allowing them to return whenever they choose. Additionally, many traditional cyber defences, such as antivirus and firewalls, cannot always protect against such attacks.

2.7.2.5 Railway reservation systems made inaccessible

In 2005, an errant virus definition distributed by an antivirus PC software company, Trend Micro, caused a large-scale disruption in the reservation division of the East Japan railway company (JR East) [17]. The software company's automatic update site in the Philippines had released a new virus definition file for its Virus Buster antivirus software which had not been adequately tested. This file was picked up by many users who either automatically or manually invoked the virus definition update function of the software. Windows XP SP2 and Windows 2003 server users who updated their definition file and rebooted their PCs after the update (as suggested by the software) saw the CPU usage go up to 100% immediately after booting, and could not control their systems. LANs became inaccessible at the JR East reservation division, so they could no longer check the status of the reservation system. Agents diverted all telephone customers to manned counters at railway stations. The company estimated the number of affected users at around 10 million.

Whilst inadequate engineering practices at Trend Micro had led to the release of the software, JR East appears to have rolled out the update to their operational systems without first verifying it on a test system.

2.7.2.6 Exposure of airport employee personal details

A third party conducting mandatory background checking of airport workers for the Australian government was hacked in 2018 [18]. This resulted in the exposure of the personal details of hundreds of aviation employees working primarily in regional areas. The information was required to grant an Aviation Security Identity Card (ASIC) and included name, date and place of birth, driver's licence, Medicare number and passport details.

This breach causes concern in terms of the invasion of privacy for workers, the potential malicious use of the information by criminals and the increased vulnerability of the aviation system.

2.7.3 System breaches

Vulnerabilities in web-connected applications have resulted in breaches of both IT and OT systems, providing hackers with control of systems and access to sensitive information. The causes include inadequate intrusion detection systems and patch management processes, along with poorly configured Web applications. The need for a holistic approach to security is emphasised by one particular incident described in this section, which employed a multi-pronged attack including spear-phishing, social engineering, a physical security breach, physical modification of a system, and finally, remote modification of information within the system.

2.7.3.1 Tram derailment

In 2008, a 14-year-old Polish teenager used a modified remote control device to interfere with the system controlling the points of the trams in the city of Łódź in Poland [19]. Having studied the system for some time, he modified a TV remote control and used it to activate switches which controlled the track points at junctions, allowing him to divert trams from their intended course at will. While using his device, four trams were derailed and 12 people were injured.

2.7.3.2 Databases compromised

In 2009, the FAA made public some incidents in aviation which had occurred between 2006 and 2009 [15]. This included cyber attacks that had compromised IT and OT components of the FAA's systems.

- In 2008, hackers took over FAA computers in Alaska and, by taking advantage of interconnected networks and a stolen administrator's password in Oklahoma, the FAA's domain controller in the Western Pacific Region was compromised by malicious code. This provided the attackers with access to 40,000 FAA user IDs, passwords and other information, thus breaching the integrity of the FAA system over a wide area, and impacting the confidentiality of a large amount of information.
- In 2009, hackers compromised an FAA public-facing Web application computer and used it as a conduit to enter an FAA internal database server containing the PII of 48,000 current and former employees.

The recommendations of the report included the deployment of adequate Intrusion Detection System (IDS) monitoring devices covering local area networks at all ATC facilities, strengthening patch management processes on Web applications, and ensuring that all Web applications were configured in compliance with government security standards.

2.7.3.3 Breach of cargo handling systems to enable drug smuggling

In 2013, police in the Netherlands and Belgium seized a tonne of cocaine, a tonne of heroin and a suitcase containing €1.3 m, after uncovering a massive drug smuggling operation [5]. Smugglers had recruited hackers to breach the cargo handling systems of shipping companies at the port of Antwerp.

The attack was carried out in a number of phases, beginning with a spear-phishing campaign, followed by malware being emailed to staff. This allowed the attackers to gain access to information remotely. The initial breach was discovered, however, and a firewall was installed to prevent further attacks. The attackers then physically broke into a building containing computer terminals with access to the target system, and installed a keyboard-video-mouse switch, which allowed them to interact with the system remotely, via the cellular phone network.

Once in the system, the smugglers were able to change the physical location of shipping containers known to contain drugs and would send their own drivers to pick them up before the legitimate haulier could collect them. The operation had been going on for 2 years prior to its detection.

In this attack, hackers were able to install malware on the target system by successfully employing social engineering in the form of a phishing campaign. Once the malware was detected and removed, additional security controls were put in place to defeat similar attacks in the future. In response, the attackers overcame physical security measures to enter the building and install equipment which provided them with remote access to the target system.

This attack highlights the importance of taking a holistic approach to security and of employing defence-in-depth to protect systems.

2.7.3.4 Breach of airline booking system

In 2018, vulnerabilities in the flight reservation system of British Airways (BA) resulted in a data breach impacting the information of more than 400,000 customers and staff over a period of 15 days in August and September. Those using the system were redirected to a fake website, and personal details including name, address and credit card details were siphoned off. The attack also impacted mobile users. BA did not detect the breach itself but was informed of the attack through a third party [20].

Under the EU's General Data Protection Regulation (GDPR) [21], the UK Information Commissioner's Office (the independent authority responsible for information rights and personal data privacy), initially fined BA the sum of £184m (1.5% of global turnover), but this was subsequently reduced to £20m due to the impact of the SARS-CoV-2 virus on the airline industry, since the issues had been addressed.

The attackers used a cross-site scripting attack, injecting 22 lines of their own code into a piece of third-party Javascript called *Modernizr* on BAs payment processing website, which exfiltrated data to another site. The vulnerability in *Modernizr* was known, but the code had not been updated since 2012, indicating a problem with risk management and patch management procedures. In addition, the attack was not detected by BA, suggesting inadequacies in monitoring and detection capabilities. The incident revealed general problems with security governance which can be solved by implementing established best practices.

2.7.4 Remote monitoring, maintenance and control

Inadequately protected systems may result in unauthorised actors being able to monitor and control them. With sufficient knowledge and resources, it is possible to reverse-engineer a system with connections to networks or information sources, identify its vulnerabilities, and create exploits which can be used to remotely enable, disable, dis-inform or control it, highlighting the importance of designing-in security at the onset of development.

Publicly available data (such as ADS-B) can be monitored and used in conjunction with other available meta-data, to infer confidential information in a number of areas, potentially impacting personal privacy, national security, diplomacy and business competitiveness. The increasing dependence of transport systems on GNSS data, in general, is driving developments to mitigate the potentially serious impact of jamming and spoofing incidents, for which means of detection, response and recovery are in their infancy.

2.7.4.1 Loss of airport communications including ATC

On 10 March 1997, the breach of a Bell Atlantic control system used for air traffic communications at Worcester airport, in Massachusetts (USA), caused a system crash that disabled the airport phone system for six hours [22].

The attacker temporarily disabled two programmable remote controllers used to integrate voice and data communications from copper-wire telephone lines for transmission over a single fibre-optic cable. The systems were connected to modems to allow remote maintenance and configuration, so were accessible with knowledge of their telephone numbers. The attacker accessed the devices and disabled both, resulting in the loss of telephone services to the control tower, airport security, the airport fire service, the weather service and aircraft operators. In addition, the attack took out the tower's main radio transmitter, a transmitter for controlling runway lights and a printer used by controllers to monitor flight progress. Telephone services to 600 homes in the vicinity were also halted.

The attack was made possible by the devices in question being permanently connected to modems for the purpose of remote monitoring and maintenance, which provided access to them with only knowledge of the modem telephone numbers. A risk assessment of the system could have revealed these vulnerabilities and potential impacts, and controls could have been put in place to mitigate the risks.

2.7.4.2 Remote access to control car systems

Chrysler Jeep Cherokee

In 2015, two researchers revealed that after a copious amount of analysis and reverse engineering, they had developed the means to wirelessly access the control systems inside a popular, standard production vehicle – a Chrysler Jeep Cherokee [23]. Their software was capable of sending commands via the car's entertainment system to the dashboard functions, steering, brakes and transmission from a laptop connected remotely via the GSM network. The researchers were able to take control of the vehicle's information display and entertainment system, turn the engine on and off, enable the windscreen wipers and activate the brakes. They could also enable surveillance of the vehicle by obtaining its GPS coordinates and speed. All of these exploits could be achieved in real-time from a great distance via the GSM network.

Access to the car was achieved by exploiting a vulnerability in the car's connected computer system, called 'Uconnect', which used the Sprint cellular network to access the Internet and allow owners to interact with their cars remotely. The attackers used the system to re-write the 'Uconnect' firmware, making it capable of sending commands through the car's internal network, the CAN bus. The attackers were also able to scan Sprint's network for other Uconnect-equipped vehicles and obtain vehicle identification information and locations.

The research revealed significant vulnerabilities in the production vehicles which, if exploited by unscrupulous attackers, could potentially result in injuries or fatalities.

For the installation of a software update to address the exposed vulnerabilities, Fiat-Chrysler provided customers with the option to either visit a dealership or to install it themselves. The latter option was via either a patch made available on a website, or provided on a USB-stick sent through the post [24]. The USB-stick approach was the source of controversy in the security community for the following reasons:

- There was no easy way to verify its authenticity.
- A legitimate update could be substituted for a malicious one.

- Attackers may have been able to extract data from the USB stick and reverse-engineer them, potentially finding vulnerabilities in the code or the update process.

Tesla Model S

In 2017, researchers from KU Leuven university in Belgium discovered a vulnerability in the third-party keyless entry system of the Tesla Model S [25]. The system used only a weak 40-bit cipher to encrypt the key-fob codes, and with only two codes from any given key fob, they were able to create a table of pre-computed keys which could be used to access and operate the associated vehicle. On presenting their findings to Tesla, they were rewarded with a $20,000 bug bounty. In a follow-up over-the-air (OTA) software update, Tesla incorporated the option of inserting a user-definable 'PIN to Drive' code, and in June 2018 introduced more robust cryptography, giving owners of older cars the opportunity to switch to the new key fobs.

The story did not end there, however, since the same KU Leuven team succeeded in cracking the revised system in 2019 [26]. Tesla was able to address the vulnerability via an OTA† update which updated vehicle firmware and allowed owners to update their key fobs inside the vehicles.

The examples above illustrate the importance of designing-in security during vehicle development and maintaining it during the life cycle. Securing the supply chain is crucial where third-party components are integrated into a system, since developing and validating solutions to vulnerabilities and rolling them out to vehicles in service can be difficult, time-consuming and expensive.

2.7.4.3 Eavesdropping

The advent of affordable software-defined radios (SDRs) has made the reception of ADS-B messages and the positional tracking of aircraft trivial. A personal SDR receiver can receive information from aircraft up to 600km away, while commercial and non-profit organisations such as Flightradar24, PlaneFinder, FlightAware, ADSBexchange, Radarbox25 and OpenSky pool the data and make it available online, adding a global dimension to the tracking of aircraft.

The data on these services are largely publicly available and easily accessible. However, it may exclude aircraft considered sensitive, such as those owned by governments or corporations. Unfiltered ADS-B data, easily obtained from several sources, can impact the privacy of aviation users when used in conjunction with publicly available aircraft meta-data.

There are several documented examples demonstrating the ease with which this information can be accessed and analysed to expose confidential information. For example, investigative journalists used such data to expose CIA rendition flights

†OTA – Over-the-air.

which occurred during the 'war on terror' [27]. Similarly, such data have been used to reveal meetings between business executives, providing clues on future mergers and acquisitions [28], and to infer meetings between governments.

Consequently, exploitation of such data can potentially be used to infer confidential information in a number of areas, including national security, diplomacy and business competitiveness. Protecting the privacy of non-commercial aviation users in the future may require further regulatory and technical developments to be made.

2.7.4.4 GNSS spoofing

A closely monitored GNSS spoofing experiment was carried out by researchers on the 'White Rose', a 213-ft superyacht, in the Mediterranean in June 2013 [29]. On board was a hand-held device costing around $2,000 to build, which was capable of generating a fake GPS signal. The researchers succeeded in shifting the ship's course 3 degrees to the north, while also convincing the yacht's GPS system that it was underwater. This experiment confirmed that civilian GPS spoofing was feasible and that it was possible to alter the course of a vessel by this means, impinging on the integrity of GPS as a navigation tool.

In the 'Pokémon Go' game, which runs on GPS-enabled smartphones, players are restricted to catching Pokémons that are present near their current location. With the capability to spoof the GPS coordinates of their smartphones, players would be able to move virtually to any location, greatly improving opportunities to enhance their collections. There are software spoofing tools available which allow this to be achieved, however, the game developer is often capable of detecting that such programmes are running and restricting game functionality as a result [30]. It is, however, possible to put together a hardware spoofer using a signal generator, a radio-frequency shielded box, and other pieces of easily obtainable equipment, providing a spoofing capability which the game is unable to detect.

GNSS spoofing has therefore become mainstream, with software tools available to the general public and specifications for spoofing hardware readily available on the Internet.

A relatively innocuous safety alert released by the U.S. Maritime Administration in June of 2017 [31] revealed that an apparent mass GPS spoofing attack, involving more than 20 ships, had occurred in the Black Sea on 22 June 2017. The incident was initially reported as GPS interference, but a further report revealed that ships were intermittently unable to obtain GPS signals near the coast of Novorossiysk, Russia. When GPS signals were received again, they provided their locations as being 25 nautical miles from their known position. Their GPS systems were telling them that they were located on land, near Gelendyhik airport.

One ship's master provided photos of his navigation displays, which included a paper chart showing his actual position, and his radar display showing various AIS ‡contacts which did not have corresponding radar returns. This was the first publicly

‡AIS – Automatic Identification System.

available, well-documented account of previously anecdotal problems with AIS and GPS in Russian waters [32]. Some factors point to this being a GPS spoofing incident:

- The effects were observed by over 20 separate vessels, implying that it was not a malfunctioning GPS unit but an external incident of some kind;
- A large number of ships in the area reported identical or very close locations. This is a symptom of a large-scale spoofing attack. In a low-level jamming attack, misleading positions reported by vessels would typically display some randomness;
- Ships reported that their positions would periodically 'jump' from the true location to the incorrect location. Again, this is typical behaviour in some spoofing experiments. For various reasons, GPS receivers may temporarily lose lock on spoof satellite signals, then reacquire the real ones, and vice versa. This causes the characteristic random flipping between two well-defined locations.

By running algorithms on data from the vessels' AIS, experts were able to identify two additional instances of mass GPS interference in 2017, which each lasted for several months. Since similar disruption patterns were observed in multiple vessels in specific areas, it appears that the issue was caused by GPS disruption rather than AIS disruption, and was therefore likely to affect everything in the area, not just ships [33].

Interestingly, all three locations examined in this analysis involve airports – Gelendzhik Airport and Sochi International Airport near the Black Sea, and St Petersburg Airport near the North Sea. It would seem that spoofed signals were being transmitted to make GPS receivers in the area appear to be located at an airport. One plausible explanation could be drone defence. Many drones, particularly those operated by casual users, have geo-fencing rules that prevent flights over airports and other restricted areas. So, if someone was trying to perform aerial surveillance of the Russian border, their drone may suddenly think it was over an airport, and take action accordingly. The action taken depends, of course, on how the drone is programmed, but often includes 'land immediately' or 'return to launch point'.

2.7.5 Unintentional acts

It is possible for an asset to be negatively impacted in the absence of a threat agent, as described earlier. This may occur when a vulnerability is exploited unintentionally. The impact of an unintentional act may be very similar to that of an intentional act, and an example is now described.

2.7.5.1 Unintentional denial of GNSS service

After the installation of a GPS Ground Based Augmentation System (GBAS) at Newark airport in 2009–2010, multiple Radio-Frequency Interference (RFI) events led to the GPS receiver losing its ability to obtain signals, impacting its availability for use in the precision approach of aircraft, and ultimately leading to the GBAS system being shut down. RFI jamming events were occurring several times per day. Investigations concluded that personal GPS jammers [34] being used by drivers on a nearby road were the cause of the problems. Personal Privacy Devices (PPDs) are used to mask user positions from GPS-based tracking systems, and are often used by truck drivers or vehicle thieves who do not wish to be tracked.

The PPDs overwhelmed GPS receivers by broadcasting directly on the GPS L1 frequency, and were difficult to detect and isolate, particularly in moving vehicles [35]. The mitigation options applied were to modify the surroundings to better protect the GBAS, to modify the GBAS system to reduce the operational impact of jammers at the ground station, and to focus on the threat source by releasing awareness material, addressing jammer sales websites and customers, and developing and deploying detection equipment. The operator of one PPD causing interference was arrested and fined and subsequently lost his job as a result of his use of the device [36].

Such events continue to occur on a regular basis, as indicated by a list of enforcement orders carried out by the US Federal Communications Division (FCC) [37], which provides information on both intentional and unintentional interference with authorised radio communications.

2.8 Responding to the challenge

2.8.1 Introduction

The transport modes investigated are in a period of rapid evolution, and each is integrating new systems with legacy systems to meet future business needs. The potential impacts of a cyber-security breach within each mode are similar, with the safety and privacy of staff, passengers and third parties raising particular concerns. As we shall see, the legal framework in different countries and regions, along with the methods, tools and standards typically applied, is also different for each mode, which raises issues regarding the feasibility of achieving the goal of integrated transport systems in the future. Such a system would require the following to be addressed:

- different levels of cyber-security maturity in each transport mode;
- differences in the legal framework in each mode and potential harmonisation;
- the achievement of a common level of cyber security across all modes – an intelligent adversary will try to exploit the weakest link to breach a system;
- the potential rationalisation of methods, tools, standards and guidance material;
- the timely and secure sharing of cyber security-related information between modes;

- the development of a trust framework between modes;
- the evolution of security awareness and culture to similar levels in all modes;
- the establishment of means of cyber-security certification of products, services and processes.

Some of these requirements are discussed in the following paragraphs.

2.8.2 Cyber-security strategies

In recognition of the evolving cyber-security threats facing transport systems, and indeed all critical infrastructures, states have been developing strategies to improve the resilience of infrastructures to acts of unlawful interference.

The first national cyber-security strategies began to appear in the early 2000s, with the United States being one of the first countries to recognise cyber security as a national strategic matter. In 2003, they published the *National Strategy to Secure Cyberspace* [38]. This was part of the overall National Strategy for Homeland Security, which was developed in response to the terrorist attacks of 11 September 2001. A number of other national strategies were subsequently developed, and those released in the EU prior to 2012 are summarised in an ENISA document [39]. The United States released an executive order for improving critical infrastructure in 2013 [40], and a national cyber-security strategy document was released in 2018 [41]. The European Commission also released a cyber-security strategy in 2013 [42], which was updated in 2020 [43], and several EU states have subsequently released updated national strategies, including the United Kingdom [44] and France [45], which subsequently published a strategic review of cyber defence in 2018 [46]. ENISA maintains an online map of EU national cyber-security strategies, indicating the status of each, and it provides a download facility for the documents [47]. Australia also published a national strategy in 2016 [48], which was updated in 2020 [49].

Transport-specific strategies have also been published, such as the United Kingdom [50] and United States [51] strategies on cyber security in aviation, and the revision of the European Union Maritime Security Strategy Action Plan [52], all of which were published in 2018. ICAO published an aviation cyber-security strategy in 2019 [53].

The frequent revision and review of cyber-security strategies are indicative of the continual evolution of the threats and the strategic importance of maintaining secure and resilient systems, nationally, regionally and globally.

2.8.3 Cyber resilience

In order to make a system cyber resilient, several factors must be addressed. A system should be protected from attacks, meaning that security incidents should be prevented from having an unwanted impact. The implementation of such 'pre-event' or preventive controls should reduce the likelihood, and potentially also the impact, of a successful attack.

However, since it is not possible to reduce the likelihood of a successful attack to zero, it is also necessary to have mechanisms in place which optimise the post-event response to a successful attack, by minimising the unwanted impact, and facilitating the recovery of the system to normal operations as quickly as possible.

In Figure 2.3 (from ISO/PAS 22399 [3]), an appropriately protected system might follow the trajectory of the solid line, with system performance initially subject to limited degradation, but recovering to normal performance levels quickly. However, with inadequate controls in place, a system might follow the trajectory of the dashed line, in which system performance is reduced to zero and recovery is slow.

Resilience is realised by performing a security risk assessment to identify which assets need to be protected and to determine how that can be achieved.

To maintain knowledge of the security posture of an organisation and its systems, it is necessary to take a systematic approach to risk management. Risks are identified and evaluated by performing security risk assessments, and risks which are above an acceptable threshold (often referred to as the risk appetite) are mitigated by the implementation of appropriate security controls. There is a variety of documented standard approaches to security risk assessment which can be applied, such as those described in ISO/IEC 27005 [54], ISA/IEC 62443-3-2 [55] and IEC 31010 [56].

Risk assessments are not static deliverables. They must be repeated in the following circumstances: if new vulnerabilities or threats become evident; if there are changes to the system or if a security-incident occurs; they might also be re-assessed on a periodic basis. Consequently, it is important to maintain awareness of new and emerging threats.

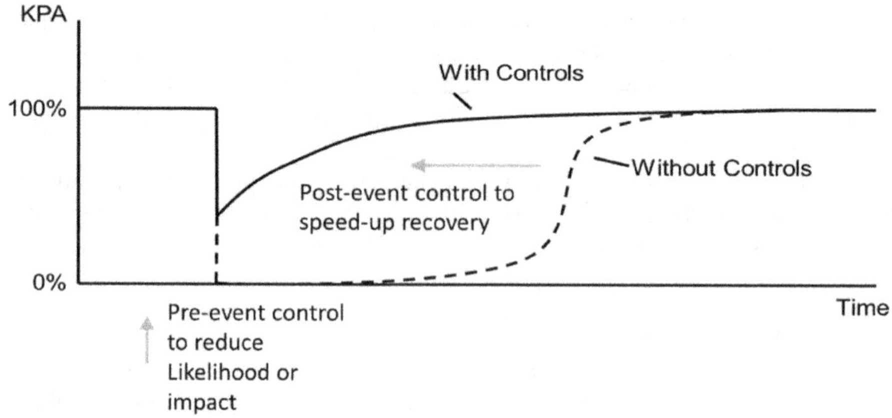

Figure 2.3 Cyber resilience

2.8.4 System-wide approach

Since transport systems comprise a large number of interacting, cooperating assets of different types, unlawful interference with one or more assets by an attacker could result in unwanted impacts, potentially extending over a large geographical area. Consequently, it makes sense to address the protection of such assets by considering the security of the system as a whole. For example, an information system considered to be cyber-secure might still be compromised through a social engineering exploit (e.g., coercion, leak of confidential information), loss of essential services or physical damage. It is therefore prudent to adopt an holistic approach to security and address the management of risk by considering all possible applicable threats.

2.8.5 Holistic view

As a result of the rapid digitalisation of transport, there can be a tendency to focus unduly on applying technical fixes to potential security problems, and to concentrate on protecting the technical infrastructure at the expense of looking at the bigger picture. The most efficient and cost-effective way to mitigate cyber-security risks may be a combination of different types of controls, perhaps including the vetting of personnel, with the support of physical, procedural and technical controls.

The incidents previously described underline the need for a holistic approach, since a multi-pronged attack may comprise a variety of means, such as spear-phishing, social engineering, malware injection, a physical breach, physical system modification, and remote monitoring and control.

2.8.6 System life cycle

It is also recommended practice to address security from the beginning of the system life cycle, from the concept stage through all subsequent stages, including development, deployment, operations and decommissioning. This involves the application of appropriate methods, tools, and guidance during the life cycle to ensure that systems are resilient in initial operations and can be maintained as such. This approach is often referred to as security by design.

At best, retro-fitting security controls to a system can be time-consuming and expensive, and it is sometimes not possible to effectively address fundamental vulnerabilities, short of re-designing system components and re-deploying them. This is particularly problematic where systems have an expected life span of 15–30 years, as is often the case with those prevalent in critical transport infrastructure.

2.8.7 Common level of security

A desirable goal in security terms is expressed in the EU's Network and Information Systems (NIS) Directive [57] as the achievement of 'a high common level of security of network and information systems in the Union'. Such an approach should facilitate the development of a harmonised level of security, and develop trust and mutual confidence in the security posture of stakeholders across all sectors providing essential services, including transport. In addition to the NIS Directive, there are two other cross-sector EU laws which support harmonisation, namely the GDPR [21], which addresses the protection of personal data and privacy rights, and the Cyber-security Act [58], which has created a cyber-security certification framework for ICT[§] products, services and processes.

2.8.8 Secure information sharing

The sharing of security-related information among stakeholders, even within the same state and sector, is not yet well established, since organisations may be reluctant to share sensitive information which might reveal vulnerabilities in their systems or procedures, and which may be detrimental if revealed to competitors or potential attackers.

The availability of means for the secure sharing of information is another issue which constrains sharing security-related data in real time and from multiple sources to provide alerts and early warnings. The introduction of transport-related security operations centres (SoCs) and information sharing and analysis centres (ISACs) is a step towards addressing these issues.

Since security is a state responsibility, the sharing of security-related information between organisations in different states may also be problematic for reasons of national security and sovereignty. Organisational and national borders in highly interconnected systems tend to be invisible to potential adversaries, who will attempt to exploit the weakest link.

In an attempt to address such issues and facilitate information sharing in European aviation, the European Aviation Safety Agency (EASA) and a number of aviation stakeholders in the European Strategic Coordination Platform (ESCP) are developing methods to facilitate the sharing of risk information between aviation stakeholders.

The establishment of a common, harmonised level of security within a single transport mode, such as aviation, is only the first step. Transport modes are expected to evolve and integrate in the future to provide society with the benefits of Mobility as a Service (MaaS) [59] requiring the closer integration of their systems. Consequently, due to issues associated with the ubiquitous 'weakest link', it will be necessary to attain a harmonised level of security across all participating transport modes, ideally prior to the integration of their systems. This could be facilitated by the use of common methods, tools and guidance material for risk management across modes. However, the section which follows on regulations, standards, and guidance material reveals that this is unlikely to be the case in the foreseeable future, given the differing security postures of transport modes, their specific needs and the level of investment in currently applied approaches.

2.8.9 Handling security incidents

The widespread implementation of transport-specific security operations centres (SoCs) facilitates the response to incidents, addressing incident detection, identification, response and recovery. The sharing of information with other SoCs and CSIRTs

[§]ICT – Information and Communications Technology.

would also contribute to the resilience of the transport community. For example, the EATM-CERT run by EUROCONTROL [60] provides support to requesting states in Europe for aviation.

2.8.10 Security culture

Cultures are based on a set of shared, underlying assumptions about reality. This means that an organisation will display tangible behaviour that derives from what the organisation assumes should be most important to it.

An effective security culture is established upon an informed awareness of the threats to, and vulnerabilities of, the organisation. All staff need to believe that where there are credible threats to the facility, security measures matter. This underlying conviction then permeates the way people do their work, and it drives their behaviour under normal and abnormal conditions.

In a facility that enjoys an effective security culture, personnel will display a firmly held belief that there are credible insider and outsider threats, and that it is their responsibility, to various degrees, to help counteract these threats and vulnerabilities.

A cultural approach to protective security involves determining the attitudes and beliefs needed to be established in an organisation, how these attitudes and beliefs manifest themselves in the behaviour of assigned personnel, and how desirable attitudes and beliefs can be transcribed into reliable working methods to produce desired outcomes, in the form of effective protections.

2.9 Regulations, standards and guidance material

2.9.1 Introduction

This section looks at some of the regulations, standards and guidance material that are applicable to different transport modes, and addresses them at global, regional and national levels (Figure 2.4). The section first considers cross-modal guidance before considering the aviation, maritime, road and rail sectors.

2.9.2 Cross modal

2.9.2.1 International standards and guidance

Ensuring security requires a systematic approach applied to people, processes and technology. International standards comprise the experience of many organisations and provide proven, accepted means for organisations to efficiently and cost-effectively establish the required processes and structures to enhance security and provide a means to support the attainment of regulatory compliance.

There are currently no generic global regulations for transport systems. However, the ISO/IEC 2700X series of standards are applied in a wide variety of industries across all sectors, in particular in the area of IT. Where industrial control systems and industrial communication networks are more prevalent in OT environments such as rail and road transport, the ISA/IEC 62443 X-X series of standards are

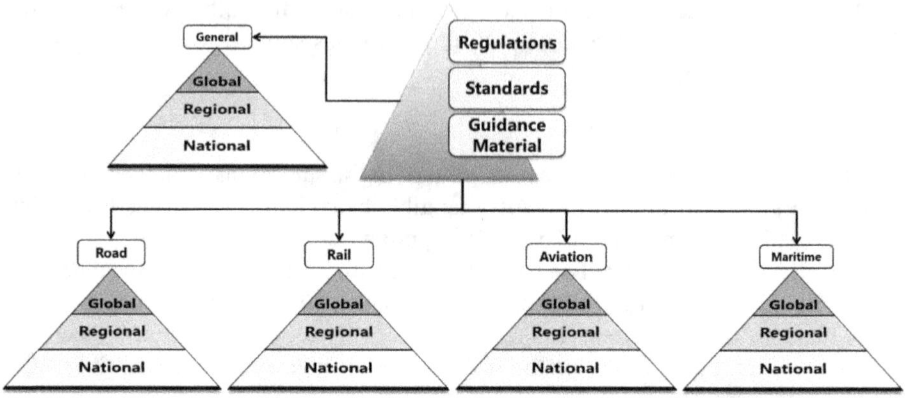

Figure 2.4 Regulations, standards and guidance material for transport modes

more commonly applied. The UITP (Union Internationale Des Transports Publics) industry body has also published general guidance for cyber security in public transport [61].

2.9.2.2 Regional regulations

A number of EU-wide regulations apply to all sectors, including transport, and have a substantial impact upon them. The most significant for transport security are now briefly described.

GDPR

The GDPR [21] is a European Union law that was implemented in 2018, which requires organisations to safeguard personal data and uphold the privacy rights of anyone in EU territory. The regulation includes seven principles of data protection that must be implemented and eight privacy rights that must be facilitated.

The GDPR replaced the 1995 Data Protection Directive, which created a country-by-country patchwork of data protection laws, and it unifies the EU under a single regime. It imposes obligations onto organisations globally if they process the personal data of EU citizens or residents, or offer goods or services to them. It also empowers member states' data protection authorities to enforce the GDPR with sanctions and fines. Violation of GDPR's privacy and security standards results in two tiers of penalties. Less severe infringements can result in a fine of up to €10 million, or 2% of an organisation's global revenue for the preceding financial year (whichever is higher). More serious infringements incur fines of either €20 million or 4% of global revenue (whichever is higher). Consequently, non-compliance can be costly, with potential penalties of tens of millions of euros. In addition, data owners have the right to seek compensation for damages.

A GDPR Enforcement Tracker website [62] provides an overview of fines and penalties which data protection authorities within the EU have imposed under the regulation. Not all fines are made public, however. A library of resources [63] is available to individuals and organisations, which provides an overview of GDPR and information on achieving compliance.

NIS Directive

The NIS Directive [57] is the first EU-wide legal framework on cyber security which addresses 'providers of essential services' in a range of sectors including energy, transport, banking, financial market infrastructures, health, water supply and distribution, and digital infrastructure. In terms of transport modes, this includes:

- In aviation, providers of essential services may include air carriers, airports and air navigation service providers (ANSPs) providing ATC services;
- In rail, infrastructure managers and railway undertakings;
- In water transport, this refers to inland, sea and coastal passenger and freight water transport companies, port management bodies and operators of vessel traffic services;
- For road transport, this refers to authorities responsible for road traffic management, including operators of intelligent transport systems.

The NIS Directive obliges EU member States to adopt a national NIS strategy and create a network of national CSIRTs, for which ENISA has the role of secretariat. It establishes security and notification requirements for digital service providers and operators of essential services. Since security is a national responsibility, each state applies its own criteria to designate providers of essential services. In support of these requirements, member states must also designate one or more National Competent Authorities (NCAs). The NCA monitors the application of the directive at national level.

A state must also designate a single point of contact to take part in a Cooperation Group, composed of state representatives, the European Commission and ENISA. ENISA's role is to provide support to states, share best practices, and provide advice and guidelines.

For example, in the United Kingdom, the Department for Transport is the competent authority for the transport sector, responsible for publishing guidance on risk management, inspecting operators of essential services, and for taking enforcement action if the directive is contravened [64]. The National Cyber Security Centre (NCSC) takes on the roles and responsibilities of the CSIRT and the single point of contact.

Notifications of security incidents with a significant impact on service provision should be provided to the NCA or CSIRT, and summary reports should be provided to the point of contact in the Cooperation Group. In addition, entities which fall outside the scope of the Directive can notify incidents of significant impact on a voluntary basis to the NCA or the CSIRT.

With some aspects of transport cyber security falling within the scope of the NIS policy, ENISA has acquired a mandate for the certification and oversight of ICT products, services and processes necessary for the functioning of providers of essential services in the air transport sector, namely air carriers, airport operators and ANSPs (Box 2.1).

Box 2.1 Transport entities and regulatory complexity

In this example, an aviation entity is used to illustrate regulatory complexity in transport. The Common Requirements Regulation (EU) No 2017/373 [10] lays down security requirements to be followed by Air Navigation Service Providers (ANSPs), with audits being carried out by the National Supervisory Authority (NSA) or the European Union Aviation Safety Agency (EASA).

However, if an ANSP is also designated as a provider of an essential service, it must comply with the NIS Directive and coordinate with the NCA* and/or the CSIRT† for the notification of incidents having a significant impact on service provision. It must also comply with the Cyber-security Act, so its products, services, and processes are subject to the European Cyber security certification framework via ENISA.

This results in a more complex governance framework for essential service providers, requiring the creation of additional roles and bodies, and resulting in additional coordination both within and outside member states. This situation is replicated in all transport modes.

*NCA – National Competent Authority (NIS Directive).
†CSIRT – Computer Security Incident Response Team (NIS Directive).

The ultimate goal in transport security is to achieve a common, minimum level of security across the system, and the regulations described seek to achieve such harmonisation in the future. However, it is worth being aware of issues which may impact the realisation of this goal:

- Since each state applies its own criteria based on NIS guidelines to designate providers of essential services, it is possible that comparable entities in different states may be designated differently;
- Each state applies its own processes and standards to comply with the Directive, and these may differ between states;
- In addition, each state determines the significance of a disruptive effect based on several cross-sectoral factors, so care should be taken to ensure that 'incidents with significant impact' are equivalent across the system.

The Cyber-security Act

The Cyber-security Act [58] created a European cyber-security certification framework for ICT products, services and processes, and reinforced ENISA by allocating it the role of developing the certification framework and coordinating the response to large-scale cross-border cyber attacks and crises. This is facilitated by ENISA also being the secretariat of the CSIRTs network.

This act also complements the NIS Directive, since an entity which is categorised by a state as being a 'provider of an essential service' is obliged to follow the requirements of the Cyber-security Act, while its products, services and processes are subject to the European cyber-security certification framework.

2.9.2.3 National standards and guidance

A variety of generic standards and guidance material are available at the national level for both IT and OT systems. A selection of references from the United States (NIST¶, NCCIC**), the United Kingdom (NCSC††), France (ANSII‡‡) and Germany (BSI§§) is provided in Table 2.1.

Table 2.1 National cross-sectorial standards and guidance material

Guidance	Content
(USA) NIST SP800-30	Risk Management Guide for IT Systems
(USA) NIST SP800-61	Computer Security Incident Handling Guide
(USA) NIST SP800-82 Rev 2	Guide to ICS Security
(USA) NIST SP800-160 Vol. 1	System Security Engineering – Considerations for a Multidisciplinary Approach in the Engineering of Trustworthy Secure Systems, November 2016
(USA) NCCIC, ICS-CERT S508C	Recommended Practice: Improving Industrial Control System Cyber security with Defence-in-Depth Strategies, September 2016
(USA) NIST Cyber-security Framework V1.1	Framework for Improving Critical Infrastructure Cyber security
(F)ANSII Cyber security for ICS	Classification Method and Key Measures
(UK) Cyber Essentials	Cyber-security Certification Scheme
(D)IT Grundschutz	IT Baseline Protection Manual (BSI)
(D)BSI-Standard 100–3	Risk Analysis Based on IT Grundschutz, Version 2.5, 2008
(D)Supplement to BSI-Standard 100–3	Application of the Elementary Threats from the IT-Grundschutz Catalogues for Performing Risk Analyses, August 2011
(USA) SANS Critical Security Controls	CIS Critical Security Controls for Effective Cyber Defense

EATM-CERT, European Air Traffic Management-Computer Emergency Response Team.

¶NIST – National Institute of Standards and Technology.

**NCCIC – National Cyber security and Communications Integration Center (US DHS).

††NCSC – National Cyber Security Centre.

‡‡ANSSI – Agence Nationale de la Sécurité des Systems d'Information.

§§BSI – Bundesamt für Sicherheit in der Informationstechnik.

2.9.3 Aviation

2.9.3.1 Global regulations, standards and guidance

Since the events of 9/11, ICAO has been extremely active in aviation security awareness, training, support, facilitation, and oversight, developing and maintaining 'Standards And Recommended Practices' (SARPs) and guidance material in these areas. Note that:

- A standard is mandatory unless a contracting state notifies ICAO that it is non-compliant, as the standard is considered necessary to facilitate and improve some aspect of international air navigation;
- A recommended practice is considered to be highly desirable to facilitate and improve some aspect of international air navigation. ICAO contracting states should endeavour to conform in accordance with the Convention.

The main ICAO documents applicable to ICAO contracting states in the area of Air Traffic Management (ATM) Security are shown in Figure 2.5 and briefly described below.

ICAO Annex 9 – security and facilitation

ICAO Annex 9 [65] embodies the SARPs and guidance material relevant to landside formalities for the clearance of aircraft and passengers, goods, and mail, i.e., for customs, immigration, public health and agriculture authorities. As such, it provides a frame of reference for planners and managers of international airport operations, describing the obligations of industry as well as the minimum facilities to be provided by governments. In addition, Annex 9 specifies methods and procedures for carrying out clearance operations in such a manner as to achieve compliance with States' laws, while enabling maximum productivity for the air transport operators, airports and government inspection agencies involved.

ICAO Annex 17 – security

Annex 17 [9] is the main document addressing SARPs for aviation security. This annex was first published in 1974 and has been amended on a regular basis. It was originally limited in scope to aircraft operators and airports. One requirement of Annex 17 was that participating states were required to establish a National Civil Aviation Security Programme (NCASP) and supporting structures. The applicability of the NCASP was initially limited to aircraft operators and airports, focusing primarily on hijacking and

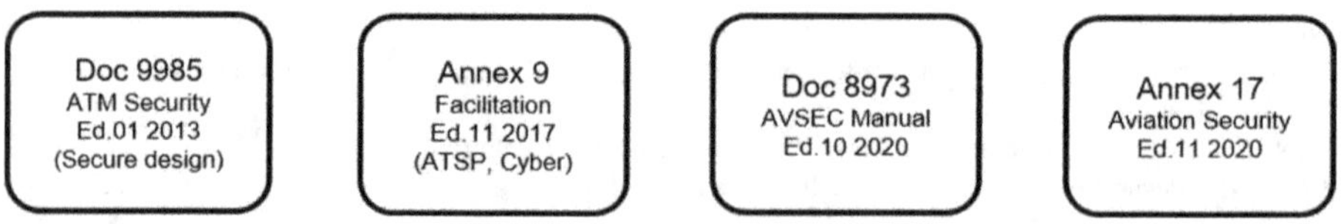

Figure 2.5 Global aviation regulations and guidance

bomb threats. However, following the use of aircraft themselves as weapons in the destruction of the New York City (USA) World Trade Centre on 11 September 2001, there was heightened awareness of the possibility of other types of security-related events, including the possibility of similar attacks, or attacks against air traffic service facilities and the navigation and surveillance infrastructure. In response to these concerns, the ICAO Council, on 17 November 2010, adopted Amendment 12 to Annex 17, requiring states to include ATM in the NCASP, and to ensure that ANSPs implement appropriate security provisions. This Amendment also addressed cyber security, and supply-chain security for the first time. Amendment 16 of 2018 included revised provisions on information sharing and cyber threats.

ICAO also provides two publications containing guidance material to assist contracting States in implementing the Annex 17 SARPs, which are briefly described.

ICAO Doc 8973 – aviation security manual

The ICAO aviation security manual [66] (often referred to as the AVSEC manual) is targeted primarily at airports and aircraft operators, and is regularly updated in line with Annex 17 amendments. For example, in response to Amendment 12 of Annex 17, the eighth edition of Doc 8973 introduced a new chapter (Chapter 18) on cyber security and additional guidance has been added since.

ICAO Doc 9985 – ATM security manual

The ATM security manual [67] complements the aviation security manual by providing specific guidance to ANSPs delivering ATM services on how to meet the Annex 17 requirements, and includes the following information:

- Guidance on security issues specific to ATM in order to assist states and ANSPs in implementing appropriate security provisions to meet
  - the requirements of the NCASP
  - national security and law-enforcement requirements
- Guidance on the protection of ATM systems, including
  - infrastructure protection
  - ATM security operations
  - (Cyber) security risk management
  - ICT controls.

The Universal Security Audit Programme (USAP) [68] is part of ICAO's 'Aviation Security Plan of Action', and provides for mandatory and regular audits of all ICAO contracting states. These audits assess a state's capability in security oversight and make recommendations for mitigating or resolving any weaknesses, with the first such audit taking place in November 2000. To promote transparency and mutual trust between states, information on the level of implementation of the critical elements of an audited state's aviation security oversight system is available to all ICAO Member States on a restricted website.

Table 2.2 Aviation-specific publications

Publication	Title
ICAO Annex 17	Safeguarding International Civil Aviation Against Acts of Unlawful Interference
ICAO Doc 8973	The Security Manual for Safeguarding Civil Aviation Against Acts of Unlawful Interference
ICAO Doc 9807	Universal Security Audit Program Continuous Monitoring Manual
ICAO Doc 9985	Air Traffic Management Security Manual
ICAO Doc 10047	Aviation Security Oversight Manual
IATA SeMS	Security Management System
IATA	Aviation Cyber-security Toolkit
CANSO	Cyber Security and Risk Assessment Guide

CANSO, Civil Air Navigation Services Organization; IATA, International Air Transport Association.

International industry organizations, such as IATA, representing airlines, and CANSO, representing ANSPs, also provide guidance material to their members to assist them in complying with ICAO SARPs.

Table 2.2 lists several global regulations, standards and guidance documents which are specific to aviation.

2.9.3.2 Regional regulations, standards and guidance

The Single European Sky (SES) initiative was launched in 2000 by the European Commission, following severe delays to flights in Europe in 1999. A High Level Group was established and, building on the recommendations in its report, the Commission drafted a legislative package at the end of 2001, which was adopted by the European Parliament and Council in March 2004 and entered into force one month later.

The goals of the SES package were to

- enhance safety and efficiency of air transport in Europe;
- reduce delays by improving the use of scarce airspace and airport resources;
- improve services and reduce cost to air transport passengers by reducing the fragmentation of the ATM in Europe;
- improve the integration of military systems into the European ATM system.

In spite of previous efforts to modernise and streamline Europe's ATM system, it remained fairly costly, but with high levels of safety. It was also hampered by heterogeneous working practices, and constrained by air route networks which, in the main, were still based on national borders rather than air traffic flows. The SES initiative attempted to put forward a legislative approach to solve these issues and enable ATM to cope with projected future traffic demand.

The legislative package adopted in 2004 comprised four basic regulations, which reinforced safety and fostered the restructuring of European airspace and air navigation services. These were complemented by more detailed implementing rules adopted by the European Commission, which, similarly to the guidance documents for ICAO Annex 17, address aviation security as two distinct streams, namely AVSEC and ATM/ANS.

The SES II Legislative Package

The introduction of Regulation (EC) No 216/2008 (EASA Basic Regulation) [69] permitted renewed legislative changes to come into effect with regard to the governance framework in aviation. The creation of EASA opened the door to new practices of certification and oversight, and provided the EU with an agent capable of exercising more direct control over aviation-related policy. However, ATM-related policy and security were not yet included in EASA's mandate, so in 2009 amendments were made to the first package of SES legislation. It was also necessary to adapt the SES legislation to achieve technical progress and to create a single safety framework in the European Community. This second legislative package included two notable regulations:

- Reg (EC) No 1070/2009 [70] addressed improving the performance of the European aviation system, and amended the four basic SES regulations of 2004
- Reg (EC) No 1108/2009 [71], extended EASA's remit to airports and ATM and Air Navigation Services (ANS), giving EASA competence in ATM-related policy and ATM security.

Aviation regulations in Europe may be considered to be represented by two separate streams with different regulatory requirements, one being the AVSEC stream, the other the ATM/ANS stream. The AVSEC stream is now described.

AVSEC

The first European AVSEC regulation (Reg (EC) 2320/2002) [72] was driven by the events of 9/11, and specified common basic standards for aviation security measures for airports (Figure 2.6). It addressed the following topics: airport security; aircraft security; passengers and cabin baggage; hold baggage; cargo, courier and express parcels; mail; air carrier mail and materials; air carrier catering stores and supplies; general aviation; staff recruitment and training; guidelines for equipment. The regulation was based on the recommendations of ECAC Doc 30 (European Civil Aviation Conference) [73] which was available at the time.

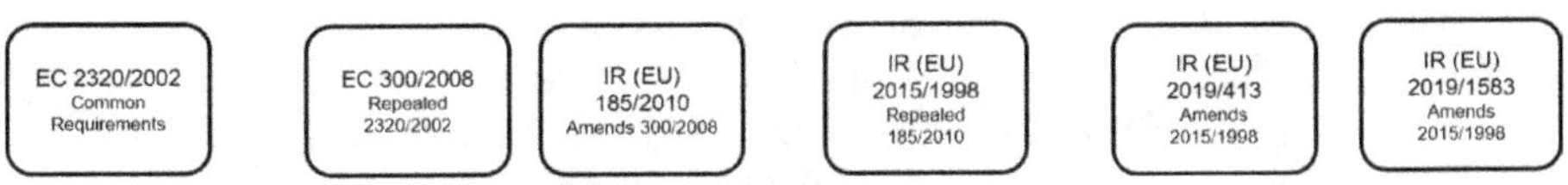

Figure 2.6 European AVSEC regulations and implementing rules

In 2008, the 2001 regulation was repealed by Reg (EC) 300/2008 [11], in order to simplify, harmonise, and clarify the rules while adding standards and mechanism for monitoring compliance. In 2010, the 2008 regulation was amended by Reg (EU) 185/2010 [74], which added measures for the screening of liquids, aerosols and gels. By 2015, Reg (EU) 185/2010 had been amended 20 times, and it was repealed by Reg (EU) 2015/1998 [75], which consolidated the 2008 act and all subsequent amendments.

Annex 17 amendments may result in new implementing rules coming into force. For example, IR (EU) 2019/1583 [76] resulted from Standard 4.9.1 (in Annex 17 Edition 10 Amendment 16), which placed new requirements on operators to identify critical information and communications systems, perform risk assessments, and implement measures to protect systems from unlawful interference.

In 2016, ENISA published a guidance document for airport decision-makers and information security professionals, policy makers, and industry, which identified the key assets of smart airports and presented a detailed analysis of threats potentially impacting smart components [77]. Smart components are defined as any networked ICT system that has a data processing capability ranging from aggregating simple data to extracting insights to support human decisions and/or triggering an automated response.

The second stream for aviation regulation in Europe (ATM/ANS) is now described.

ATM/ANS

Published in 2001, Reg (EC) 2096/2005 [78] 'The Common Requirements', was the first regulation to place security requirements on Air Navigation Service Providers (ANSPs) and Air Navigation Services (ANS) within Europe. Each ANSP was required to establish a security management system to ensure:

- the security of its facilities and personnel to prevent unlawful interference with service provision;
- the security of operational data received, produced or employed, maintaining its confidentiality.

The security management system required security risk assessments to be carried out, risks to be mitigated, monitoring to be in place, and continuous review and improvement to be established. The requirements also extended to the detection of security breaches, their containment, and means of response and recovery.

An update to the 'Common Requirements' (Figure 2.7) was published in 2011 [79], which permitted the option of an integrated management system (IMS) for safety, quality and security. This regulation was subsequently repealed in January 2020 by Reg. (EU) No 2017/373 [10], which added explicit requirements on information and cyber-security threats, and broadened the scope of security management requirements as follows:

Figure 2.7 European ATM regulations

- By adding the obligation for ANSPs to 'protect their systems, constituents in use, and data and prevent compromising the network against information and cyber-security threats which may have an unlawful interference with the provision of their service' – these topics were only implicit in the previous regulations;
- By extending the scope of the regulation to encompass the Network Manager, and air traffic flow managers.

Regulation (EU) 73/2010 [80] laid down requirements on the quality of aeronautical data and aeronautical information in terms of accuracy, resolution and integrity. However, this was repealed by Reg (EU) 2020/469 [81], which also amends parts of IR (EU) 2017/373 by placing requirements to ensure that data are not corrupted, to ensure the implementation of integrity assurance processes to mitigate potential data integrity risks, and to ensure that transferred data is subject to a suitable authentication process.

Cross-modal regulations

The National Supervisory Authority (NSA) performs the oversight task on aviation entities to ensure compliance with aviation-specific regulations. However, a number of European legal instruments also apply to aviation entities. For example, an entity which is designated as a 'provider of an essential service' will be subject to GDPR [21], the NIS Directive [57] and the Cyber-security Act [58]. This creates the following potential issues:

- Multiple applicable regulations may contain similar, overlapping security requirements;
- Evidence of compliance with security requirements may have to be provided to separate authorities;
- Security incidents may have to be reported to multiple authorities.

For the short term at least, IR (EU) 2019/1583 [76] (see Figure 2.6) aims to address such issues in the AVSEC stream by allowing for coordination by a single appropriate authority, which may address compliance with multiple security requirements by permitting alternative means of compliance using elements contained in other EU or national legislation. Longer term solutions are currently being developed by EASA in cooperation with aviation stakeholders.

The new EASA basic regulation

More recently, IR (EU) 2018/1139 [82] on 'Common Rules in the Field of Civil Aviation and Establishing EASA' repealed Reg (EC) No 552/2004 (Interoperability) [83], with one of its objectives being to contribute to establishing and maintaining a high uniform level of civil aviation security. Article 88, 'Interdependencies between civil aviation safety and security', recognised the interdependencies between safety, security and cyber security. The regulation modified EASA's mandate, allowing it to act on security and cyber-security issues, as long as safety-related matters are involved.

The need for rationalising and harmonising the complex regulatory framework for aviation in Europe resulted in EASA creating the European Strategic Coordination Platform (ESCP) with stakeholders from all aviation domains, to address cyber-security strategy, risk assessment methods, the sharing of risk information between organisations, and to address gaps and overlaps in the regulatory framework. The ESCP developed a Strategy for Cyber Security in Aviation [84] to make aviation an evolutionary cyber-resilient system, adopting a 'built-in' security approach and addressing security from a system's conception through its development, deployment and operations. In support of this development, EASA created two 'Rulemaking Tasks' (RMT), namely RMT.0648, to introduce new cyber-security provisions in aircraft certification specifications, and RMT.0720, to address cyber-security risks in other aviation domains, including ATM/ANS and aerodromes. The simplification of the regulatory framework will lead to the development of comprehensive guidance material and acceptable means of compliance (AMCs) for the security certification of aviation systems.

ECAC

The European Civil Aviation Conference (ECAC) is Europe's largest and longest standing aviation organisation consisting of 44 member states. Its mission is the promotion of the continued development of a safe, secure, efficient and sustainable European air transport system. ECAC plays a key role in supporting its member states as they address issues affecting the European civil aviation sector. Aviation security measures are described in ECAC Doc 30, Part II [73], including a policy statement adopted by all ECAC member states. It includes provisions for securing airports, aircraft, passengers, cabin baggage, hold baggage, cargo and mail, in-flight supplies and airport supplies. Provisions on in-flight security, ATM and cyber security and the management of threats and hijackings are also included.

EUROCAE

Several aviation-specific standards and guidance material have recently been developed by the European Organisation for Civil Aviation Equipment (EUROCAE¶¶) (see Table 2.3), addressing the security of aircraft and ATM systems, including

¶¶EUROCAE – European Organisation for Civil Aviation Equipment.

Table 2.3 Aviation-specific publications

Guidance	Content
EUROCAE ED201	Aeronautical information system security (AISS) framework guidance
EUROCAE ED202A	Airworthiness security process specification
EUROCAE ED203A	Airworthiness security methods and considerations
EUROCAE ED204A	Information security guidance for continuing airworthiness
EUROCAE ED205A	Process standard for security certification and declaration of ATM ANS ground systems security aspects for certification and declaration
CEN EN 16495	Air Traffic Management – information security for organisations supporting civil aviation operations
ICAO EUR Doc 22 AFS Security Guidelines	Guidelines for security requirements in the AFS network and COM centres in the EUR region
ECAC Doc 30 Part II	Policy statement on aviation security measures
EUROCONTROL	Security risk management toolkit
SESAR	ATM Security Guidance Material
ENISA	Securing smart airports
EASA Easy Access e-Rules for ATM/ANS	Rules, links to acceptable means of compliance, and guidance

communications, navigation and surveillance systems. This work has been carried out in cooperation with the RTCA*** in the United States.

EUROCONTROL

EUROCONTROL††† collaborates with several organisations to support the development of cyber-security regulations, standards and guidance material. At ICAO, it contributes to the Aviation Security (AVSEC) Threat and Risk Working Group (TRWG) and the Secretariat Study Group on Cyber Security (SSGC), and is an observer on the AVSEC Panel.

In Europe, it participates in the Stakeholders Advisory Group on Aviation Security (SAGAS) at the European Commission (EC), and participates in EASA's ESCP. It also participates in the ECAC Security Forum, the Guidance Material Task Force (GMTF) and the study group on cyber threats to aviation. EUROCONTROL also takes part in EUROCAE working groups, including WG-72 (Aeronautical Systems Security), developing security certification standards for both ground and

***RTCA – Radio Technical Commission for Aeronautics.

†††EUROCONTROL – The European Organisation for the Safety of Air Navigation.

airborne systems. It led the development of EUROCAE ED205A, which is elaborating ED205 (see Table 2.3), a process to assess the security of ATM/ANS ground systems.

EUROCONTROL is the home of the EATM-CERT, which supports EUROCONTROL services and products and ATM stakeholders in protecting themselves against cyber threats that could impact operational IT assets and data. To provide these services, the organisation collaborates with national and international ATM stakeholders, ATM manufacturers, sectoral and national CERTs, EASA, ENISA, ISACs, Europol and others. It is also involved in cyber-security awareness building and training, delivering courses at EUROCONTROL's Aviation Learning Centre in Luxembourg, and providing tailored cyber-security workshops to ANSPs, civil aviation authorities, and national supervisory authorities in member states.

2.9.3.3 National regulations, standards and guidance

Contracting states of ICAO must comply with the ICAO SARPs described earlier. Given the large number of government departments and agencies likely to be involved in aviation security activities within a state, ongoing coordination between key players is needed. This is achieved by means of a National Civil Aviation Security Committee (NCASC) or similar arrangements. The NCASC comprises senior government officials and senior representatives of the aviation industry, the latter acting as consultants to government. Ideally, NCASC meetings should take place at least twice a year.

Each state must also develop a 'National Civil Aviation Security Training Programme' (NCASTP) to provide a framework for the selection and training of staff involved in aviation security by stipulating the various responsibilities for performing this task. The NCASTP also spells out the respective training-related responsibilities of all entities involved in, or overseeing, security measures described in the NCASP.

States should also monitor the various entities to ensure that their security policies and standards are being adequately implemented, and this is achieved by means of a National Civil Aviation Security Quality Control Programme (NCASQCP).

ICAO conducts audits under the Universal Security Audit Programme (USAP) [68], and classified results are available to authorised users in the ICAO USAP website, sorted both by region and by audit area. In addition, graphical representations of the implementation status of the critical elements, as well as significant security concerns identified, are available. States audited under the second cycle of the USAP are provided on the USAP secure website, to which access is limited to authorised personnel and Member States. ICAO adheres to the principle of maintaining strict confidentiality of all state-specific information derived from USAP audits. If such information is requested by another state, however, states should ideally share and advise ICAO that sharing has taken place.

Table 2.4 Aviation-specific guidance material

Guidance	Content
Academie de L'Air et de L'Espace, Dossier #45, Jan 2019	Cyber threats targeting air transport
Aviation Cyber-security Guidelines	CAA and Ministry of Transport and Communications, Qatar, April 2019

EATM-CERT, European Air Traffic Management-Computer Emergency Response Team.

States may also have to comply with regional requirements. For example, EU states must comply with the SES regulations described earlier, as well as cross-sectorial requirements, such as those prescribed by the NIS Directive.

At the national level, there appear to be few publicly available cyber-security publications for aviation, perhaps due to the sensitivity of such material. However, Table 2.4 provides examples of useful guidance from France and Qatar.

2.9.3.4 Observations

In aviation, cyber-security regulations have historically evolved in two distinct streams, one for the AVSEC community and one for ATM/ANS.

The guidance material associated with each stream has also evolved separately, and guidance documents have also been developed by aviation stakeholder groups such as IATA[‡‡‡] and CANSO[§§§] to interpret the regulations in such a way as to address the particular needs of their communities.

Europe-wide legal instruments such as the NIS Directive, GDPR, and the Cyber-security Act have resulted in a more complex governance framework for aviation cyber security, but several initiatives are ongoing to address this.

2.9.4 Maritime

2.9.4.1 Global standards

The International Maritime Organisation (IMO) is the United Nation's regulatory body responsible for the safety of life at sea (SOLAS) and environmental protection. The IMO has adopted a large number of conventions and regulations since its inception in 1959, covering a broad range of maritime topics. These topics include collision prevention, emissions, crew training and ship/port security. The events of 9/11 in New York in 2001 exposed gaps in aviation security triggering a review of aviation security practices. The events also raised the question of the vulnerability of ships, and of the possibility of shipping being used as a vector of terrorist activity. Consequently, in November 2001, IMO Assembly resolution A.924(22)[¶¶¶] was made

[‡‡‡] IATA – International Air Transport Association.

[§§§] CANSO – Civil Air Navigation Services Organisation.

[¶¶¶] On the review of measures and procedures to prevent acts of terrorism which threaten the security of passengers and crews and the safety of ships.

to reduce risks to vessels and their cargo, passengers, crews and port personnel, and to enhance the security of ships and ports to minimise the possibility of shipping becoming a target of international terrorism.

The International Convention for SOLAS is an international maritime treaty which sets minimum safety standards in the construction, equipment and operation of merchant ships. The convention requires signatory flag states to ensure that ships flagged by them comply with at least these standards. As a result of the adoption of resolution A.924(22), a Diplomatic Conference on Maritime Security (the 2002 SOLAS Conference) was held, at which a number of amendments to SOLAS were adopted. The most far-reaching amendment enshrined the new 'International Ship and Port Facility Security' (ISPS) Code. Part A of the ISPS Code contained detailed, mandatory security-related requirements for governments, port authorities, and shipping companies, while Part B contained recommendations on how to implement those requirements.

The first edition of the *Guide to Maritime Security and the ISPS Code* [85] was published in 2012 to assist member states in implementing a comprehensive technical cooperation programme to support them in achieving compliance with the Code. The threats considered in the ISPS Code are primarily physical in nature, so gaps remained regarding the topic of cyber security.

Alongside the ISPS Code, the IMO ratified the *International Safety Management (ISM) Code* [86]. Investigations into several accidents revealed major errors from poor management standards in shipping. The ISM Code establishes safety-management objectives and the requirement for a safety management system (SMS). The SMS outlines who within a company is responsible for operating and ensuring compliance with safety standards of a vessel. The ISPS and ISM codes include security requirements under which the IMO will target maritime cyber security.

The last decade has seen an increased discussion on maritime cyber security by the international community. Whilst the latest IMO resolution MSC.428(98) (Maritime Safety Committee) is a broad attempt to include cyber security in management practices, the industry is still a long way from mandatory cyber-security regulations. However, as can be seen in Figure 2.8 and Table 2.5, there have been advancements in maritime cyber-security guidance in the last few years.

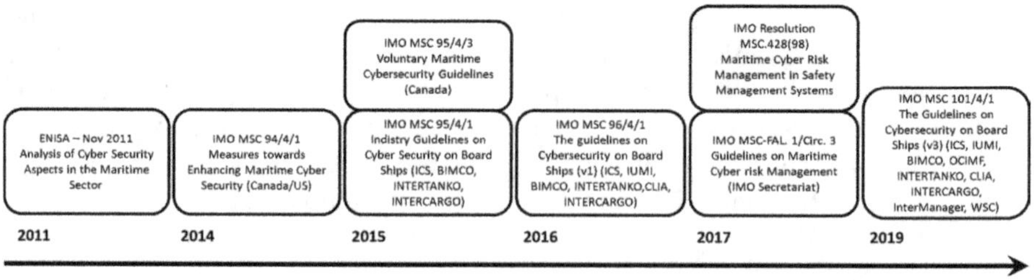

Figure 2.8 Global maritime regulations and standards

Table 2.5 Global maritime regulations and guidance material

Guidance	Content
IMO MSC 94/4/1, 2014	Measures towards enhancing maritime cyber security
IMO MSC 95/4/1, 2015	Industrial guidelines on cyber security on board ships
IMO MSC 95/4/3, 2015	Voluntary maritime cyber-security guidelines
IMO MSC 96/4/1, 2016	The Guidelines on Cyber Security on Board Ships (v1)
IMO MSC-FAL 1/Circ. 3, 2017	Guidelines on Maritime Cyber Risk Management
IMO MSC 101/4/1, 2019	The Guidelines on Cyber Security on Board Ships
IMO Resolution 428(98), 2017	Maritime Cyber Risk Management in Safety Management Systems

IMO MSC 94/4/1 – measures towards enhancing maritime cyber security

This document [87], submitted by Canada and the United States, marks the start of maritime cyber-security discussions within the IMO. While providing high-level recommendations, the document highlights that the increased dependence on cyber technology and the lack of expertise in its security have led the industry to be vulnerable to attack. The document proposes that the IMO, through the Maritime Safety Committee, should create voluntary cyber-security guidelines that provide information regarding potential vulnerabilities and risks within the industry. It also proposes that these guidelines should be designed to support the objectives of the ISPS Code, and address cyber security within ports.

IMO MSC 95/4/3 – voluntary maritime cyber-security guidelines [88]

At the following meeting of the Maritime Safety Committee, Canada submitted a second document appealing to the international community for voluntary guidelines addressing maritime cyber security. The document argues that for effective cyber-risk management, it must be implemented through the ISPS Code and SOLAS. The document highlights that amending these core international frameworks is a lengthy process, thus the industry requires interim guidance on potential cyber risks and mitigation practices. The document also appealed for the creation of plain-language voluntary guidance and to ensure that there is the right balance between maritime security and the facilitation of international maritime trade.

IMO MSC 95/4/1 – industry guidelines on cyber security on board ships [89]

Submitted by an agglomeration of commercial interests with consultative status within the IMO, this document argued that to mitigate maritime cyber-security risks requires a risk management approach. An approach of identifying, prioritising and responding to threats comes at the price of potentially modifying business processes. MSC willingly delayed further discussion of cyber-security issues until the first version of industry guidelines were published the following year, as they represent the views of the commercial interests within the industry.

IMO MSC 96/4/1 – the guidelines on cyber security on board ships – version 1

The following year, a collaboration of shipowners, cyber-security specialists, communications experts, producers of satellite and radio communications equipment, and insurance experts produced the industry guidelines on cyber security onboard ships [90]. The recommendations within the document focus on the distinctive issues of cyber security on board ships, while acknowledging that not all the content will be applicable for every ship. It provides a guide to established procedures, plans and instructions for key onboard operations that organisations can use as a blueprint to develop a better cyber-security posture.

The document provides a high-level structure to the proposed guidelines, dividing the document into eight key themes:

1. Awareness and education for all stakeholders
2. A generic risk-based framework drawing on existing standards and guidelines augmented by current intelligence and best practice
3. Cyber systems addressing their integrity, confidentiality and availability
4. Establish clear guidelines on the management of key information in order to retain operational cyber capability
5. How to integrate elements of both physical and software security to ensure safety and business continuity
6. Importance of identifying and mitigating third-party interfaces that could compromise cyber security
7. The consideration of cyber-security monitoring systems and network management
8. Development of contingency plans

IMO MSC-FAL. 1/Circ. 3 – guidelines on maritime cyber risk management

In 2016, the interim guidelines on maritime Cyber Risk Management were proposed (MSC. 1/Circ. 1526), and were superseded in 2017 by MSC-FAL. 1/Circ. 3. These guidelines [91] mark the first cyber-security guidance published by the IMO secretariat. The document argues that it is difficult to address cyber security through technical means only. Therefore, a number of potential control options should be developed and form part of an organisation's existing risk management practices. The document also highlights the importance of IT, OT and their interaction.

IMO resolution MSC.428(98)

This MSC resolution [92] marks the formal acknowledgement by the IMO that cyber risk management should be included within a ship's safety management systems. The International Management Code for the 'Safe operation of Ships and for Pollution Prevention' stipulates that an organisation is responsible for ensuring its management practices promote a safe working environment on board a vessel. As the safe operation of a vessel is now reliant upon digital infrastructure, these systems now fall under the remit of the ISM safety management practices. The resolution encourages administrations to ensure that cyber risks are appropriately addressed within the company's Document of Compliance by 1 January 2021.

IMO MSC 101/4/1 – the guidelines on cyber security on board ships – version 3

Replacing both Version 1 (2016) and Version 2 (2017) guidelines, this document [93] represents the third iteration of what are known within the maritime community as the 'BIMCO guidelines'. Much like the previous two versions, this document acts as a 'living' reflection on developments in both threats and protection measures. This version provides the authoritative guide on compliance with IMO resolution MSC.428(98), and the implementation of cyber risk management into an organisation's security frameworks. It also provides more use case examples, highlighting some of the types of cyber threats that the maritime community face.

What the future holds…

The future of international maritime regulation is currently unclear. This lack of clarity is due to several key issues within the fragmented governance framework of the maritime sector. First, the IMO is very well versed in regulating and producing guidance for vessels; however, it has few capabilities in ports and associated infrastructures. Second, whilst Resolution MSC.428(98) highlights the obligation to include cyber risk management within the ISM Codes remit, it does little to stipulate how, or what this means. Therefore, it is open to an organisation's interpretation to decide what a cyber threat is, and how to mitigate it.

Furthermore, there has been little discussion within the IMO on how these cyber risk management practices are to be inspected and enforced. Many risks within the maritime space have seen little change over the last few decades, thus there are internationally agreed formal and informal practices and procedures to mitigate those risks. However, with the infinite number of differing integrations of cyber-enabled technology creating a standard set of mitigations is challenging. Therefore, where an awareness training course may be appropriate for one organisation or crew, it may not be appropriate for another.

Many of the major classification societies are now offering some form of 'cyber' notation on a ship's safety certificate. This notation signifies that the vessel has met a certain standard of cyber security as defined within the societies' rules. If a ship gains this notation, it could be used as a way of proving compliance to MSC.428(98). However, they are currently voluntary, so do not offer much in the way of enforcement. There are benefits for operators gaining this notation as it could reduce their insurance premiums, or provide a competitive edge in the market.

Arguably, adequate uptake of cyber risk management, and subsequent cyber risk assessment, into an organisation's security management systems should lead to increased breadth of knowledge about the key risks posed by cyber-enabled technology. This information could then be used to develop guidance and regulation that targets threats generic to the whole sector, as well as address those found only in specific subsets, such as LNG tankers or cruise ships. This information [94] could be used to develop a 'Cyber Code'. Like the Polar Code****, this would be a collection of regulations and practices underpinned by IMO Conventions to provide a framework for the enforcement of requirements to specific systems/practices.

Regardless of approach, one element that will most certainly be present in the future of maritime cyber-security regulation is the community. Successful and effective cyber-security regulation requires the community to discuss, create, and then enforce it. Therefore, through the local, regional, national, and international groupings of maritime communities, regulation and practices will be formed that address the dynamic issue of maritime cyber security.

2.9.4.2 Regional regulations and standards

ENISA's analysis of cyber-security aspects in the maritime sector [95] marked one of the first documents by a supra-national organisation highlighting the importance of maritime cyber security. The document highlights the lack of holism in maritime cyber-security practices, which is ingrained within the fragmented governance structure of the industry. The document argues that due to the limited examples of successful cyber incidents, there is generally a low awareness of the threats that insecure technology poses to the industry. The document recommends that awareness-raising campaigns should start the process of moving towards the development of standards and enforceable regulation. These standards and regulations should be designed to ensure they help the industry achieve a good level of cyber security.

Additional applicable publications in the European region are shown in Table 2.6.

2.9.4.3 National regulations and standards

At the national level, there appear to be few cyber-security publications for maritime, with the bulk of available publications addressing physical security, such as those in Table 2.7.

Table 2.6 Regional maritime regulations and guidance material

Publication	Content
Reg. (EC) No 725/2004	On enhancing ship and port facility security
Directive 2005/65/EC	On enhancing port security

BIMCO, Baltic and International Maritime Council.

****International code for ships operating in polar waters (Polar Code).

Table 2.7 Maritime-specific regulations

Regulation	Content
The Ship and Port Facility (Security) Regulations 2004 (UK) SI 1495	Merchant shipping, maritime security
The Ship and Port Facility (Security) (Amendment) Regulations 2005 (UK) SI 1434	Merchant shipping, maritime security

2.9.4.4 Observations

In the maritime space, the IMO has delivered regulations and guidance material for vessels, but there are gaps with respect to ports and associated infrastructures. In addition, while there exist obligations with respect to cyber risk management, there is minimal information available on how this is to be achieved.

In terms of governance, there are gaps in the area of oversight and how risk management practices are to be inspected and enforced.

2.9.5 Rail

2.9.5.1 Global standards and guidance

The International Union of Railways, UIC,[††††] is the worldwide railway organisation, with goals to develop the coherence of the rail system at a global level, develop strategies and initiatives to improve business performance and increase rail transport investment, and to execute and manage projects and other activities on non-commercial issues, including research, development and technical efficiency.

The UIC provides high-level guidance for rail network security [96] (see Figure 2.9), which attempts to frame the required security policies and measures in relation to a spectrum of differing demands and needs, both nationally and internationally. The document includes a risk assessment and planning framework, information on threats, and describes the essence of a Railway Security Management System.

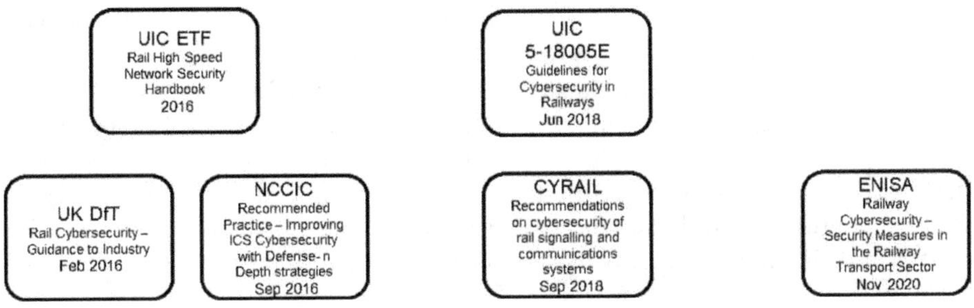

Figure 2.9 Rail guidance material

[††††]https://uic.org/.

Table 2.8 Global rail guidance

Publication	Content
UIC ETF	Rail High Speed Network Security Handbook, 2016
UIC 5-18005E	Guidelines for Cyber security in Railways, June 2018
ISA/IEC 62443 series	Security for Industrial Automation and Control Systems

Guidelines for Cyber Security in Railways was published by the UIC in 2018 [97], and it addresses areas such as signalling and telecommunications in some detail, taking into account system design in evaluating security requirements. The guidelines use best practices applied in other industries, including aeronautics and energy. The document applied the ISMS [‡‡‡‡] approach described in ISO/IEC 27001 [8].

Due to the ubiquity of IACS[§§§§] in rail, the ISA/IEC 62443 group of standards (Table 2.8) is widely referenced and applied within the sector, providing a flexible framework to address and mitigate current and future security vulnerabilities in this area.

2.9.5.2 Regional regulations and standards

The Shift2Rail Joint Undertaking (S2RJU) in Europe supports railway research and innovation in Europe and published recommendations on cyber security of rail signalling and communications systems which was delivered by the project 'Cyber security in the RAILway sector' (CYRAIL[¶¶¶¶]) [98]. This guidance has also been endorsed by the UIC.

The NIS Directive [57] applies to rail, meaning that EU member states must transpose the requirements of the directive into national legislation applicable to the rail transport mode. In Europe, ENISA has a mandate to assist member states in effectively preventing and responding to cyber attacks in the region, including meeting the NIS Directive requirements. In this role, ENISA conducted a study in 2020 to assess the status of security measures in the rail industry [99]. While cyber-security basics were implemented, measures requiring cyber-security expertise had a lower level of implementation, particularly those involving OT and legacy systems. The outdated or obsolete nature of many OT systems and their physical spread across the network also adds to the difficulty of being in control of cyber security. A lack of cyber-security expertise is a major concern, since the sector is undergoing digital transformation which increases the need. Railway suppliers with disparate technical standards and cyber-security capabilities add to the complexity, as does low cyber-security awareness and the absence of a security culture. ENISA has teamed with railway stakeholders to address these issues, and the document proposes a set of

[‡‡‡‡] ISMS – Information Security Management System.

[§§§§] IACS – Industrial Automation and Control Systems.

[¶¶¶¶] www.cyrail.eu.

Table 2.9 Regional rail guidance material

Publication	Content
(EU) Cyber security in the RAILway sector (CYRAIL)	Recommendations on cyber security of rail signalling and communication systems [98], Shift2Rail, September 2018
ENISA, Railway Cyber security	Security Measures in the Railway Transport Sector, November 2020
CENELEC Technical Specification 50701	Railway Applications – Cyber security (In development)

29 'Minimum Security Measures' classified in four domains, namely 'Protection', 'Defence', 'Resilience' and 'Governance and Ecosystem', the implementation of which is being monitored by member states.

The ISA/IEC 62443 series of standards, widely applied in Industrial Automation and Control Systems (IACS) in general, are becoming the de facto standard in Europe. However, some adaptations are desirable for the rail environment. As a result of working in collaboration with S2RJU, CENELEC***** is developing Technical Specification 50701 (Railway Applications – Cyber security) [100], which is based on the ISA/IEC 62443 standards but is tailored for the rail industry. The ISA/IEC 62443 standards describe four levels of network and system criticality, while the upcoming CENELEC TS-50701 proposes five levels of criticality, on the basis of the increased complexity of rail networks relative to conventional automation networks.

Table 2.9 lists the guidance material described above which is available for the rail sector within the European region.

2.9.5.3 National regulations, standards and guidance

In Europe, national governments have sovereignty over security matters in rail; however, the requirements of the NIS Directive have been transposed into the national laws of EU member states and apply to the rail industry. Examples of nationally developed standards and guidance material for Industrial Control Systems (ICS) and rail cyber security are provided in Table 2.10.

The UK government's high-level guidance document for Rail Cyber security [101] (see Figure 2.10) is designed to support the rail industry in reducing its vulnerability to cyber attack. It describes good practice principles and a general approach to cyber security but does not provide detailed instructions. It describes protecting infrastructure and rolling-stock systems, and incorporates advice for railway-specific systems.

*****CENELEC – European Committee for Electrotechnical Standardization.

Table 2.10 National rail publications

Publication	Content
(UK) Rail Cyber Security – Guidance to Industry	UK Department for Transport, February 2016
(F) Cyber security for Industrial Control Systems	Managing Cyber security for Industrial Control Systems, ANSSI, June 2012
(F) Cyber security for Industrial Control Systems	Classification Methods and Key Measures, Version 1.0, ANSSI, January 2014
(F) Cyber security for Industrial Control Systems	Detailed Measures, Version 1.0, ANSSI, January 2014
(F) Cyber security for Industrial Control Systems	Use Case, Version 1.0, ANSSI, June 2012
(USA) NCCIC, ICS-CERT S508C	*Recommended Practice: Improving Industrial Control System Cyber security with Defense-in-Depth Strategies*, September 2016
(USA) APTA SS-CCS-RP-001–10 Recommended Practice	Securing Control and Communications Systems in Transit Environments, *Part I: Elements, Organisation and Risk/Assessment Management,* July 2010
(USA) APTA SS-CCS-RP-002–13 Recommended Practice	Securing Control and Communications Systems in Rail Transit Environments, *Part II: Defining a Security Zone Architecture for Rail Transit and Protecting Critical Zones,* June 2013
(USA) APTA SS-CCS-WP-003–15 White Paper	Securing Control and Communications Systems in Rail Transit Environments, *Part IIIa: Attack Modelling Security Analysis White Paper,* April 2015
(USA) APTA SS-CCS-RP-004–16 Recommended Practice	Securing Control and Communications Systems in Rail Transit Environments, *Part IIIb: Protecting the Operationally Critical Security Zone,* October 2016

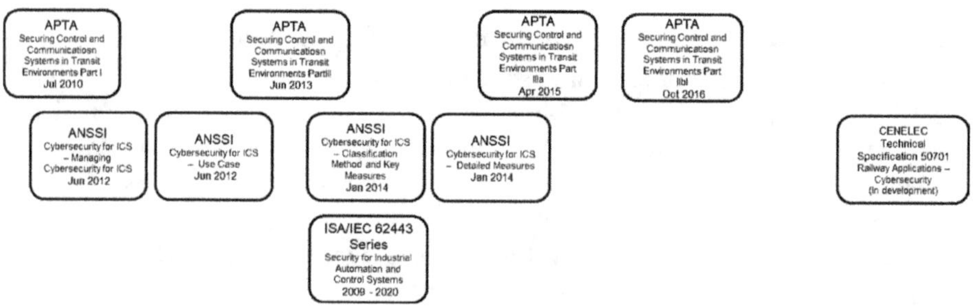

Figure 2.10 Rail standards

In France, ANSSI is responsible for the state's digital security strategy [45] and its enforcement, and has published a series of four guidance documents (see Figure 2.10) on the topic of 'Cyber security for Industrial Control Systems', the first of which is a high level introduction to the topic [102]. A supplementary use case document [103] presents situations that pose a risk for businesses and provides pertinent recommendations to secure the systems. 'Classification Method and Key measures' [104] is a general standard which recognises three classes of ICS, and provides cyber-security recommendations depending on this classification. Class three relates to an ICS in which the impact of an attack is critical, which is the class allocated to railway switch automation. The 'Detailed Measures' document [105] provides technical and organisational controls to be applied for the protection of ICS.

In the United States, the NCCIC[†††††] document 'Recommended Practice: Improving Industrial Control System Cyber security with Defence-in-Depth Strategies' [106] is a general ICS guide, which includes rail in its target industries. It comprises a collection of recommendations for ICS security programmes on topics including risk management, security controls and technologies, physical security, and training and awareness. It also describes attack scenarios and the limitations of security technologies, providing specific recommendations.

Also in the United States, APTA[‡‡‡‡‡] has published a number of rail transit system standards which address the securing of the control and communications systems. The first document [107] describes transit system elements, how to organise a control system security program, and risk assessment and management. The next document in the series [108] addresses topics which include defence-in-depth, risk zones, minimum controls and implementing controls in safety-critical zones. A white paper on attack modelling security analysis is provided [109], along with recommended practices for protecting the operationally critical security zone [110].

2.9.5.4 Observations

There is currently little in the way of global standards and regulations for the rail industry, but there are many standards and guidance documents available for IACS in general, and some have been specifically adapted for rail.

The first regional standard specifically developed for rail cyber security is CENELEC TS-50701 (Railway Applications – Cyber Security), which is based on the ISA/IEC 62443 standards for IACS.

2.9.6 Road

In road transport, there has been substantial activity in the development of policy and guidance material on ensuring the cyber security of Connected and Automated Vehicles (CAVs). A small number of standards are available, the most important of which was published in 2021. The first CAV-specific cyber-security regulations were published in 2020.

The following sections describe developments in policy and guidance material, standards and regulation for CAVs.

[†††††]NCCIC – National Cyber security and Communications Integration Centre.
[‡‡‡‡‡]APTA – American Public Transportation Association.

2.9.6.1 Guidance material

US DOT

Between 2016 and 2020, the US Department of Transportation (DOT) published four policy and guidance documents (see Table 2.11) in the area of cyber security for automated vehicles (Figure 2.11).

The DOT NHTSA[§§§§§] first published non-binding guidance for industry to improve automotive vehicle cyber security in 2016, with the publication of 'Cyber-security best practices for modern vehicles' [111]. This presented voluntary best practices based on the use of the NIST Cyber-security Framework [112], and promoted a risk-based approach to the protection of safety-critical vehicle control systems and PII. It addressed the timely detection and rapid response to vehicle cyber-security incidents in the field, and recommended designing-in measures to facilitate rapid recovery from incidents.

The following year, the document 'A Vision for Safety: Automated Driving Systems 2.0' [113] was published, which again recommended the use of the NIST Cyber-security Framework and other established cyber-security guidance to support the safe testing and integration of automated driving systems.

In 2018, AV3.0 [114] introduced guiding principles for automated vehicle innovation and extended the scope to all surface transportation modes. It described the US DOT's strategy to address existing barriers to potential safety benefits and

Table 2.11 National road publications

Publication	Content
USDOT AV1.0, 2016	Cyber-security Best Practices for Modern Vehicles
USDOT AV2.0, 2017	A Vision for Safety: Automated Driving Systems
USDOT AV3.0, 2018	Preparing for the Future of Automation
USDOT AV4.0, 2020	Ensuring American Leadership in AV Technology
UK DfT, 2017	The Key Principles of Cyber Security for Connected and Automated Vehicles

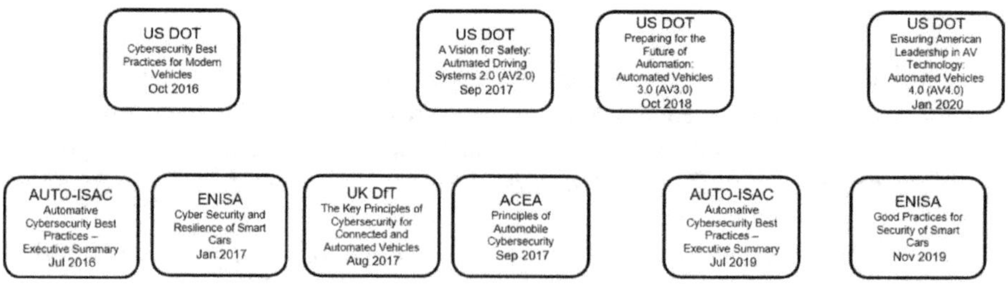

Figure 2.11 Guidance material

§§§§§NHTSA – National Highway Traffic Safety Association.

progress. Concerns about the safety, security and privacy of automated technology result in cyber security being a key topic throughout the document.

In January 2020, AV4.0 'Ensuring American Leadership in Automated Vehicle Technologies - AV4.0' was published [115]. This describes the approach of the US government to support stakeholder collaboration in automated vehicle development, and to deter, detect, protect, respond and safely recover from known and evolving risks. Topics addressed include security, cyber security, privacy and data security. Modernising regulations and ensuring the use of consistent policies, standards, and best practices, alongside conducting research to promote a layered approach to cyber security across all domains of the transportation system.

The DOT guidance documents clearly recognise the importance of risk-based approaches, designing-in security, and incident management and information sharing between stakeholders to address safety, privacy and data security concerns. The multi-mode nature of transport in the future is another key topic.

AUTO-ISAC

The Automotive Information Sharing and Analysis Center (Auto-ISAC) was formed in August 2015 by car manufacturers to establish a global information sharing community to address vehicle cyber security risks. It operates a central hub for sharing, tracking, and analysing intelligence about cyber threats, vulnerabilities, and incidents related to connected vehicles. The Auto-ISAC released a Best Practices Executive Summary in July 2016, which provided a high-level overview of the available best practices. A new version was published in 2019 [116] (Figure 2.11). The Best Practices provide implementation guidance for the seven key cyber-security functional areas identified in the Executive Summary. The Auto-ISAC may periodically add a guide for a new functional area to address the evolving vehicle cyber risk landscape. The functional areas currently addressed are the following:

1. incident response
2. collaboration and engagement with appropriate third parties
3. governance
4. risk assessment and management
5. awareness and training
6. threat detection, monitoring and analysis
7. security development life cycle.

ENISA

In 2016, ENISA performed a study on smart car security issues which resulted in the document 'Cyber Security and Resilience of Smart Cars' [117]. The objective of the study was to identify good practices that ensure the security of smart cars against cyber threats, in particular to ensure that vehicle security would also guarantee safety. The study lists the sensitive assets present in smart cars, as well as the corresponding threats, risks, mitigation factors and possible security measures to implement. To obtain this information, experts in the fields and areas related with smart cars were contacted to gather their know-how. These exchanges led to the publication of the report 'Good Practices for the Security of Smart Cars' in 2019 [118], which addressed policy and standards, organisational measures, and Security functions.

The report took stock of existing standardisation, legislative and policy initiatives for automated vehicles, to serve as a reference point to promote cyber security for connected and automated cars across Europe, raising awareness of relevant threats and risks. The focus of the report is on 'cyber security for safety', and it defines good practices for the security of connected and (semi-) autonomous vehicles, providing added-value features in order to enhance the car user experience and to improve safety.

The study provides the following:

- High level reference model of connected and autonomous vehicles;
- Detailed asset and threat taxonomy for the connected and autonomous vehicles ecosystem;
- Concrete and actionable good practices to improve the cyber-security posture of connected and autonomous vehicles;
- Mapping to existing legislative, standardisation and policy initiatives to foster harmonisation.

UK DfT and CPNI

In 2017, in order to provide all stakeholders with a consistent set of high-level guidelines, the UK DfT[¶¶¶¶¶] collaborated with CPNI[******] to produce 'The Key Principles of Cyber Security for Connected and Automated Vehicles' [119], for use in the automotive sector, CAV and Intelligent Transport System ecosystems, and their supply chains. The targeted parties include those in the manufacturing supply chain, from designers and engineers, to retailers and senior-level executives. The eight key principles are the following:

1. Organisational security is owned, governed and promoted at board level;
2. Security risks are assessed and managed appropriately and proportionately, including those specific to the supply chain;
3. Organisations need product aftercare and incident response to ensure systems are secure over their lifetime;
4. All organisations, including subcontractors, suppliers, and potential third parties, work together to enhance the security of the system;
5. Systems are designed using a defence-in-depth approach;
6. The security of all software is managed throughout its lifetime;

[¶¶¶¶¶]DfT – Department for Transport.

[******]CPNI – Centre for the Protection of National Infrastructure.

7. The storage and transmission of data are secure and can be controlled;
8. The system is designed to be resilient to attacks and respond appropriately when its defences or sensors fail.

The principles require security to be owned and driven at board level, to be risk-based and designed-in, for the security of software and transmitted data to be ensured, for the system to be resilient to attack, and for external stakeholders to be involved in achieving these goals.

The European Automobile Manufacturers Association

In 2017, European OEMs published another set of cyber-security principles through the European Automobile Manufacturers Association (ACEA), entitled 'Principles of Automobile Cyber security' [120]. ACEA identified six key principles to enhance the protection of connected and automated vehicles against cyber threats:

1. cultivating a cyber-security culture;
2. adopting a cyber-security life cycle for vehicle development;
3. assessing security functions through testing phases: self-auditing and testing;
4. managing a security update policy;
5. providing incident response and recovery
6. improving information sharing amongst industry actors.

All ACEA members endorsed the principles, which are broad-based and support the implementation of a holistic approach to cyber security within the automotive industry.

2.9.6.2 Standards

Due to the rapid pace of development in road transport, domain-specific standards and recommended practices are still evolving. Due to the mix of technologies employed in vehicle systems and similarities to industrial control systems, the standards shown in Table 2.12 could be applicable to road transport and CAVs. Table 2.13 provides an overview of additional road-specific standards which are now described.

Motor vehicle and equipment manufacturers, suppliers, and other industry stakeholders have been active in their efforts to contribute to improving the security posture of motor vehicles. The Society of Automotive Engineers standard, SAE J3061 (Cyber-security Guidebook for Cyber-Physical Vehicle Systems [121]), published in January 2016, is the first standard to explicitly address automotive cyber security. It provides a set of high-level cyber-security principles and guidance for cyber-physical vehicle systems (see Figure 2.12).

Subsequently, the British Standards Institution (BSI) published the Publicly Available Specification (PAS), PAS 1885:2018 ('The Fundamental Principles of Automotive Cyber security') [122] (Figure 2.12). This standard provides high-level guidance on protecting vehicles and vehicle systems from cyber threats across the whole automotive life cycle, from design to de-commissioning. It is intended to be

Table 2.12 Example international standards used in transport

Standard	Title/Content
ISO/IEC 27000	Overview and vocabulary
ISO/IEC 27001	Information security management systems (ISMS)
ISO/IEC 27002	Information security controls
ISO/IEC 27003	ISMS implementation guidance
ISO/IEC 27004	Monitoring, measurement, analysis, evaluation
ISO/IEC 27005	Security risk management
ISO/IEC 27010	Information security management for inter-sector and inter-organisational communications
ISO/IEC 27018	Handling PPI/SPI (Privacy) – Protection of PII in public clouds
ISO/IEC 27034	Guidance to ensure software delivers' necessary levels of security in support of an organisation's security management system
ISO/IEC 27035–1	Security incident management – Principles of incident management
ISO/IEC 27035–2	Security incident management – Guidelines for incident response
ISO/PAS 23399 Societal security	Guideline for incident preparedness and operational continuity management
ISO/IEC 29101	Privacy architecture framework
ISO/IEC 29119	Software testing standard
ISA/IEC 62443 Security for industrial automation and control systems	
ISA/IEC 62443-1-1:2009 Industrial communication networks – Network and system security	Part 1–1: Terminology, concepts, and models
ISA/IEC 62443-2-1:2010 Industrial communication networks – Network and system security	Part 2–1: Establishing an industrial automation and control system security program
ISA/IEC 62443-2-3:2015 Security for industrial automation and control systems	Part 2–3: Patch management in the IACS environment
ISA/IEC 62443-2-4:2015 Security for industrial automation and control systems	Part 2–4: Security program requirements for IACS service providers
ISA/IEC 62443-3-1:2009 Industrial communication networks – Network and system security	Part 3–1: Security technologies for industrial automation and control systems
ISA/IEC 62443-3-2:2020 Security for industrial automation and control systems	Part 3–2: Security risk assessment for system design
ISA/IEC 62443-3-3:2013 Industrial communication networks – Network and system security	Part 3–3: System security requirements and security levels

(Continues)

Table 2.12 Continued

Standard	Title/Content
ISA/IEC 62443-4-1:2018 Security for industrial automation and control systems	Part 4–1: Secure product development life cycle requirements
ISA/IEC 62443-4-2:2019 Security for industrial automation and control systems	Part 4–2: Technical security requirements for IACS components

CPNI, Centre for the Protection of National Infrastructure; DfT, Department for Transport.

Table 2.13 Road standards

Guidance	Title
SAE J3061	Cyber-security guidebook for cyber-physical vehicle systems
SAE J3101	Hardware protected security for ground vehicle applications
PAS 1885:2018	The fundamental principles of automotive cyber security
PAS 11281:2018	Connected automotive ecosystems – impact of security on safety – code of practice
ISO/SAE 21434	Road vehicles – cyber-security engineering (in development)

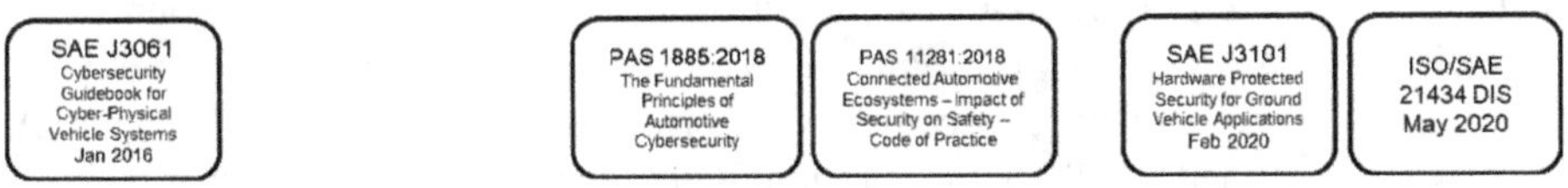

Figure 2.12 Road standards

read in conjunction with the guidance document 'Key Principles of Cyber Security for Connected and Automated Vehicles' [119], published in 2017.

This was followed by the publication of PAS 11281:2018 ('Connected automotive ecosystems – Impact of security on safety – Code of practice') [123] (Figure 2.12), which provides recommendations for managing security risks that might lead to a compromise of safety in a connected automotive ecosystem. It aims to assist organisations to ensure that security-related risks in their products, services or activities do not pose an unacceptable safety risk in the physical world. The scope of the document covers potential risks to single systems through to multiple systems, and considers interdependencies and vulnerabilities. It also addresses the link between cyber security and safety. The recommendations cover the entire connected automotive ecosystem and its constituent systems throughout their lifetimes, including manufacturing, supply chain and maintenance activities. The PAS is intended to be used by manufacturers, operators, and maintainers of products, systems and services, and is complementary to BSI PAS 1885:2018 [122].

In 2020, SAE J3101 ('Hardware Protected Security for Ground Vehicles') [124] (Figure 2.12) was released. A hardware-protected security environment provides a platform to implement access control by enabling secure authentication, authorisation and access enforcement. This standard defines common use cases of such an environment and establishes common requirements associated with those use cases. The standard provides OEMs and chip makers with a common reference that facilitates the communication of security requirements. Use case examples include the following: creating a new key fob; re-flashing ECU[††††††] firmware; reading/exporting PII[‡‡‡‡‡‡] from an ECU.

Although the publication of the first standard on automotive cyber security (SAE J3061) was an important step forward, the automotive industry recognised during the development of the safety standard ISO 26262:2018 (Road Vehicles – Functional Safety) [125] that there was a need for a comparable cyber-security standard. Consequently, it was decided to limit the consideration of cyber security in ISO 26262:2018, so that communications channels are required between safety and cyber-security engineering, with the aim being to coordinate the development of these disciplines and exchange requirements at system, software and hardware level. An annex in the standard provides background to the interaction between safety and security.

ISO and SAE subsequently collaborated on the development of the new cyber-security standard, ISO/SAE 21434 ('Road Vehicles – Cyber security Engineering') [126] (Figure 2.12) which superseded SAE J3601 and addresses the issues which became apparent during the development of ISO 26262:2018. ISO/SAE 21434 focuses on automotive cyber-security engineering by specifying requirements and providing recommendations for cyber-security risk management for road vehicles (including their components, software and interfaces) throughout their life cycle. A framework is defined which includes requirements for a cyber-security process, and a common language for communicating and managing cyber-security risk among stakeholders.

2.9.6.3 Regulations

Although there have been a number of initiatives in the areas of policy, guidance and standards for automotive cyber security, regulations have only been in development fairly recently. This situation was changed in 2020 when the World Forum for Harmonization of Vehicle Regulations, under UNECE[§§§§§§], released two regulations (Figure 2.13 and Table 2.14) which apply to passenger cars, vans, trucks and buses. One regulation relates to cyber security [127], while the other relates to software updates [128].

[††††††] ECU – Electronic Control Unit.
[‡‡‡‡‡‡] PII – Personally Identifiable Information.
[§§§§§§] UNECE – The United Nations Economic Commission for Europe.

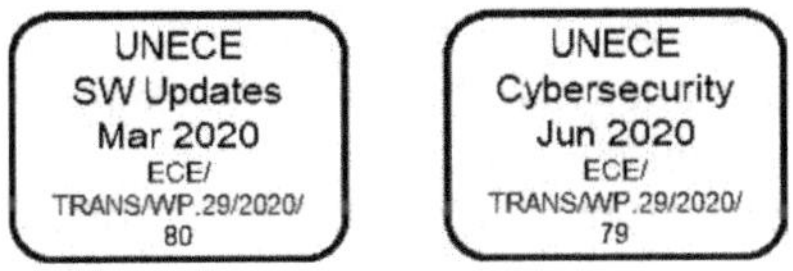

Figure 2.13 Road regulations

Table 2.14 Regulations

Publication	Content
ECE/TRANS/ WP.29/2020/79	Proposal for a new UN Regulation on uniform provisions concerning the approval of vehicles with regard to cyber security and cyber-security management system
ECE/TRANS/ WP.29/2020/80	Proposal for a new UN Regulation on uniform provisions concerning the approval of vehicles with regard to software update and software updates management system

Under the UNECE regulations, vehicle manufacturers must demonstrate that their vehicles meet requirements relating to cyber security and software updates for the life of the vehicle. Manufacturers must certify regulatory compliance with each regulation, and successfully complete a government audit of compliance in order to sell the vehicles in implementing countries. Both UNECE regulations provide requirements and a framework for manufacturers to reference when creating their compliant processes, some of which we set out below.

The Cyber-security Regulation [127] requires vehicle manufacturers to

- Create a cyber-security management system that applies to the development, production, and post-production phases of a vehicle's life cycle;
- Identify both critical risks and mitigation measures;
- Verify the effectiveness of mitigation measures;
- Monitor and respond to cyber threats, including through use of vehicle data and logs and data forensics;
- Provide yearly reports on any cyber-security risks found through the required monitoring and the ongoing effectiveness of the risk mitigation measures in place.

The Software Updates Regulation [128] deals with cyber-securing the process of Software updates, specifically OTA updates. It requires vehicle manufacturers to

- create a software update management system;
- assess the interdependence of systems in need of an update and whether the update affects safety or safe driving;

- identify which vehicles need a system update and inform users of the updates;
- document the need for changes, the vehicle systems affected by the changes, whether the changes require additional government approvals, and the verification and validation procedures performed on the software update;
- secure the software update and the delivery system during its development and deployment;
- ensure OTA updates can be completed safely, are performed when the vehicle has adequate power, and do not prevent functionality if the update fails.

Several countries have already announced their domestic timetables to implement these regulations. In the EU, all new vehicle types must comply with them by July 2022; by July 2024, the requirements will apply to all new vehicles. Japan will apply the regulations once they enter into force in January 2021. South Korea will implement them as guidelines to be released in 2020, and will formally implement them at a later date.

Given the widespread use of UN Regulations in the automotive sector globally, it is anticipated that other countries will also adopt them. Once implemented, any manufacturer that sells vehicles in the implementing countries must comply with the regulatory requirements. Since addressing supply chain security is required for compliance, the regulations are expected to influence manufacturers in countries in which they are not yet applicable, such as the United States.

UNECE is trying to appease as many stakeholders as possible and avoid being prescriptive by leaving much to interpretation and providing few actual technical cyber-security guidelines. However, the requirements have much in common with the contents of the SAE J3061 and ISO/SAE 21434 standards, which have influenced the UN documents.

2.9.6.4 Observations

Until 2016, there was very little in the way of regulations, standards or guidance material for road vehicle cyber security. Since then, however, a variety of publications have appeared on policy, principles and good practices from governments, standards organisations, and stakeholder groups, underlining the increasing importance of the topic and the rapid pace of development in vehicle automation.

In 2021, a key standard in road vehicle cyber-security engineering was released [126], and the first regional regulations related to the topic were published by UNECE.

2.10 Conclusions

This chapter has attempted to document the security landscape of the current transport system by summarising threats, describing historical security incidents and their underlying causes, and revealing the initiatives of governments, transport stakeholders, standards organisations, and others to address the issues. It should be clear that making the transport system resilient to cyber threats requires concerted action from stakeholder organisations at all levels, including addressing management policy and structure, processes, the training of personnel, system development and procurement, and system monitoring and control. Some key takeaways are presented below.

- The regulatory framework for cyber security continues to evolve at pace in all transport modes, with global, regional and national bodies regularly releasing new material and associated amendments in response to the constantly changing landscape. It is necessary to maintain awareness of such developments to ensure regulatory compliance. However, the rate of emergence of new cyber-security threats far exceeds the rate at which regulations and standards can be developed and adopted.
- The regulatory framework for each transport mode has evolved separately, so there is minimal commonality. In Europe, the goal of improving and harmonising security across states and sectors has led to the introduction of legal instruments such as GDPR, the NIS Directive, and the Cyber-security Act, promoting the legal right to personal and data privacy, and supporting the establishment of means of cyber-security certification of products, services and processes. These have made the regulatory environment more complex in some cases, although initiatives are in place to address these issues in the near future.
- It is important to secure information of all types, including that associated with privacy, systems (e.g., hardware and software configurations and status), sensitive procedures, and the sharing of security-sensitive information between stakeholders. The capability to securely store, share information and update systems (particularly OTA with new vehicle types) is key to maintaining transport system resilience.
- In all modes a structured, holistic approach to security is being mandated, requiring security to be endorsed at board-level and effectively resourced. Security management systems are being implemented in transport operators and throughout the supply chain (for both hardware and software) to ensure the development of security awareness among staff, and to support the development of a security culture.
- A whole life-cycle approach to security is now required, incorporating risk assessment and risk management as an integral part of system development, deployment, operations, and maintenance, through to decommissioning. Such an approach ensures that security is designed-in to systems and reduces the likelihood that potentially costly and time-consuming re-designs or retro-fits become necessary, resulting in unwanted impacts on deployment, operations and resilience.
- Regulations mandate a structured approach to managing cyber-security incidents, requiring measures to be in place to protect systems, detect incidents, identify them, respond rapidly and recover to normal operations as quickly as possible. The existence of appropriate protections and the prompt implementation of procedures by suitably trained personnel minimises down-time before service provision can be resumed.

- The potential impact of cyber security on safety is being recognised within and across modes, with explicit references to such interdependencies commonly found in guidance material, underlining the need for experts from both disciplines to coordinate throughout the life cycle.
- The level of cyber-security maturity varies between transport modes. Aviation, for instance, has been addressing the issues for an extended period, while initiatives in road transport, though substantial, are fairly recent.
- The applicable regulations, standards, and guidance material are also different for each mode, although there is evidence of convergence in their overall goals.
- The legal right to personal and data privacy compels transport stakeholders to ensure the security of customer-facing systems in addition to those providing transport services. Non-compliance can be financially costly and reputationally damaging.
- The need for smarter mobility will require a new generation of intelligent transportation services which deliver efficient, seamless connectivity in the planning and execution of a multi-mode journey.
- A pre-requisite for the secure introduction of multi-mode transport is the achievement of a common level of cyber security across modes for organisations and the components which must interface with each other for the secure and timely sharing of data. The development of a trust framework between transport modes could support this, perhaps by mutual recognition of the equivalence of certain standards or certifications, along with the development of security awareness and culture to similar levels. In the longer term, the rationalisation of methods, tools, standards and guidance material may provide a way forward.

2.11 Forthcoming Developments

To assist cyber-security experts to navigate through the regulatory framework and applicable standards for their specific transport mode, tailored standards are either available or in development which meet the individual requirements of each mode, and which use terminology which is familiar to practitioners. Examples of these are CENELEC TS-50701 for rail, ISO/SAE 21434 for automotive, EUROCAE documents for aviation, and the BIMCO guidelines for shipping. Such standards will continue to evolve to provide the necessary guidelines for each mode to develop and mature its security posture.

Higher level publications aimed at transport in general and including other critical infrastructures are also evident. For example, the *EU Transport cybersecurity toolkit [129]* is a repository of tips and recommended practices to enhance cybersecurity and cyber-resilience in the transport sector, while the *US DoT Strategic Plan 2022-2026 [130]* states strategic objectives in critical infrastructure cybersecurity, including resilient supply chains and enterprise cyber risks.

In order to encourage the harmonization of cybersecurity standards industry-wide, cross-sectorial regulations and standards will also continue to evolve. In the USA, *Executive order 14028 [131]* addresses the improvement of cybersecurity at the Federal level, including standardizing the government's playbook for responding to cybersecurity vulnerabilities and incidents.

In Europe there has recently been an update of the NIS Directive [57], commonly referred to as 'NIS2'. The 'NIS 2' Directive came into force on 16 January 2023 and must be transposed into EU Member States' National Laws by October 2024 [132, 133]. NIS 2 enacts measures to achieve a high common level of cybersecurity across the European Union, with a view to improving the functioning of the internal market. Key aspects of NIS2 are:

- The scope of the Directive is now expanded to cover new sectors based on their criticality for the economy and society, including all medium and large companies of these sectors, and smaller entities which may have a high risk profile.
- The distinction between Operators of Essential Services and Digital Service Providers is removed and instead classified as entities which are essential or important.
- Critically, the management bodies of these entities can be held liable if they do not take appropriate and proportionate technical, operational, and organisational measures to manage risks, and to prevent or minimise the impact of incidents on recipients of their services and on other services.
- Member States of the EU are required to adopt national cybersecurity strategies and designate competent authorities, cyber crisis management authorities, and single points of contact on cybersecurity and computer security incident response teams (CSIRTs). The strategies shall include policies in a number of areas, including the following:
    - supply chain cybersecurity for ICT products and services,
    - the inclusion and specification of cybersecurity-related requirements for ICT products and services in public procurement,
    - managing vulnerabilities,
    - promoting the development and integration of advanced technologies to implement risk-management measures, and
    - promoting and developing education and training on skills, awareness and R&D amongst other things.

The imminent introduction and rapid growth of new vehicles and services is also anticipated. These will include connected, autonomous vehicles such as air taxis, goods delivery drones, self-driving road vehicles, and autonomous trains and ships. These will add to the complexity of the transport system and will support the implementation of Mobility as a Service (MaaS), facilitated by the exploitation of cloud-based services, network connectivity, and smart analytics. MaaS will contribute to efficiency improvements at the multi-mode level. Such developments will accentuate the need for security and resilience to be achieved and maintained system-wide.

Climate change is driving developments towards the goal of low emission mobility in several ways [134].

For example, road vehicle electrification is accelerating, as shown by the proliferation of battery electric vehicles (BEVs), and the introduction of hydrogen fuel cell powered cars, vans, and buses [135] [136] on the roads. In aviation, electric aircraft supporting UAM are in development, and some are in the process of being certified for operations, one example being the Joby S4 eVTOL[¶¶¶¶¶] [137]. Commercial airlines have begun to use Sustainable Aviation Fuels (SAF) [138], while the development of electric and hybrid-electric passenger aircraft is ongoing [139] [140].

In the maritime space, electric ships are becoming increasingly common, and the first fully electric, battery-powered autonomous cargo ship set sail at the end of 2021 [141]. High-power hydrogen fuel cell units are also in development [142] for application onboard a wide range of vessels, such as ferries and inland tow-boats, and could allow cruise vessels to either run entirely on fuel cell power, or switch to it when operating in environmentally sensitive areas. Fuel cells powered by ammonia [143] are also under investigation for shipping, and ammonia has advantages over hydrogen in relation to temperature and pressure requirements for storage [131] [144]. Ammonia is consequently considerably easier to transport than hydrogen, potentially making it viable for powering large ships over long distances. The first ship powered by such a fuel cell, the 'Viking Energy', is planned to set sail in the second half of 2023.

Trains powered by hydrogen fuel cells can convert existing, non-electrified infrastructure into zero-emission rail lines, and the world's first hydrogen-powered passenger trains came into service in Germany in 2022 [145].

The importance of cybersecurity in the energy transition may not be immediately obvious, however, the brief summary above suggests that in the near future we will begin to see the proliferation of electric charging stations and hydrogen/ammonia fueling stations on roads, at airports, at seaports, and on rail networks, each with their own particular distribution infrastructures, systems, people, cultures, and supply chains.

The increased diversity and complexity resulting from these developments will require the development of appropriate regulations, standards, and guidance material to secure them, and will further increase the available attack surface, potentially exposing new security vulnerabilities which must be addressed to maintain transport system resilience.

The application of Artificial Intelligence (AI) will become increasingly important in all transport modes. AI systems will provide traffic forecasts, optimise vehicle routing, scheduling and traffic management, maintain vehicle separation, pilot autonomous vehicles transporting goods and passengers, and ensure the timely maintenance of vehicles and supporting systems to maintain safety and maximise reliability. However, the high level of computational functionality and connectivity required to exploit the benefits of AI will also enlarge the attack surface and increase the likelihood of physical and cyber-attacks. To address these issues, a number of initiatives are ongoing in the transport sector [146] [147] [148]. The *Fly AI Report* [149] developed by the European Aviation AI High Level Group provides a comprehensive overview of current and expected uses of AI in aviation, including an analysis of how AI may support aviation cyber-resilience. EASA's AI roadmap [24] discusses the implications of AI on aviation and the challenges which it brings to the sector, and addresses the notion of the trustworthiness of AI.

[¶¶¶¶¶] eVTOL - electric Vertical Take-Off and Landing

Developments in machine- and deep-learning will increase the resilience of systems and services in, for example, malware detection, network analysis, message filtering, and the provision of support to human operators [147]. In the Security Operations Centres of the future, AI may contribute to the automation of cyber-defence in areas such as vulnerability identification, predicting the evolution of threats, and adaptively mitigating attacks in real-time. However, AI can also be used to attack systems [150], and will be used by malicious actors to facilitate the exploitation of the attack surface.

The transport system security landscape is evolving rapidly in several areas. In order to carry out the tasks required to ensure that it remains resilient, a sufficient number of suitably qualified cybersecurity experts is required at all stages of the system life-cycle, and also in governance and oversight functions. However, there is already a significant shortage of such skills globally [151] [152]. For example, between 2013 and 2021, the number of vacant cybersecurity posts worldwide grew by 350%, from 1 million to 3.5 million, and it is predicted that the same number of vacancies will still exist in 2025. This shortage is further exacerbated in the transport sector by the more complex nature of transport systems and operations compared to the majority of ICT systems. This results in experienced practitioners being very difficult to find, and employing new, transport-naive cybersecurity experts requires substantial investment in their development. Consequently, progress in transport cybersecurity in the near future is likely to be severely limited throughout the transport sector by the shortage of available, appropriate talent.

References

[1] ENISA CSIRTs Relations Team. *CSIRTS by country – interactive map* [online]. European Union Agency for Cybersecurity (ENISA). Available from https://www.enisa.europa.eu/topics/csirts-in-europe/csirt-inventory/certs-by-country-interactive-map [Accessed Mar 2021].

[2] International Organization for Standardization. 'ISO Guide 73:2009, risk management – vocabulary'. n.d.

[3] International Organization for Standardization. 'ISO/PAS 22399:2007'. *Societal Security – Guideline for Incident Preparedness and Operational Continuity Management*. n.d.

[4] Insider Threat: What Do You Do When The Call Is Coming From Inside The House. Transport Security International Magazine [online]. February 2016. Available from https://www.tsi-mag.com/insider-what-do-you-do-when-the-call-is-coming-from-inside-the-house/ [Accessed April 2021].

[5] Bateman T., BBC News. Police warning after drug traffickers' cyber-attack [online]. 2013. Available from https://www.bbc.com/news/world-europe-24539417 [Accessed March 2021].

[6] Gruschka N., Jensen M. 'Attack surfaces: aa taxonomy for attacks on cloud services'. *2010 IEEE 3rd International Conference on Cloud Computing*; 2010 Jul 5. pp. 276–79.

[7] Gupta S., Winstead J. 'Using attack graphs to design systems'. *IEEE Security & Privacy Magazine*. 2007;5(4):80–3.

[8] International Organization for Standardization. 'ISO/IEC 27001:2013'. *Information Security Management Systems Information Security Management Systems – Requirements*. n.d.

[9] International Civil Aviation Organization. 'ICAO annex 17 to the convention on international civil aviation, safeguarding international civil aviation against acts of unlawful interference'. *Amendment 17, 11th Edition*. 2020.

[10] 'Commission implementing regulation (EU) 2017/373'. *Laying Down Common Requirements for Providers of Air Traffic Management / Air Navigation Services and Other Air Traffic Management Network Functions and their Oversight*. 1st March 2017.

[11] European Parliament, Council of the European Union. *Regulation (EC) 300/2008, on common rules in the field of civil aviation security*. The Publications Office of the European Union; 2008.

[12] Risidata. *Sobig Virus Strikes CSX Train Signaling System*. 2003. Available from https://www.risidata.com/Database/Detail/sobig-virus-strikes-csx-train-signalling-system [Accessed March 2021].

[13] Messmer E. 'Worm outbreak saturates networks'. *Network World*. August 25, 2003.

[14] *Virus Shuts Down AC Jazz Airline Flight Planning Computer* [online]. Risidata. 2003. Available from https://www.risidata.com/Database/Detail/virus-shuts-down-ac-jazz-airline-flight-planning-computer [Accessed March 2021].

[15] Federal Aviation Administration (FAA). *Review of web applications security and intrusion detection in air traffic control systems, federal aviation administration*. Number: FI-2009-049. 2009.

[16] Walsh W. *Phishing Scam Targeted 75 US Airports* [online]. Information Week. 2014. Available from http://www.informationweek.com/government/cybersecurity/phishing-scam-targeted-75-us-airports/d/d-id/1278762 [Accessed March 2021].

[17] *Bug in Anti-Virus Software Hits LANs at JR East* [online]. Japan Times. 2005. Available from https://www.japantimes.co.jp/news/2005/04/24/national/bug-in-antivirus-software-hits-lans-at-jr-east-some-media/ [Accessed March 2021].

[18] Pash C. *Cyber Security is Being Tightened at Australian Airports After an Identity Card Data Hack* [online]. Business Insider Australia. 2018. Available from https://www.businessinsider.com.au/identity-card-data-hack-data-breach-australian-airports-2018-7 [Accessed March 2021].

[19] Leyden J. *Polish Teen Derails Tram After Hacking Train Network* [online]. The Register. 2008. Available from https://www.theregister.co.uk/2008/01/11/tram_hack/ [Accessed March 2021].

[20] Newman L.H., WIRED. *How Hackers Slipped by British Airway's Defences*. 2018. Available from https://www.wired.com/story/british-airways-hack-details/ [Accessed March 2021].

[21] 'Regulation (EU) 2016/679, … on the protection of natural persons with regard to the processing of personal data and on the free movement of such data'. *General Data Protection Regulation*. April 2016.

[22] Rindskopf A. *Juvenile* computer hacker cuts off FAA tower [online]. Irational.org. 1998. Available from http://www.irational.org/APD/CCIPS/juvenilepld.htm [Accessed March 2021].

[23] Bradbury D. *How to* hack a Jeep Cherokee – but don't try this at home, kids [online]. Naked Security. 2017. Available from https://nakedsecurity.sophos.com/2017/05/10/how-to-hack-a-jeep-cherokee-but-dont-try-this-at-home-kids/ [Accessed March 2021].

[24] Paganini P. *Fiat-Chrysler distributes the fix for flawed Jeep via mailed USB* [online]. Security Affairs. 2015. Available from http://securityaffairs.co/wordpress/39899/hacking/fiat-chrysler-mailed-usb-fix.html [Accessed March 2021].

[25] Greenberg A. *Hackers can steal a Tesla Model S in seconds by cloning its keyfob* [online]. Wired. 2018. Available from https://www.wired.com/story/hackers-steal-tesla-model-s-seconds-key-fob/ [Accessed March 2021].

[26] Winder D. *Tesla has facepalm moment as hackers defeat 'Fixed' Model S security* [online]. Forbes. 2019. Available from https://www.forbes.com/sites/daveywinder/2019/08/30/tesla-has-facepalm-moment-as-hackers-defeat-fixed-model-s-security/#5a77c6104135 [Accessed March 2021].

[27] *The Rendition Project* [online]. The Rendition Project. 2001. Available from https://www.therenditionproject.org.uk/index.html [Accessed March 2021].

[28] Strohmeier M., Smith M., Lenders V., Martinovic I. 'The real first class? Inferring confidential corporate mergers and government relations from air traffic communication'. *IEEE European Symposium on Security and Privacy*; 2018.

[29] Psiaki M., Humphreys T. *Protecting GPS from Spoofers Is Critical to the Future of Navigation* [online]. IEEE Spectrum. 2016. Available from https://spectrum.ieee.org/telecom/security/protecting-gps-from-spoofers-is-critical-to-the-future-of-navigation [Accessed March 2021].

[30] Gartenberg C. *This Pokémon Go GPS Hack Is the most impressive yet* [online]. The Verge. 2016. Available from https://www.theverge.com/circuitbreaker/2016/7/28/12311290/pokemon-go-cheat-gps-signal-spoofing [Accessed March 2021].

[31] 2018-004A – Eastern Mediterranean Sea - Possible GPS interference [online]. US Maritime Administration. 2018. Available from https://www.maritime.dot.gov/content/2018-004a-eastern-mediterranean-sea-possible-gps-interference [Accessed March 2021].

[32] Goward D. *Mass GPS Spoofing Attack in Black Sea?* [online]. The Maritime Executive. 2017. Available from https://www.maritime-executive.com/editorials/mass-gps-spoofing-attack-in-black-sea [Accessed March 2021].

[33] Virginia A. *GPS Spoofing Patterns Discovered* [online]. Resilient Navigation and Timing Foundation. 2017. Available from https://rntfnd.org/wp-content/uploads/GPS-Spoofing-Patterns-Press-Release.1-26-Sep-17-RNT-Foundation.pdf [Accessed Mar 2021].

[34] Gabrowski J.C. *Personal Privacy Jammers: Locating Jersey PPDs Jamming GBAS Safety-of-Life Signals* [online]. GPS World. 2012. Available from

https://www.gpsworld.com/personal-privacy-jammers-12837/ [Accessed March 2021].

[35] FAA Navigation Team. *GPS Privacy Jammers and RFI at Newark* [online]. FAA. 2011. Available from http://laas.tc.faa.gov/documents/Misc/GBAS%20RFI%202011%20Public%20Version%20Final.pdf [Accessed March 2021].

[36] Inside GNSS. *FCC Fines Operator of GPS Jammer That Affected Newark Airport GBAS* [online]. Inside GNSS. 2013. Available from https://insidegnss.com/fcc-fines-operator-of-gps-jammer-that-affected-newark-airport-gbas/ [Accessed March 2021].

[37] Federal Communications Commission. *Jammer Enforcement* [online]. FCA. 2020. Available from https://www.fcc.gov/general/jammer-enforcement [Accessed March 2021].

[38] US Department of Homeland Security. *The National Strategy to Secure Cyberspace* [online]. 2003. Available from https://georgewbush-whitehouse.archives.gov/pcipb/.

[39] *National Cyber Security Strategies – setting the course for national efforts to strengthen security in cyberspace* [online]. ENISA. 2012. Available from https://www.enisa.europa.eu/topics/national-cyber-security-strategies [Accessed March 2021].

[40] Obama B. *Executive Order – Improving Critical Infrastructure Cybersecurity,* [online]. The Whitehouse. 2013. Available from https://obamawhitehouse.archives.gov/the-press-office/2013/02/12/executive-order-improving-critical-infrastructure-cybersecurity [Accessed March 2021].

[41] *National Cyber Security Strategy of the USA* [online]. The Whitehouse. 2018. Available from https://trumpwhitehouse.archives.gov/wp-content/uploads/2018/09/National-Cyber-Strategy.pdf [Accessed March 2021].

[42] European Commission. *Cyber security strategy of the European Union: an open, safe and secure cyberspace*. The Publications Office of the European Union; 2013.

[43] European Commission. *The EU's Cyber Security Strategy for the Digital Decade* [online]. 2020. Available from https://ec.europa.eu/digital-single-market/en/news/eus-cybersecurity-strategy-digital-decade [Accessed March 2021].

[44] UK HM Government. *National Cyber Security Strategy, 2016-2021* [online]. 2016. Available from https://www.gov.uk/government/publications/national-cyber-security-strategy-2016-to-2021 [Accessed March 2021].

[45] ANSSI. *Stratégie Nationale pour la Sécurité de Numérique* [online]. 2015. Available from https://www.ssi.gouv.fr/actualite/la-strategie-nationale-pour-la-securite-du-numerique-une-reponse-aux-nouveaux-enjeux-des-usages-numeriques/ [Accessed March 2021].

[46] 'Sécretariat Général de la Défense et de la Sécurité national (SGDSN'. *Revue stratégique de cyberdéfense*. 2018, SGDSN.

[47] ENISA. *ENISA map of National Cyber Security Strategies* [online]. 2021. Available from https://www.enisa.europa.eu/topics/national-cyber-security-strategies/ncss-map [Accessed Mar 2021].

[48] Department of the Prime Minister and Cabinet. *Australia's cyber security strategy, enabling innovation, growth & prosperity*. Commonwealth of Australia; 2016.

[49] Department of Home Affairs. *Australia's cyber security strategy 2020*. Commonwealth of Australia; 2020.

[50] UK Department for Transport. *Aviation cyber security strategy*. UK Government; 2018.

[51] 'The Whitehouse'. *National Strategy for Aviation Security of the USA (United States Whitehouse Office), 2018*.

[52] Council of the European Union. *Council conclusions on the revision of the European Union Maritime Security Strategy (EUMSS) Action Plan*. Outcome of Proceedings, 10494/18; 26 June 2018.

[53] International Civil Aviation Organization. 'Security and facilitation strategic objective' in *Aviation cybersecurity strategy*. International Civil Aviation Organization; 2019.

[54] International Organization for Standardization. *ISA/IEC 27005:2018*. Information Security Risk Management.

[55] International Organization for Standardization. *ISA/IEC 62443-3-2:2020, Security for industrial automation and control systems - Part 3-2: Security Risk Assessment for System Design*.

[56] International Organization for Standardization. *ISO/IEC 31010:2019, Risk management – Risk assessment techniques*.

[57] *Directive (EU) 2016/1148 concerning measures for a high common level of security of network and information systems across the union*. The Publications Office of the European Union; 2016.

[58] *Regulation (EU) 2019/881 of the European Parliament and of the Council on ENISA (the European Union agency for cyber security) and on information and communications technology cyber security certification*. The Publications Office of the European Union; 2019.

[59] United Nations - Inland Transport Committee. *Transport trends and economics – mobility as a service*. United Nations Economic Commission for Europe; 2020.

[60] EUROCONTROL. *EUROCONTROL Air Traffic Management Computer Emergency Response Team (EATM-CERT)* [online]. 2021. Available from https://www.eurocontrol.int/service/european-air-traffic-management-computer-emergency-response-team [Accessed March 2021].

[61] *Action Points – Cyber security in Public Transport*, February 2017, UITP (*Union Internationale des Transports Publics – International Organisation of Public Transport*) [online]. Available from https://cms.uitp.org/wp/wp-content/uploads/2021/02/Action_Points_Cyber_Security-v2.pdf [Accessed April 2021].

[62] CMS-Law. *GDPR Enforcement Tracker* [online]. CMS Law, 2021. Available from https://www.enforcementtracker.com/ [Accessed March 2021].

[63] Proton Technologies AG. Proton Technologies AG. Complete Guide to GDPR Compliance [online]. 2020. Available from https://gdpr.eu/ [Accessed March 2021].

[64] UK Department for Transport. *Implementation of the NIS Directive – DfT Guidance version 1.1*. London: UK DfT; 2018.

[65] 'ICAO annex 9 to the convention on international civil aviation' in *Facilitation*. International Civil Aviation Organization; 2017.

[66] ICAO Doc. 8973 (Restricted). *Security manual for safeguarding civil aviation against acts of unlawful interference*. 12th edition. International Civil Aviation Organization; 2020.

[67] *Air traffic management security manual*. 1st Edition. International Civil Aviation Organization; 2013.

[68] ICAO Doc 9807. *Universal security audit programme continuous monitoring manual*. 3rd Edition. International Civil Aviation Organization; 2021.

[69] *Regulation (EC) 216/2008, on common rules in the field of civil aviation and establishing a European Aviation Safety Agency*. Official Journal of the European Union; 2008.

[70] *Regulation (EC) 1070/2009, amending regulations (EC) NO 549/2004, (EC) NO 550/2004, (EC) NO 551/2004 and (EC) NO 552/2004 in order to improve the performance and sustainability of the European aviation system*. Official Journal of the European Union; 2009.

[71] *Regulation (EC) 1108/2009, amending regulation (EC) NO 216/2008 in the field of aerodromes, air traffic management and air navigation services and repealing directive 2006/23/EC*. Official Journal of the European Union; 2009.

[72] *Regulation (EC) NO 2320/2002, establishing common rules in the field of civil aviation security*. Official Journal of the European Union; 2002.

[73] ECAC. *Doc 30 Part II - restricted*; 2018.

[74] *Implementing regulation (EU) 185/2010, laying down detailed measures for the implementation of the common basic standards on aviation security*. Official Journal of the European Union; 2010.

[75] *Implementing regulation (EU) 2015/998, laying down detailed measures for the implementation of the common basic standards on aviation security*. Official Journal of the European Union; 2015.

[76] *Implementing regulation (EU) 2019/1583, ... laying down detailed measures for the implementation of the common basic standards on aviation security, as regards cyber-security measures*. Official Journal of the European Union; 2019.

[77] European Union Agency For Network And Information Security (ENISA). *Securing smart airports*. European Union Agency For Network And Information Security (ENISA); 2016. Available from https://www.enisa.europa.eu/publications/securing-smart-airports

[78] Official Journal of the European Union. 'Commission regulation (EC) 2096/2005, laying down common requirements for the provision of air navigation services'. 20th December 2005.

[79] Commission Implementing Regulation (EU) 1035/2011. 'Laying down common requirements for the provision of air navigation services'. 17th October 2011.

[80] Official Journal of the European Union. 'Regulation (EU) 73/2010, laying down requirements on the quality of aeronautical data and aeronautical information for the single European sky'. 26th January 2010.

[81] Official Journal of the European Union. 'Commission implementing regulation (EU) 2020/469, … as regards requirements for ATM/ANS services, …'. February 2020.

[82] Official Journal of the European Union. 'Regulation (EU) 2018/1139, on common rules in the field of civil aviation and establishing a European Union aviation safety agency'. 4th July 2018

[83] Official Journal of the European Union. 'Regulation (EC) 552/2004, on the interoperability of the European air traffic management network (the interoperability regulation)'. 10th March 2004

[84] EASA European Strategic Coordination Platform *Strategy for cybersecurity in aviation*. Official Journal of the European Union; 10th September 2019.

[85] International Maritime Organization, IMO IA116E. *Guide to maritime security and the ISPS code*; 2012 Edition.

[86] International Maritime Organization. *IMO international safety management (ISM) code*.

[87] International Maritime Organization, IMO MSC 94/4/1. *Measures towards enhancing maritime cyber security*.

[88] International Maritime Organization, IMO MSC 95/4/3. *Voluntary maritime cyber security guidelines*.

[89] International Maritime Organization, IMO MSC 95/4/1. *Industry guidelines on cyber security on board ships*. 2015.

[90] International Maritime Organization, IMO MSC 96/4/1. *The guidelines on cyber security on board ships*.

[91] International Maritime Organization, IMO MSC-FAL 1/Circ. 3. *Guidelines on maritime cyber risk management*.

[92] International Maritime Organization, IMO Resolution MSC.428(98). *Maritime cyber risk management in safety management systems*.

[93] International Maritime Organization, IMO MSC 101/4/1. *The guidelines on cyber security on board ships*.

[94] Hopcraft R., Martin K.M. 'Effective maritime cyber-security regulation – the case for a cyber code'. *Journal of the Indian Ocean Region*. 2018;14(3):354–66.

[95] ENISA. 'Analysis of cyber security aspects in the maritime sector'. 2011.

[96] International Union of Railways (UIC). *High Speed Network Security Handbook*. Paris: UIC; 2016.

[97] International Union of Railways (UIC). *Guidelines for Cyber Security in Railways - UIC 5-18005*. Paris: UIC; 2018.

[98] CYRAIL project. 'CYRAIL recommendations on cyber security of rail signalling and communications systems'. *Horizon*. 2018;**2020**.

[99] ENISA. *Railway cyber security – security measures in the railway transport sector*. European Union Agency For Network And Information Security (ENISA); 2020.

[100] CENELEC. *Technical specification 50701 - railway applications – cyber security*. European Committee for Electrotechnical Standardization (CENELEC); 2021.

[101] UK Department for Transport. *Rail cyber security – guidance to industry (London)*. UK Government; 2016.

[102] Agence nationale de la sécurité des systèmes d'information. 'Cyber security for ICS – managing cyber security for ICS'. 2012.Version 1.0.

[103] Agence nationale de la sécurité des systèmes d'information. 'Cyber security for ICS – use case'. 2012. Version 1.0.

[104] Agence nationale de la sécurité des systèmes d'information. 'Cyber security for ICS - classification method and key measures'. 2014. Version 1.0.

[105] Agence nationale de la sécurité des systèmes d'information. 'Cyber security for ICS - detailed measures'. 2014. version 1.0.

[106] National Cybersecurity and Communications Integration Center. 'ICS-CERT S508C'. *Recommended Practice: Improving Industrial Control System Cyber Security with Defense-in-Depth Strategies*. 2016.

[107] American Public Transportation Association. 'Securing control and communications systems in transit environments, part I: elements, organisation and risk/assessment management, APTA SS-CCS-RP-001-10'. *Recommended Practice*. 2010.

[108] American Public Transportation Association. 'Securing control and communications systems in transit environments, part II: defining a security zone architecture for rail transit and protecting critical zones, APTA SS-CCS-RP-002-13'. *Recommended Practice*. 2013.

[109] American Public Transportation Association. 'Securing control and communications systems in transit environments, part IIIa: attack modelling security analysis, APTA SS-CCS-WP-003-15'. *White Paper*. 2015.

[110] American Public Transportation Association. 'Securing control and communications systems in transit environments, part IIIb: protecting the operationally critical security zone, APTA SS-CCS-RP-004-16'. *Recommended Practice*. 2016.

[111] NHTSA. Cyber-security best practices for modern vehicles [online]. 2016. Available from https://www.nhtsa.gov/sites/nhtsa.dot.gov/files/documents/812333_cybersecurityformodernvehicles.pdf [Accessed April 2021].

[112] NIST Cyber-security Framework, version 1.1 [online]. NIST. April 2018. Available from https://www.nist.gov/cyberframework [Accessed April 2021].

[113] NHTSA. A vision for safety: automated driving systems 2.0 [online]. 2017. Available from https://www.nhtsa.gov/sites/nhtsa.dot.gov/files/documents/13069a-ads2.0_090617_v9a_tag.pdf [Accessed April 2021].

[114] US Department of Transportation. Preparing for the Future of Transportation: Automated Vehicles 3.0 [online]. 2018. Available from https://www.transportation.gov/av/3/preparing-future-transportation-automated-vehicles-3 [Accessed April 2021].

[115] Ensuring American Leadership in Automated Vehicle Technologies: Automated Vehicles 4.0 [online]. 2020. Available from https://www.nhtsa.gov/vehicle-manufacturers/automated-driving-systems#automated-driving-systems-av-40 [Accessed April 2021].

[116] Automotive Information Sharing and Analysis Center (Auto-ISAC). "Automotive Cyber-security Best Practices – Executive summary", v1.1 [online]. 2019. Available from https://www.automotiveisac.com/best-practices/ [Accessed April 2021].

[117] European Union Agency For Network And Information Security (ENISA). *Cyber security and resilience of smart cars*. ENISA; 2017.

[118] European Union Agency For Network And Information Security (ENISA). *Good practices for security of smart cars*. ENISA; 2019.

[119] DfT U.K., CPNI U.K. *The Key Principles of Vehicle Cyber Security for Connected and Automated Vehicles*. London: UK HM Government; 2017.

[120] European Automobile Manufacturers Association (ACEA). *ACEA-Principles-of-Automobile-Cyber Security*. Brussels: ACEA; 2017.

[121] *SAE standard J 3061: cyber-security guidebook for cyber-physical vehicle systems*. 2016. Available from http://standards.sae.org/wip/j3061/

[122] The British Standards Institution. 'BSI PAS 1885:2018'. *The Fundamental Principles of Automotive Cyber Security The Fundamental Principles of Automotive Cyber Security – Specification*. 2018.

[123] The British Standards Institution. *PAS 11281:2018 connected automotive ecosystems – impact of security on safety – code of practice*.

[124] Society of Automotive Engineers. 'SAE J3101'. *Hardware Protected Security for Ground Vehicles*. 2020.

[125] International Organization for Standardization. *ISO 26262:2018, road vehicles – functional safety*.

[126] International Organization for Standardization. *ISO/SAE 21434:2021 - road vehicles – cybersecurity engineering*.

[127] 'Proposal for a new un regulation on uniform provisions concerning the approval of vehicles with regards to cyber security and cyber security management system'. *Unece, Ece/trans/wp.29/2020/79*. 2020.

[128] United Nations Economic Commission for Europe. 'Proposal for a new un regulation on uniform provisions concerning the approval of vehicles with regards to software update and software updates management system'. *UNECE, ECE/TRANS/WP.29/2020/80*. 2020.

[129] *Cybersecurity toolkit*. European Commission, DG for Mobility and Transport;

[130] *Strategic plan FY 2022-2026*. Washington: US Department of Transportation;

[131] 'Executive office of the president. executive order 14028, improving the nation's cybersecurity'. 2021.

[132] *Directive (EU) 2022/2555 on measures for a high common level of cybersecurity across the union, amending regulation (EU) no 910/2014 and directive (EU) 2018/1972, and repealing directive (EU) 2016/1148*. The Publications Office of the European Union; 2016.

[133] ENISA. *NIS Directive*. 2022. Available from https://www.enisa.europa.eu/topics/cybersecurity-policy/nis-directive-new [Accessed 25 Jan 2023].

[134] *Fly the green deal - report of the advisory council for aviation research and innovation in Europe (ACARE)*. European Commission, Brussels; 2022 Jun.

[135] *Hydrogen cars – all models at a glance* [H2Live]. Available from https://h2.live/en/wasserstoffautos/

[136] Buses F.C.E. 'Towards clean public transport with hydrogen'. [online] *Fuel Cell Electric Buses*. 2022. Available from https://fuelcellbuses.eu/

[137] Warwick G. 'Joby agrees on revised EVTOL certification basis with FAA'. [online] *Aviation Week*. 2022. Available from https://aviationweek.com/air-transport/joby-agrees-revised-evtol-certification-basis-faa

[138] ICAO 'Sustainable aviation fuels'. [online] *ICAO Website*. 2022. Available from https://www.icao.int/environmental-protection/pages/SAF.aspx

[139] AIRBUS *Electric flight* [online]. 2022. Available from https://www.airbus.com/en/innovation/zero-emission/electric-flight

[140] Gates D. 'First all-new, electric commuter airplane takes flight at moses lake'. [online] *The Seattle Times*. 2022. Available from https://www.seattletimes.com/business/boeing-aerospace/first-u-s-all-electric-airplane-takes-flight-at-moses-lake/

[141] Resilience 'Making waves: electric ships are sailing ahead'. [online] *Resilience*. 2022. Available from https://www.resilience.org/stories/2022-07-28/making-waves-electric-ships-are-sailing-ahead/

[142] Habibic A. 'High-power fuel cell concept for larger ships gets DNV's blessing'. [online] *Offshore Energy*. 2022. Available from https://www.offshore-energy.biz/high-power-fuel-cell-concept-for-larger-ships-gets-dnvs-blessing/

[143] Rokke N. *Ammonia: A sustainable fuel option for shipping* [[online]. Forbes]. 2021 Oct. Available from https://www.forbes.com/sites/nilsrokke/2021/10/05/ammonia-a-sustainable-fuel-option-for-shipping/

[144] Fraunhofer Institute 'The world's first high-temperature ammonia-powered fuel cell for shipping'. [online] *Microengineering and Microsystems*. 2021. Available from https://www.fraunhofer.de/en/press/research-news/2021/march-2021/worlds-first-hightemperature-ammonia-powered-fuel-cell-for-shipping.html

[145] Buckley J. 'The world's first hydrogen-powered passenger trains are here'. [online] *CNN*. 2022. Available from https://edition.cnn.com/travel/article/coradia-ilint-hydrogen-trains/index.html

[146] *Artificial intelligence - A European perspective*. Luxembourg: European Commission Joint Research Centre; 2018.

[147] *Artificial intelligence roadmap - A human-centric approach to AI in aviation*. 1.0. Cologne: European Union Aviation Safety Agency;

[148] *Cybersecurity challenges in the uptake of artificial intelligence in autonomous driving*. Luxembourg: ENISA;

[149] *The fly AI report - demystifying and accelerating AI in aviation/ATM*. European Aviation Artificial Intelligence High Level Group;

[150] Brundage et al *The malicious use of artificial intelligence: forecasting, prevention, and mitigation*. Future of Humanity Institute, University of Oxford; 2018 Feb.

[151] Lake S. 'Companies are desperate for cybersecurity workers - more than 700k positions need to be filled'. [online] *Fortune Education*. 2022. Available from https://fortune.com/education/business/articles/2022/06/30/companies-are-desperate-for-cybersecurity-workers-more-than-700k-positions-need-to-be-filled/

[152] Pollet M. 'Looming talent shortage will limit cybersecurity efforts in Europe, French agency warns'. [online] *EURACTIV*. 2022. Available from https://www.euractiv.com/section/cybersecurity/news/looming-talent-shortage-will-limit-cybersecurity-efforts-in-europe-french-agency-warns/

Chapter 3

Introduction to risk management

Martin Hawley, Paul Thomas, Amy Ertan, and Mauricio Antonio Duarte Lara

3.1 Introduction

3.1.1 Overview

Most readers will be familiar with project, programme or enterprise risk management. Our aim in the chapter is to promote a general understanding of risk in the context of cybersecurity, and how it is approached practically. We draw on ISO 27005 [1] but are generally not prescriptive; we reference several standards and frameworks available to organisations looking to develop a mature approach to cyber risk. A large part of the chapter is a walk-through of the risk management process, where we address the management of risks from remote keyless systems (RKS) in cars, focusing on the main steps and some of the issues that arise. We also analyse risks both quantitatively and qualitatively to highlight the relative merits of the approaches.

3.1.2 Risk management

From a business perspective, risk management is about understanding the organisation's exposure to loss, compared with its appetite to risk. It helps the organisation manage how it will react to events that were not planned, but can be foreseen in the business' evolution. Enterprise risk management is the predominant process through which organisations manage risk, providing a structure that combines all risk management activities into one integrated framework that enables the identification of interdependencies [2], and cyber risk should be a part of this. The goal of enterprise risk management is to obtain a balance between capitalising on opportunities while minimising vulnerabilities and losses. Risk management is tasked with the ongoing identification, assessment and control of risks.

With a fundamental reliance on information technology infrastructure, cybersecurity risk management is essential. Effective cybersecurity risk management requires an organisation to understand its current and target 'cybersecurity posture'. For example, an organisation should at least be aware of the assets it should protect, the potential threats to those assets and the defences currently in use to protect each asset.

Risk management has been a recognised management tool since the early twentieth century. However, the level of maturity in different organisations varies widely. All too often, risk analysis and assessment is regarded as a 'tick-box' exercise, or may be done superficially, with only loose coupling to decision-making. Much has been written on this, for example, by Hubbard [3], who has criticised approaches to risk assessment in general, and there are particular concerns about the effectiveness of risk management in information system projects [4, 5].

3.1.3 Risk and decision-making

3.1.3.1 What is risk?

Risk is a human construct or 'metric', albeit one that has been appreciatively and widely used in technology programmes since the 1950s. In general, we regard risk as having two main components:

- the impact of an event should it occur
- the likelihood that it will occur.

Hence risk is a function of 'impact and likelihood' [6]. Two other aspects, however, are central to risk – the scenario for which risk is being estimated and the body of knowledge of any scenario, which will likely be incomplete [7]. These aspects will be returned to later in the chapter as we present a worked example.

The 'risk equation' is usually described as 'Risk = Impact × Likelihood'. This risk equation also illustrates that risk is widely used quantitatively. However, it is also used with semi-quantitative and qualitative 'risk scales'. In the transport sector, cybersecurity risk assessment has tended to use qualitative methods, but as we will see later in the chapter, there are good reasons for quantifying risk and it is less daunting than might be imagined, even for complex systems.

Risk is primarily about making choices before events make them for us. Moving to more familiar territory than cybersecurity, consider a risk of disruption to a sporting fixture by rain. The impact is the cost of delay or cancellation, borne by the spectators, players and organisers. The risk is derived from the estimated impact and probability* of rain on the day, with the risk scenario defined by the location of the event, the event schedule and the effects of different intensities of rain.

For example, a 33% chance of light rain for a short period may be readily manageable and incur delays, whereas a 66% chance of long periods of heavy rain may risk cancellation. In these two scenarios, the costs may range from a few thousand to tens of thousands. Let us say €5,000 and €50,000, with probabilities of 0.33 and 0.66, equating to quantified risks of €1,650 and €33,000.

*Note that there are distinct mathematical definitions of likelihood and probability. In this chapter, we refer to likelihood in a general way and probability when we are referring to the chance of an event happening as a number between 0 and 1 (or 0% and 100%).

These two scenarios also demonstrate how different mitigation options may be appropriate for each risk. The lower risk scenario may require extra staff costs but also an opportunity for more bar sales. The higher risk scenario, if manifested, may lead to a decision to cancel, with refunds due to spectators. There is clearly plenty of scope to assess different scenarios and resulting decisions.

3.1.3.2 Risk embodies knowledge

What should be noted from the preceding example and Reference [7] is that risk calculations require not only likely impact and probability but also the body of knowledge about the risk in question, which may change over time. We also link potential opportunities to risk, in this example more bar sales, although we appreciate this may not be a common finding in cybersecurity risk assessments.

3.1.4 Dealing with the extremes of impact and probability

One interesting point to note is that the metric of risk as 'impact × probability' assigns similar values to risks that are 'high impact/low probability' and 'low impact/high probability'. How these issues are dealt with often depends on the specific industry. Invariably managers will have limited resources and will be faced with difficult decisions on what to do with risks. Risks also need to be combined in some way to effectively consider the overall risks to the operation. Risk curves, shown in Figure 3.1, are a way of visualising the range of risks as they plot frequency or probability† of event occurrence against the impact. At the extremes of the curves, higher frequency but low impact events may be tolerable, but low frequency high impact events may not be. Because of the wide range of values plotted (10^2–10^8) they are normally plotted on a log scale as shown in the right hand side of Figure 3.1.

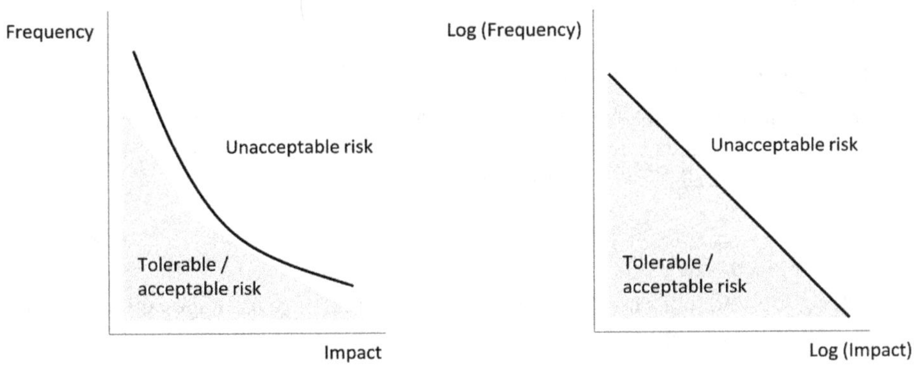

Figure 3.1 Risk curves

†As mentioned previously, frequency is an historic measure and is often used as an estimate of future probability of occurrence.

3.1.5 Taking decisions from risk assessment

Once a risk has been established, it leads to decisions about how to mitigate it. These decisions are guided by the following broad choices:

- Tolerate – accept the risk (its likelihood and potential consequences) and take no further action.
- Treat – apply a mitigation or 'control' to the risk to reduce its value.
- Transfer – make the risk 'someone else's problem'.
- Terminate – stop the activity or project that the risk is associated with.

The above options emphasise how risk assessment is ultimately a tool to support decision-making. Risks should also be 'Tracked' or monitored to identify any change (especially increase) in the estimated level of risk for any given scenario. Risk assessments always need to be kept under review and updated, and may be refreshed at certain intervals in line with best practice frameworks or guidelines. In particular, any changes to the system should trigger an update of the risk assessment, as well as the identification of new threats and vulnerabilities. Organisations may be inclined to patch vulnerabilities as they occur but not update the risk assessment. While this may be expedient in the short term, the concern is that the risk assessment becomes invalidated, resulting in poor risk management overall.

Another fundamental point to note is that risk measurement of all forms is based at least partially on human judgement [8]. This unavoidably raises challenges relating to subjectivity, personal biases and how the perspectives and assumptions held by colleagues shape their approach to risk. Individuals doing the same risk assessment would likely come up with different answers, so a process of review and working group approaches are often necessary, especially in safety-critical contexts.

3.1.6 The language of risk

There are a number of concepts that relate to risk, which is often mistakenly referred to as a sense of 'chance', but as a management concept, it includes the element of impact or cost as well. The key concepts are as follows.

3.1.6.1 Probability

This is mostly used in a statistical context, as a decimal or percentage between 0 and 1; where 0 signifies impossibility and 1 denotes certainty.

3.1.6.2 Likelihood

In general usage, this is the same as probability, or we may speak of a low, medium or high likelihood when doing qualitative risk assessments. There are also statistical definitions of likelihood which we do not cover in this chapter.

3.1.6.3 Frequency

This refers to the measured frequency of past events in order to estimate the probability of future events. For example, we may estimate that an event could occur once a year or once every 10 years based on historic data.

3.1.6.4 Uncertainty

This is used in general conversation to express a degree of probability that an event will occur. It is also regarded qualitatively, as something that is unmeasurable, as defined by Knight [9]. So called 'Knightian uncertainty' refers to aspects of risk that are unknown. Thus, a risk may be estimated and bounded by error bars, but there still exists some unquantifiable amount that is uncertain, or there may remain risks that have not been identified and are therefore unmeasured. In this sense, there is a difference between risk, which is measurable, and uncertainty, which is not.‡

3.1.7 Approaches to risk management

Risk approaches also vary according to different disciplines:

- high level approaches to economic risks;
- more detail for project management (overruns, overspend, project quality etc.);
- high technical detail for safety assessment

In the transport sector, cybersecurity risk management has been evolving from a high level approach to a more detailed technical approach, akin to the detailed technical approaches for safety and requiring similar levels of resource. This is a very important trend for safety critical domains as it is hard to argue a safety case if the detailed security work is not done.

One expression often seen in cybersecurity is the need for organisations to consider their 'crown jewels' and ensure that these are well protected. At the corporate level, cybersecurity risk may even be defined as a single line item. This high-level approach means that the organisation may be well aware of the general risk of cyber attack but treat it as a 'black-box' challenge. Resources will then be applied to mitigate the risk, but the organisation will not have uncovered a deeper understanding of how the components of cyber risk impact their operations. This generalised approach is shown in Figure 3.2. The figure also shows how the technical approach identifies specific risks, which may then have specific controls applied to them, for example, Risk A is mitigated by controls X and Y.

‡Knight's work was in the field of economics and he considered that if uncertainty could be measured, there would be no reward for decision owners when taking risks, such as an entrepreneur/owner investing in a new factory. This also distinguishes between the decision-maker, typically a manager and the risk owner, the shareholder who bears the cost or enjoys the proceeds of the risk outcome.

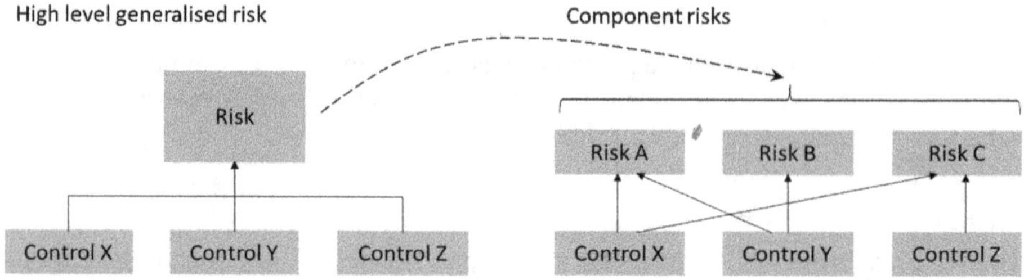

Figure 3.2 Generalised risk and technical approaches to risk assessment

3.1.7.1 Generalised approach to risk management

In the high level, generalised approach to risk, we are establishing a broad risk of 'cyber attack', and then applying a set of controls to mitigate attacks: access controls, authentication, firewalls, backups etc. The risk of cyber attack could be estimated from reports of other organisations, taking ranges of impact and frequency. This may help inform a budget for the controls, an estimate of the likely reduction in successful attacks and thereby residual risk. This is similar to how we might treat other corporate risks, such as currency changes and corruption. It is essentially treating the risk as a black box and then applying mitigations (controls) to reduce it. Such an approach is appropriate for essential baseline or 'cyber-hygiene' controls, but beyond this more detailed risk assessment is needed for critical systems.

3.1.7.2 Technical approach to risk management

In the safety domain, and increasingly for cybersecurity, risk analysis is done at a more detailed level. This requires looking at hazards (in safety) or threats (in security) to assets. The estimated effect of a safety hazard or security threat event on operations defines the impact and, coupled with estimated event frequency/likelihood, leads to a first pass estimate of risk. Risk is not static, however, so new knowledge can be added on the likelihood to support a re-assessment of the risk, using techniques such as Bayesian networks [3].

The level of detail for the analysis is an important consideration. The idea is to first identify the essential operational functions or components and the critical information and assets that support them. For example, an airport's operational database or its baggage handling system would be regarded as central to operations. From here, the system can be broken down into components, which may require an iterative approach to not be overwhelmed with detail.

There is clearly a link between the generalised and technical approaches to risk assessment as the generalised risk can be built up from the component risks. Generalised and technical approaches are not mutually exclusive and both have application in cybersecurity.

3.2 Cybersecurity risk management

3.2.1 Introduction

The concepts discussed in the preceding paragraphs are applicable to risk in general: how it is defined, used and described within an organisation. However, cybersecurity risk presents an additional set of challenges.

First, risk management relies on several assumptions concerning the ability to measure certain variables. Amongst others, one may hope to correctly identify vulnerabilities and the correct status of untreated vulnerabilities, measure a threat level and identify possible losses as the result of an attack. [10]. Unlike traditional domains, many of the assets, threats, vulnerabilities and risks are intangible, and therefore extremely difficult to measure – particularly if an organisation does not know where to look. This difficulty in effective measurement can lead to inaccurate risk assessment findings and, as a result, inadequate organisational responses.

Related to the previous point, there is a knowledge deficit concerning cyber risks. Cyber incidents can go unreported, hampering future efforts in threat and vulnerability identification [11]. Furthermore, organisations have difficulties in obtaining expert knowledge to help them.

Finally, cyber is a domain characterised by its dynamism. New technologies are constantly introduced, challenging businesses to constantly adapt. For existing systems, it is more of a question of when, not if, new vulnerabilities will be found. State actors and cyber criminals are also motivated to be constantly searching for new exploits. This dynamism makes cyber risk assessment a more difficult endeavour.

All these challenges underline the importance of implementing cyber risk management as a continuous activity.

3.2.2 Cybersecurity risk concepts

In information security, there are some specific risk concepts which bear further explanation, as their usage is not as straightforward as it might seem. We define these terms below, but note that various standards and frameworks, such as National Institute of Standards and Technology (NIST) [12], International Standards Organisation (ISO) 27005 [1] and Control Objectives for Information and Related Technology (COBIT) [13], have their own definitions.

3.2.2.1 Assets

An asset is something that helps an organisation accomplish its business objectives and has value for the organisation. Dubois *et al.* [14] describe two types of assets:

- Business asset: an intangible asset (function, service, process or information) required to fulfil the business objectives of an organisation. For example, a satellite navigation service.
- Information system asset: where system assets are 'tangible assets' that enable business assets, such as the satellites, ground stations and communication

infrastructure that supports the satellite navigation service. It is the information system assets that have vulnerabilities, which may be exploitable by threats aiming to impair business assets.

Similarly, ISO 27005 describes 'primary' and 'supporting' assets:

- Primary asset: Intangible function, service, process or information that is part of the information system within the scope of the project and has value to the system.
- Supporting asset: entities which enable the primary assets. Supporting assets possess the vulnerabilities that are exploitable by threats aiming to impair primary assets.

Assets possess certain security properties or objectives, which characterise the security needs of confidentiality, integrity and availability:

- Confidentiality: the information should be disclosed only to the intended and authorised persons, entities and processes and no one else.
- Integrity: information should be accurate and complete. If modifications and deletions are made by unauthorised parties, they should be detectable.
- Availability: information should be available to the authorised persons, entities and processes when it is needed.

3.2.2.2 Risk

As discussed previously, risk is the combination of likelihood and impact. The likelihood is of a threat being carried out by an 'attacker' or 'threat actor' and the impact concerns the effect of the attack, such as disruption to operations.

3.2.2.3 Threat

This is the method of attack, to be carried out by a threat actor against an information system asset. Dubois describes a threat as a possible cause of an unwanted incident which may result in an impact to the organisation. In this sense, lightning strikes which disable a control centre could be considered a threat. NIST [15] defines a threat as 'any circumstance or event with the potential to adversely impact organisational operations and assets, individuals, other organisations or the Nation through an information system via unauthorised access, destruction, disclosure, or modification of information, and/or denial of service'.

Recalling our earlier sporting event example, we would normally speak of the 'chance' of rain, but for the sporting event, it would be more appropriate to think of rain as a threat with a likelihood of occurring and an impact, which leads to a risk for the day in question. In the information security context, a simple example of a threat would be the theft of a laptop. Theft is the threat, potentially enacted by a thief, and we can imagine a clear impact and likelihood leading to a risk.

3.2.2.4 Threat actor

A threat actor could be a person or software, defined as something that can potentially cause harm to system assets. In our laptop example, the threat agent is the thief. NIST describes threat actor as synonymous with 'Threat Source', and the NIST definition combines the intent and method[§] that exploit a vulnerability.

3.2.2.5 Attack method

The attack method is the means by which a threat agent executes a threat. This could be a thief breaking and entering, or stealing an unattended bag on a train. Attack methods can be quite elaborate and are discussed later in the book.

3.2.2.6 Vulnerability

Vulnerability is a security weakness of an information system asset, which can be exploited by an attacker via a threat. For example, the security weakness of an unattended laptop is that it is mobile and can easily be moved. Without any additional mitigation, such as hard disk encryption, the data on the laptop are vulnerable to being accessed by an attacker.

3.2.2.7 Impact

Impact is the extent to which a loss of confidentiality, availability or integrity of an asset affects the operation or achievement of business objectives. For example, loss of availability or integrity of a satellite navigation service could lead to extended journey times.

3.2.2.8 Risk evaluation

Risk evaluation involves deciding what to do with the risks analysed: to tolerate (accept) a risk, treat (mitigate) it, transfer it (to another party) or to terminate (avoid) it. These are further detailed as follows:

Tolerate the risk (risk acceptance)

To tolerate or accept a risk, it must be low in the context of the benefits of the operation and may also be costly to treat.

For example, radiotelephony communications between pilots and controllers [16] are openly broadcast. Modernisation to digital communications supports encryption, but there is no regulatory requirement to do so. Hence the communications may remain open, influenced by costs and bandwidth issues; that is, there may

[§]NIST also refers to a 'situation and method' that exploit a vulnerability.

be insufficient bandwidth for encryption without costly upgrades to communications infrastructure. This in turn may limit the operational concepts that can be deployed, such as issuing and responding to safety critical clearances, or influence the level of risk that is tolerated.

Treat the risk (risk mitigation)

In deciding to treat a risk we are aiming to reduce the likelihood and/or impact.

For example, risk mitigation could include the installation of technical controls, security monitoring and the implementation of business continuity and disaster recovery plans. Risk mitigation may be achieved through multiple controls, thereby creating 'defence in depth'.

Transfer the risk

To transfer risk, the organisation contracts a third party to take on the risk.

In a cybersecurity context, the mechanism for risk transference would be cyber insurance, where a premium would be adjusted according to the risks and the security posture of the business wanting to transfer the risk. Transferring risk needs to be thought about carefully as it may be that it is liability only that is really being transferred, as operators may still carry the risks of loss of, for example, reputation or market share.

Transferring of risk brings another aspect of risk into focus, in that risk is personal to those experiencing it. Passengers suffering delay from transport system failure are unlikely to be recompensed for their entire loss. The costs of delay in the European air traffic system [17] impact both airlines and passengers, with passengers compensated for significant delays by airlines. Airlines, however, may not be compensated by failures outside of their control, such as when the European Network Manager experienced an outage in 2018 [18].

Terminate the risk (risk avoidance)

If the risk exceeds the benefit of continuing with a project, then an organisation may adapt to avoid any chance of the risk being realised. This could mean choosing not to do an activity, deploy a system or provide a service. In the authors' experience, terminating risks is uncommon, and it is more likely that the operational concept or service is scaled back rather than entirely dropped.

3.2.2.9 Security control

The concept of a 'security control' or 'control' is specific to security and cybersecurity. It refers to a risk mitigation, which can be any measure, process or technical system applied to protect a given asset. For example, deploying a firewall on a network boundary, vetting staff, deploying antivirus etc. Controls are covered in detail in Chapter 6.

3.2.3 Cybersecurity risk management standards

3.2.3.1 Introduction

There are a wide variety of methods to manage risk, including those described in international standards such as ISO 27005 [1], and others that were introduced in Chapter 2. Most organisations rely on some form of cybersecurity framework to assess cyber resilience, or as a starting point for a bespoke framework [19]. There are advantages with a common risk management framework, as this supports sharing best practices within and between different sectors. For instance, EUROCAE Document ED-201 [20] asserts that 'co-operative and collaborative information security is more effective and efficient than independent and isolated information security'.

3.2.3.2 ISO 27005

Cybersecurity risk management is simply risk management applied to the cyber domain. In this section, we give an overview of the process based on the ISO 27005 standard, which offers guidance for information security risk management.

The ISO 27005 standard follows from ISO 31000, which is a more generalised approach to enterprise risk management. This is intended to ensure that information security risk management is consistent with more general enterprise risk management approaches.

ISO 27005 does not specify a specific risk management methodology [21] but embeds a continuous information risk management process with five components: context establishment; risk assessment (including risk identification, analysis and evaluation); risk treatment; communication and consultation; risk monitoring and review [22]. Figure 3.3 presents a simplified structure of the framework and its interactions.

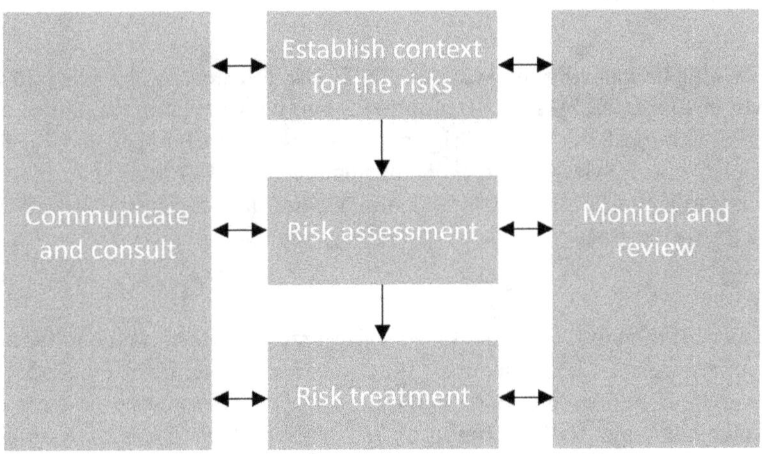

Figure 3.3 ISO 27005 risk management process

The information security risk management process begins with 'context establishment', as security management does not take place in a vacuum. To do this effectively, professionals need to have a wide understanding of the operations and business, including its goals, stakeholder requirements and applicable regulations. It is also necessary to define the scope of the risk management exercise, its schedule, and the staff and budget available.

Risk assessment follows from context establishment and is split into three substantial steps: risk identification, risk analysis and risk evaluation:

- With risk identification, the task is to identify the risks that affect the assets.
- Risk analysis is conducted to determine the specifics of the risk.
- Risk evaluation determines the measures to be adopted for each risk: tolerate, treat, transfer and terminate.

The risk management process does not end with the application of an appropriate treatment but is a continuous endeavour requiring effective monitoring and communication of risk. Risk monitoring ensures that any change in probability or impact of a risk being realised is captured as quickly as possible, allowing for security controls to be updated to reflect changing risk profiles. Communicating the organisation's risks therefore plays a role in updating the risk picture.

3.2.3.3 Other risk frameworks

There are a variety of risk management frameworks, with the IET/ISO (International Standards Organisation) and NIST among the most widely recognised. The choice of standard may depend on the sector, the nature of the assets to be protected and the organisation's capabilities (in terms of organisation size, security team size and budget). For smaller organisations where lighter frameworks are needed, there are alternatives such as OCTAVE-S. Other examples are given in the following sections.

RMF – NIST

The Risk Management Framework, or RMF, is published by the US NIST [23]. The RMF is supported by a NIST-supplied list of controls and mitigations such as Special Publication 800–53 (Security and Privacy Controls) [24] and SP 800–82 (Industrial Control Systems) [25]. While mandated in US federal organisations, NIST standards can be applied to any organisation and are in the public domain (making them freely accessible).

COBIT

COBIT is a framework of 'best' IT security practices, designed to manage and document enterprise IT and security functions for an organisation [26]. COBIT has the goal of addressing risk management, IT performance and regulatory compliance concepts, with mapping of IT security goals to business objectives. COBIT is based on five key principles for governance of enterprise IT: meeting stakeholder needs; covering the enterprise end to end; applying a single, integrated framework; enabling a holistic approach and separating governance from management.

ITIL

Formally the Information Technology Infrastructure Library (ITIL) is a joint venture between the UK government and a private firm and concentrates on how a security function maps to an organisation's business goals [27]. ITIL is mapped to the ISO 27000 family of standards¶.

OCTAVE Allegro

The Operationally Critical Threat, Asset, and Vulnerability Evaluation (OCTAVE) – 'OCTAVE [28] Allegro' is an asset-focused method that follows through the establishment of qualitative risk measurement criteria, risk profiling, real-world threats and impacts, and risk prioritisation [29].

3.2.3.4 Comparing risk frameworks

Risk methodologies have differences in terms of scope and focus. For example, a comparative analysis between ISO 27005 and OCTAVE Allegro revealed ISO 27005 having a strong focus on threat, vulnerability and control assessments compared to OCTAVE Allegro. In contrast, OCTAVE Allegro addresses broader topics such as organisational direction and impact area prioritisation [30]. OCTAVE Allegro focuses on information assets; the NIST RFM offers an extremely structured approach; ISO 27005 covers the security risk assessment process in a more generic manner, while COBIT holds a heavy focus on governance.

A common factor in risk frameworks is the idea of a hierarchy, in which a risk analysis or risk methodology forms a sub-step within a risk assessment, which then fits into a risk management process or programme [31].

3.2.3.5 Supply chain risk

An organisation must look beyond their own infrastructure to gain a complete picture of their risk profile. Any third-party vendor or supplier providing hardware, software or firmware for an organisation can increase risk. Supply chain risk refers to the risk throughout the supply chain, including service providers, such as the email provider, or delivery service, for an organisation. Supply chain risks come in many different forms. For example, a compromised supply chain actor can be used as a means to target another organisation, or can have another organisation's data revealed as part of a breach.

¶The ISO 27000 family includes 27001 on an Information Security Management System, 27002 on a code of practice for information security management and ISO 27005 on Information Security Risk Management.

There are different approaches to supplier risk management. These may involve asking suppliers to self-report, meet requirements listed in an agreed framework or set of guidelines, or provide the results of an internal or external audit. The level of supplier assurance required will be commensurate to the level of risk presented should the supplier be compromised. The more potential a supplier has to present a higher-level risk to an organisation, the more enhanced the supplier assurance process is likely to be. For example, there may be less stringent checks on an organisation that only provides paperclips, compared to an organisation which provides critical software containing confidential company data. We can therefore say that effective supplier assurance will take a risk-based approach to supplier management [32].

Under the European Union's General Data Protection Regulation [33] (GDPR), an organisation is accountable for the security of their data even when it is in the hands of supply chain actors, which creates additional supply chain risk. GDPR Article 28 stipulates that the data controller** holds primary responsibility for compliance with the regulation [34]. This responsibility overrides any contracts with data processors (who would form part of the supply chain for the data controller). Should data be exposed by data processors, it is the data controller who faces liability through sanctions and corrective measures, including administrative fines. To avoid liability, the data controller must be able to prove that they are in no way responsible for the event that led to the damage. This may be achieved by demonstrating sufficient diligence through an effective supplier assurance procedure.

In addition, an organisation may choose to exercise other risk management activities, such as working with suppliers to increase their security levels (risk treatment), avoiding less secure suppliers (terminate risk) and insurance (risk transfer). Risk transfer may also include the shifting of liability through contracts that specify security expectations.

There are many challenges relating to risk management across the supply chain. As well as inherited risks and liability, accurately tracking and auditing suppliers can be a lengthy exercise, particularly in large organisations. A supply chain often has many suppliers, and in a complete-knowledge scenario, an organisation would have visibility of a supplier's suppliers. The lack of transparency across the supply chain is well recognised and the potential mitigations include the use of blockchain as a mechanism to provide traceability with low administrative cost [35].

3.2.4 Analysing cybersecurity risk

With any risk framework, the analysis of cyber risk can be done in a *quantitative, qualitative or semi-quantitative way*. The argument for using a qualitative method is that it is relatively quick, and the likelihood of an attack is difficult to know anyway.

**The data controller is the ultimate owner and processor of the data, who determines the purposes and means of the processing of personal data.

As will be seen in the rest of the chapter, this approach is hard to defend. The choice of method will depend on the maturity of cybersecurity risk management within an organisation. Given that the transport sector already uses semi-quantitative methods for safety, which is also a difficult area to analyse, it is likely that cybersecurity risk evaluation will become at least semi-quantified.

Using quantitative methods has the advantage of providing a better granularity when assessing different aspects of a risk. This granularity supports easier risk prioritisation and more accurate cost-benefit analysis of risk treatments. Taken together, they help with decision-making around the key areas to address. However, the results of quantitative methods are harder to present. They often require additional interpretation and explanation, especially for non-technical audiences. In addition, the use of numerical values might hide the fact that the assessments are still the product of a subjective analysis [36], according to the assumptions made, but where these are transparent there is the opportunity to model different scenarios.

Examples of quantitative methods include the 'Annual Loss Expected' (ALE) method, Jack Freud's and Jack Jones' 'FAIR method' [37] and Hubbard and Seiersen's *How to Measure Anything in Cybersecurity Risk* book [4]. Underlying the latter two methodologies is the estimate of impact, or loss, on the occurrence of a cyber attack and the frequency of such a loss. By converting the frequency of loss into an annual equivalent, we arrive at the ALE method. The steps are as follows:

1. Estimate the impact of a loss event, such as theft of a laptop. This will depend on the information stored on the laptop and whether it gives access to additional information, or control access to corporate servers. A Ponemon Institute survey in 2009 estimated the average cost of losing a laptop to be €35k and a staggering 265 laptops per organisation were lost or missing over a 1-year period [38].
2. Estimate the frequency of any such event, here 265 per year. Further data in the Ponemon study show that the average loss ratio over a laptop's useful life was 8%.
3. Multiply (1) by (2) to get the annual loss expectancy, so €9.4 million (M) per year multiplying the above figures, and an average of €4.7M per organisation reported in the study.

Qualitative assessment methods are in wider use than quantitative approaches and assess risk with *non-numerical levels* [36]. The major advantage of qualitative methods is the ease in which the results can be communicated through the organisation. Nevertheless, the limited range of values can hinder risk prioritisation and comparison as there is a possibility of 'result collision'. Depending on the criteria, some risks with different values might end up with a similar outcome. Furthermore, qualitative methods are more prone to the influence of subjective experience when designating values for the same risk. For those reasons, care must be exercised when defining values with meaningful examples [36].

Semi-quantitative methods are a hybrid of qualitative and quantitative features. The use of 'bins' or 'scales' supports risk communication as they translate easily into qualitative terms [36]. With the right amount of granularity in the bins and scales, the prioritisation of the results is better than in purely qualitative methods.

3.2.5 Resourcing cybersecurity risk management

In practice, the majority of resources are expended on treating risks in the first instance and addressing residual risks (the risks remaining after treatment) through risk transfer to insurers or suppliers, or risk acceptance by the business. These actions also have cost impacts and expenditure on treatment, and transference is a key decision-making area.

Without sufficient investment in cybersecurity, there is also the potential for significant fines from regulators. For example, in October 2020, the UK Information Commissioner issued a £20M fine to British Airways for 'failing to protect the personal and financial details of more than 400,000 of its customers' [39]. At the time of writing, the action against British Airways was the largest enforced fine to date by the UK Data Protection Authority. Given the £20M fine was explicitly lower than the £183M intended fine (with the decrease taking COVID-19 economic challenges into account), it is likely that penalties will continue to grow in size, signalling the capability to hold organisations financially accountable to data protection regulation [40].

It is worth noting that the size of penalty fines has increased considerably under the EU GDPR. With two-tiered levels of fines, the first level may be up to €10M or 2% of an organisation's revenue for the previous year (whichever is the higher total). The second tier, for major infringements, can fine up to €20M or 4% or an organisation's annual revenue for the previous year, whichever is higher [41].

There is no hard and fast rule on how much an organisation should invest in cybersecurity, and certainly no quantitative consensus. Prominent academic approaches include the Gordon-Loeb model, which broadly advises the investment dedicated to protecting a particular asset should not exceed 37% of the worth of that asset [42]. This would not apply in safety critical contexts but gives insight into the decision-making process concerning business systems. For critical national infrastructure such as transport, the amount invested in cyber defence may readily exceed the financial value of the asset, with safety coming first.

Of course, risk analysis is not the only (and often, not the main) factor determining security spending. Financial constraints impact the extent to which security resources may be secured, with many organisations facing a reducing budget for security and/or cybersecurity more specifically. Budgets will usually be allocated by business decision-makers, including board members such as the Chief Financial Officer and Chief Executive Officer. As such, an engaged senior leadership that understands the implications of cybersecurity risk management may be more likely to allocate resources towards cybersecurity investment. Risk communication therefore becomes a key mechanism for a successful risk approach.

3.3 Walk-through of risk management

3.3.1 Introduction

In the following paragraphs, we provide a high level walk-through of a risk assessment to familiarise readers with how the process works. The walk-through is based on the ISO 27005 framework described in Section 3.2.3.2, aiming to provide some insight into the information used and practicalities of the process.

We have chosen a transport-relevant example which will be familiar to most readers. Our example is a vehicle's remote keyless system RKS, which is present in the majority of modern cars. The functions of an RKS include unlocking a car's doors from a distance (remote keyless entry) and starting the engine without inserting the car key (remote keyless ignition). This system is also known as a 'Passive Keyless System'. The authors have no specialist knowledge of the area and the example is purely to illustrate the process. We initially take a qualitative approach to risk evaluation and later add the quantitative approach.

3.3.2 Establishing the context

The overall context of the assessment is that the manufacturer of the RKS wants to be assured that their system provides an adequate level of security for their customers. The *internal context* may include the following:

- The manufacturer's business objectives.
- The security policy for the manufacturer's products. This may include requirements on conducting and reviewing risk assessments, such as requiring independent review and testing.
- The risk assessment methodology, which is likely to be tied to an international framework standard such as ISO 27005. This may include requirements on who conducts the product risk assessments, such as competent security design experts.
- Products may be marketed with different levels of 'acceptable risk' for different clients (e.g. for civil or military users).

The *external context* may include:

- Regulatory requirements to implement security.
- Industry standards for security.
- Customer expectations for the security of the product.
- The operational context for the product, where and how it is used and by who.

The above items, and more, may need to be documented as part of the risk assessment, with standard items documented in the organisation's 'risk management manual'.

The *scope* of the assessment may include the following:

- The systems included in the assessment, for example, identified on a system architecture diagram.
- Interfaces to, and dependencies on, systems that are not included in the scope.

Bearing in mind that controls are applied to tangible assets, such as adding a firewall to a network, the 'business' or 'primary' assets are a valuable step in sub-dividing the overall problem. A primary asset could be described as a service or process, such as 'unlocking and starting the car', or information, such as the key information held by the key fob. These non-tangible assets can then be ascribed to the physical assets that support them, which includes the software needed. This leads us to two primary assets to consider:

1. service or process of unlocking and starting the car
2. key information

Where it exists, an organisation's Enterprise Architecture is potentially a good source of information for identifying business/primary assets. Enterprise architecting can be an involved process, however, so where it does not already exist, it will be much faster to identify primary assets from workshops with relevant experts. Primary assets can be captured on whiteboards and then later in drawing software. Workshops are invaluable in identifying primary assets, especially when involving a wide range of operational, engineering and business experts.

In defining the scope, some consideration also needs to be given to the granularity of the primary assets. In Figure 3.4, we show primary assets broken down into two or three components, by separating 'unlocking' and 'starting' in the second example. A more detailed breakdown of primary assets may ease the identification of supporting assets in the next step.

Some caution is needed against becoming too granular as the mapping of primary and supporting asset could become one to one, unnecessarily increasing the work of identifying impacts.

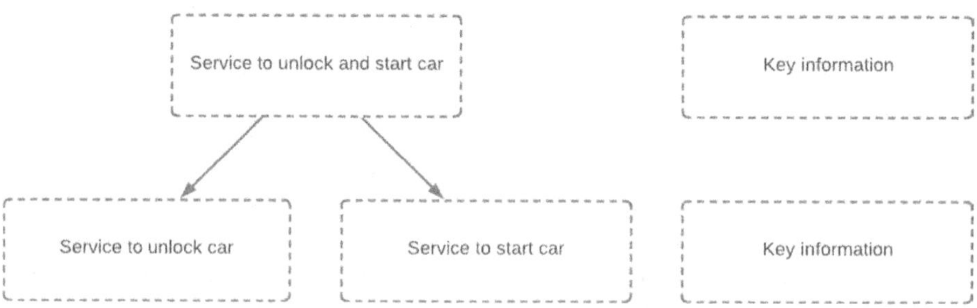

Figure 3.4 Defining primary assets

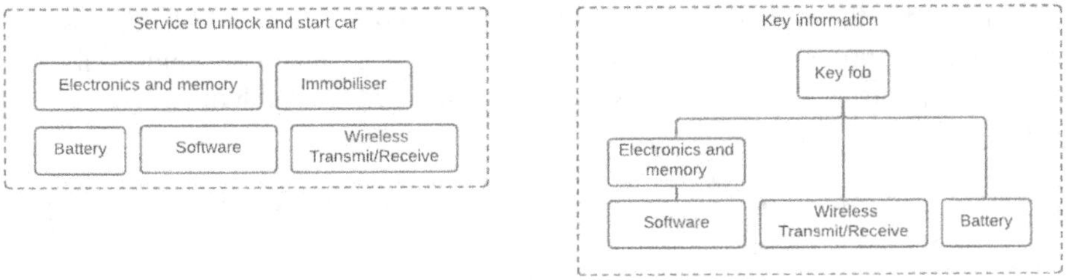

Figure 3.5 Establishing tangible supporting assets

Following the identification of the business or primary assets, the next stage is to identify the tangible assets supporting them. In ISO 27005, these are known as 'supporting' assets and specifically include software. For the RKS, the obvious tangible assets are the software and hardware of the car key fob held by the driver, the car receiver unit and the power supplies to them (Figure 3.5). The RKS car receiver communicates wirelessly with the key fob, which holds a chip with the data that ties a key to a specific car.

In identifying supporting assets, there is also a question of granularity, but this time the reference point is to the controls that might be applied. For example, if the same controls would be applied to the key fob as to its components (say chip, battery and transceiver), then there may be no need to break down the supporting asset list further. The above figure shows the key fob subdivided into 'electronics and memory', software, wireless transmit/receive and a battery.

Could this subdivision be too much detail for a small device? The answer may depend on the organisation's methodology, component supply chain and the need to control component costs so that they are not overspecified. For example, it may be useful to separate out the software to apply specific development and test controls to. These may be standard for the organisation and form part of the outsourcing agreement. The main reason for separating out components is to create structure for the risk assessment, so that the coverage of risks can be maximised. If we did not separate the software out of the key fob, it might be missed in the identification phase and security requirements not placed on the supplier. These principles apply at different levels, so that while a manufacture may be interested in the component level, service providers may breakdown to system and subsystem level only.

3.3.2.1 Security criteria/objectives

A further aspect of the context is the security criteria[††] of interest. This is very likely to be the standard set of 'confidentiality', 'integrity' and 'availability', also known as 'CIA', meaning that we want to protect the CIA of the business or primary asset in question.

[††]These security criteria are also referred to as security attributes, objectives, requirements etc.

The CIA criteria are then 'inherited' by the supporting assets. Hence to protect the confidentiality of the key information, we need to protect the confidentiality of all of its supporting assets. The CIA criteria can be used to help define the impact of a security breach, using them as a set of questions:

- What will be the impact if we lose the **Confidentiality** of the key information?
- What will be the impact if we lose the **Integrity** of the key information?
- What will be the impact if we lose the **Availability** of the key information?

Note that additional security criteria could be defined in the risk assessment process such as non-repudiation or accountability [43], depending on the needs of the organisation.

Confidentiality of the RKS key data is vital to ensure only the legitimate drivers of the car can access it. If confidentiality is breached, unauthorised parties could steal the car. The integrity of the data must be guaranteed or else the driver will not be able to access the car due to corrupt or missing data. Finally, the availability of the data is a must. If the RKS key data are not available when the RKS needs them, the driver will not be able to take advantage of the RKS features.

3.3.2.2 Estimating the impact of loss of CIA on each primary asset

Table 3.1 is an example impact scale created for this example. An organisation will normally establish such look-up tables while developing its risk assessment methodology and five levels are more common.

An observation of the above scale is that it includes a characterisation of the recovery. This is not necessarily needed, but in cases where there is an operational service, the recovery profile is very relevant. More details on incident preparedness and operational continuity management may be found in References [44, 45].

Table 3.1 Example qualitative definitions of impact

Impact	Level	Description
Insignificant	1	The impact is almost negligible and the recovery from it is likely to be easy
Minor	2	The impact is small and recovery is likely to be straightforward
Moderate	3	The impact is significant and recovery to normal operations takes several days. For organisations, several occurrences over time could potentially put the company out of business
Major	4	The impact is initially high and recovery could take weeks. At an organisation level, the company could go out of business

Vehicle owner perspective of impact

Table 3.2 evaluates the impact using both the above scale and a quantitative estimate for comparison. We start with the perspective of a car owner before moving to the manufacturer's perspective.

Table 3.2 is the authors' interpretation and readers may take a different approach to what the CIA impacts mean in each circumstance and come up with different impacts. This part of the assessment is similar to cost-benefit analysis, but rather working out the impact of taking something out of the transport system rather than adding to it.

One of the challenging aspects of the above impact evaluation is to determine what the loss of CIA means. For example, we have taken loss of confidentiality of the 'service to unlock the car' as meaning that it leads to loss of the car. However, we could alternatively consider that loss of confidentiality means that details of the service are known, which could be an intellectual property issue for the manufacturer, and also lead to vehicle theft. We have also considered that the loss of integrity of the service results in an attacker gaining access to the vehicle, but this could be

Table 3.2 Primary assets impact of loss of CIA – owner's perspective

	Impact with loss of CIA for the case of an individual car		
Primary asset	**Confidentiality**	**Availability**	**Integrity**
Service to unlock the car	If the confidentiality of the service is lost to an attacker, this could be exploited by theft of the car	The user will be unable to enter the vehicle. This incurs a vehicle recovery and repair	Any modification or misuse of data could lead to the user being unable to enter the vehicle, incurring a recovery and repair cost or an attacker gaining access and stealing the vehicle
Impact level (value)	3 (€20k)	2 (€1k)	2 (€1k)–3 (€20k)
Service to start the car	Attacker may be able to start the car and steal it, assuming they have also gained entry to the vehicle	User unable to start, incurring vehicle recovery and repair	As for availability
Impact level (value)	3 (€20k)	2 (€1k)	2 (€1k)
Key information	Attacker could exploit the key information leading to the theft of the vehicle	User unable to start, incurring vehicle recovery and repair	As for availability
Impact level (value)	3 (€20k)	2 (€1k)	2 (€1k)

more narrowly interpreted as an act of mischief needing repair rather than loss of the vehicle. Workshops and rounds of review are therefore very helpful in carrying out this and other steps in risk assessment.

The primary asset impacts defined above will subsequently be 'inherited' by the supporting assets that relate to each primary asset. Hence the loss of 'service to start the car' is inherited by each of the supporting assets shown in. Given this, the above impacts may be consolidated in some way for the next stages. The SESAR Security Risk Assessment Methodology [46] takes the maximum value of C, I or A for each primary asset, which reduces the impact for each asset, as shown in Table 3.3.

The consolidated information should not, however, result in the loss of the different CIA impacts. Different security controls may serve in different ways, mitigating for C, I or A separately or in combination. For example, staff confidentiality agreements may be effective to reduce the loss of confidentiality around a system component, but would not help the integrity of that component, which is a design issue. In the SESAR risk assessment methodology mentioned, this issue was part of the requirements for a database support tool [47], supporting the selection of controls for each risk.

Vehicle manufacturer perspective of impact

Moving on from the single car, in Table 3.4, we address the car manufacturer's perspective, as the producer of tens to hundreds of thousands of cars per year. The main considerations we have made here are that the manufacturer may bear the costs of product recall and there may be a loss of brand value leading to fewer car sales and less net profit.

In Table 3.5, we summarise the results for the manufacturer's perspective by taking the maximum CIA values across each column.

This example already shows the value of quantifying impacts. Assuming there were no cybersecurity controls, the impact equates to a cost of €310 per car per year (3.1M divided by 100,000), indicating that strong cybersecurity which might cost less than €50 per car would be well worth the investment, unless the likelihood of cyber attack is low already.

If we had done a qualitative analysis only, however, our definition informs us only that the maximum impact was a '4', and therefore 'high'. We also do not see the granularity of impact between the assets, although there is little practical use in distinguishing between the primary assets with the impacts of €28M or €31M.

Table 3.3 Primary assets maximum impact – owner's perspective

Primary asset	Maximum impact level (cost)
Service to unlock the car	3 (€20k)
Service to start the car	3 (€20k)
Key information	3 (€20k)

Table 3.4 Primary assets impact of loss of CIA

	Impact with loss of CIA for the case of many cars		
Primary asset	**Confidentiality**	**Availability**	**Integrity**
Service to unlock the car	Inability to unlock known to others with no impact. Affects the brand image of the manufacturer, potentially leading to 5% fewer sales over 1 year while the problem is solved plus recall and repair costs	The user will be unable to enter the vehicle. This incurs a vehicle recovery and repair cost and more significantly, brand damage as for confidentiality impact	Any modification or deletion of data within the primary asset could lead to the user being unable to enter the vehicle, incurring a recovery and repair cost. Additional impact of car thefts may lead to increased brand damage (8% of sales lost) and insurance costs for owners
Impact level (value)	4 (€28M: 100,000 cars recalled at a cost of 200 per fix (20M) plus profit reduction by 8M, assuming 20k revenue per car, 8% net profit margin and 5% reduction in sales)	As for confidentiality	4 (€31M: 100,000 cars recalled at a cost of 200 per fix (20M) plus profit reduction by 11M, assuming 20k revenue per car, 8% net profit margin and 7% reduction in sales)
Service to start the car	Inability to unlock known to others with no impact	User unable to start, incurring vehicle recovery and repair costs	As for availability
Impact level (value)	4 (€28M as above)	As for confidentiality	As for confidentiality
Key information	Attacker could exploit the key information leading to the theft of the vehicle	User unable to start, incurring vehicle recovery and repair costs	As for 'Service to unlock the car'
Impact level (value)	4 (€28M as above)	As for confidentiality	4 (€31M as above)

Table 3.5 Primary assets maximum impact – manufacturer's perspective

Primary asset	**Maximum impact level (cost)**
Service to unlock the car	4 (€31M)
Service to start the car	4 (€28M)
Key information	4 (€31M)

There is, however, a dependency with these estimates, which relates to the number of cars stolen. If the manufacturer's cars are stolen at around the average rate of theft, then there may be little damage to brand and no need to recall the model. We return to this point later on in the discussion of risk and controls to be applied.

3.3.2.3 Impact categorisation

In the preceding discussion, we did not differentiate between different types of impact, but this may be a helpful step. The SESAR risk assessment methodology [48] defines several impact areas: 'personnel' (safety of people), 'capacity', 'performance', 'economic', 'branding', 'regulatory' and 'environment'. The impact of loss of CIA on a primary asset can be evaluated across all of these areas. For example, a successful attack could lead to injury, reduce capacity, have an economic cost, harm the brand of an organisation and lead to regulatory action. In practice, there are dependencies between these categories and they could be simplified.

3.3.3 Risk assessment

After having identified the assets in scope and the impact of losing their confidentiality, integrity and availability, we next consider scenarios of how this might happen and the likelihood of those scenarios.

3.3.3.1 Risk identification

Risk identification is described in ISO 27005 [1] as considering the threats to an asset, how those threats may arise, and the potential consequences. ISO 27005 also refers to 'incident scenarios', which describe 'a threat exploiting a vulnerability or a set of vulnerabilities'. We have drawn on the SESAR JU Security Risk Assessment Methodology [49] to expand on this approach, which describes 'threat scenarios' which target vulnerabilities present in the system. In Table 3.6, we identify some relevant threats to the RKS supporting assets and the scenarios in which they could arise.

For Scenario 2, it is most likely that the data transmitted are encrypted, which means that the attacker has to decipher it or replay the encrypted data for the attacks. Older systems required the driver to depress a button on the key fob to initiate the transmission, to which the car unit then responds. When these were first introduced, the replay attack was effective but was subsequently mitigated by the addition of rolling codes to each transmission, which could be used only once, with any replayed code ignored. This mitigation was not perfect, however, as attackers could simultaneously jam the key fob signals so they were not received by the car unit and store the signals for later use [50].

Threat Scenario 3 is the well-known 'relay' attack. The security design of RKS systems is for bi-directional transmission between the car unit and the key fob. The car unit regularly transmits a signal to the key fob and, if the key fob is nearby, responds by unlocking the car [51].

Table 3.6 Threat scenarios

Supporting asset	Key fob
Impact inherited by system asset	€31M
Threat Scenario 1	**Theft of key fob** A thief obtains the key fob, makes a copy of the data and replaces it back into the driver's possession. If the data are not stored in an encrypted format, this leads to the negation of the RKS key data and theft of the car at a later time
Threat Scenario 2	**Interception of wireless signal from key fob** A thief intercepts and copies the wireless transmission between the key fob and the car unit. The attacker then replays this transmission to the car unit when the key fob is no longer nearby, resulting in theft of the car
Threat Scenario 3	**Relay of wireless transmission from key fob** A thief relays the key transmission from the car unit to the key fob, using a high gain antenna, gaining sufficient access to open and start the car

Threat catalogues

Creating threat scenarios can be aided by reference to catalogues of generic threats and vulnerabilities and examples are provided in ISO 27005 [1] and, for example, NIST's 'Mobile Threat Catalogue' [52]. 'Threat' and 'vulnerability' are inextricably linked and often referred to interchangeably. This can be somewhat confusing, so we make the following observations:

- Vulnerabilities and threats are referred to in either general or more specific ways:
  - General: a threat of 'electromagnetic radiation' and a vulnerability to electromagnetic radiation.
  - Specific: a threat of SQL (structured query language) injection and vulnerability to SQL injection. More specifically, the Common Vulnerabilities and Exposures database [53] references a variety of SQL vulnerabilities, such as CVE-2021–3278 which is described as 'Local Service Search Engine Management System 1.0 has a vulnerability through authentication bypass using SQL injection. Using this vulnerability, an attacker can bypass the login page'.
- At the more general level, a vulnerability can be exploited by one or more threats, and a threat can exploit one or more vulnerabilities, that is, there is a many-to-many relationship between threats and vulnerabilities.

3.3.3.2 Risk analysis

Once the threats have been identified, we next articulate them as risks. Building on the threat scenarios and the resulting impact, the remaining question is the likelihood of the threat being actioned. For simplicity, we start with a qualitative risk assessment and scales of likelihood, with a table of definitions created for this example in Table 3.7. We have also chosen a four-level scale for likelihood but note that five-level scales are common.

In creating this table, it was apparent that it is quite difficult to specify a qualitative description without including a quantitative reference. Also apparent is that the above does not neatly join each scale, so there is a gap between the occurrence of rare (5+ years) and likely (1–3 years). Our risk table therefore needs some attention to avoid these gaps. The likelihood is further qualified in a second example in Table 3.8.

We can be more specific about likelihood, however, as car thefts are recorded crimes. Eurostat reports [54] an average of about 520k per year for the EU. With around 230M cars in the EU [55], the average rate of theft is 0.0023 per year in the EU. For our car manufacturer producing 100k cars, around 210 per year may be expected to be stolen in total. Further data suggest 92% of cars are taken without using the keys, in relay attacks [56].

This analysis could be extended to include, for example:

- How desirable the type of car it is. For Mercedes cars, there were reported [57] 122 thefts per 10,000 vehicles, a rate of 0.011 per year or 5 times the average.
- The probability of theft, based on factors such as type, model, colour, or location. Bayesian network analysis can be used to model this probability, which is not as daunting as it may sound, thanks to commercial software tools such as Bayes Server [58].

Given that 92% of attacks were 'keyless' and also believed to be relay attacks, we assume that threat scenarios 1 and 2 are less likely and assign proportions of 2%

Table 3.7 Example 1: qualitative definitions of likelihood

Likelihood	**Definition**
Rare	There are few similar events on record
Possible	The event has happened several times
Likely	The event has occurred a significant number of times
Certain	The event has occurred a very high number of times

Table 3.8 Example 2: qualitative definitions of likelihood

Likelihood	**Definition**
Rare or unlikely	The event rarely happens with few similar events on record, maybe once every 5+ years
Likely	The event occurs a few times every 1–3 years
Highly likely	The event occurs many times per year
Very likely	The event occurs on a weekly or monthly basis

to each of them. Scenario 2 is also keyless, so we estimate that the probability of each type of attack is as follows:

- 2% theft of key fob
- 2% interception of transmission
- 90% relay of transmission
- 6% other scenarios, not necessarily cyber

The 'other' category is the uncertainty we have, separate to the risks we know about. Over time, we may expect to uncover new attack methods, adding to the list above and changing the relative likelihood. For instance, relay attacks have grown substantially in the last few years: 66% in 2016, 88% in 2019 to 92% in 2019.

The probability of the theft of a single car to any particular attack is P(theft) × P(attack type), that is, 0.00226 × 0.9 = 0.00203 for the relay threat scenario. We are phrasing this as 'attack' rather than 'threat scenario' because the data are the historic frequency of attack, which is not necessarily the same as the future likelihood of a threat scenario.

In Table 3.9, we show the quantitative likelihoods (rounded) for the car and manufacturer perspective. Because the manufacturer is concerned with multiple car losses, the likelihood is estimated using the binomial theorem, where we are asking what the probability is that *X* cars are stolen. This is a more involved approach for which we provide an explanation in Box 3.1, but note that the approach should be well within the capabilities of organisations' risk experts.

From the car owner's perspective, the quantitative likelihood is for one car, so it is quite unlikely for a car owner to lose their car, after all the average rate of car thefts is 0.0023 per year.

It might be tempting to ignore some threat scenarios at this stage because they are relatively small, but this is because the relay attack has become easy for thieves. The cost of building the equipment needed by thieves is reported to be £257 [61].

Table 3.9 Likelihood for threat scenarios

	Car owner		Manufacturer	
Threat scenario	**Qualitative per car**	**Quantitative per car**	**Qualitative**	**Quantitative (expectation value)**
1. Theft of key fob	Unlikely	4.5×10^{-5}	Likely	53% (5)
2. Interception of wireless signal	Unlikely	4.5×10^{-5}	Very likely	53% (5)
3. Relay of wireless transmission	Likely	2.0×10^{-3}	Very likely	49% (203)

Box 3.1 Estimation of manufacturer's likelihood

The likelihood of any one person having their car stolen was estimated as the average number of cars stolen divided by the total number of cars in circulation (520k/2.3M) = 0.00230. We can further multiply this by 0.9 for the probability of the car being stolen by the relay attack = 203.

For the manufacturer, they could expect to find 203 of their particular model stolen by a relay attack. This raises the question on which value to use as the likelihood in Table 3.9, which is the probability that the manufacturer will see 203 cars stolen from relay attacks.

To answer this, we use the Binomial Theorem, for which good descriptions can be found in Reference [59] and in more detail in Reference [60]. To calculate the probability of a particular number of successes, the binomial formula is:

Equation 1: Binomial distribution

$$P\left(X=x\right) = {}^{n}C_{x}p^{x}\left(1-p\right)^{n-x} \tag{3.1}$$

The formula relates to the number of trials n, the number of successes, x and the probability of success p. Here a 'success' is the number of cars stolen by relay attacks. The expression ${}^{n}C_{x}$ is the number of combinations of x items chosen from n, also referred to as 'n choose x':

Equation 2: The number of combinations of size x from n

$$\frac{n!}{x!\left(n-x\right)!} \tag{3.2}$$

The inputs to the formula are:

- p, the probability of success of theft with 90% relay attacks (0.00226 × 0.9) = 0.00203
- n, the number of trials = 100,000
- x, the independent variable for the number of successes, which is an input from 0 to n

By plotting the range x = 0 to x = n (100,000 in our problem) we get the following graph:

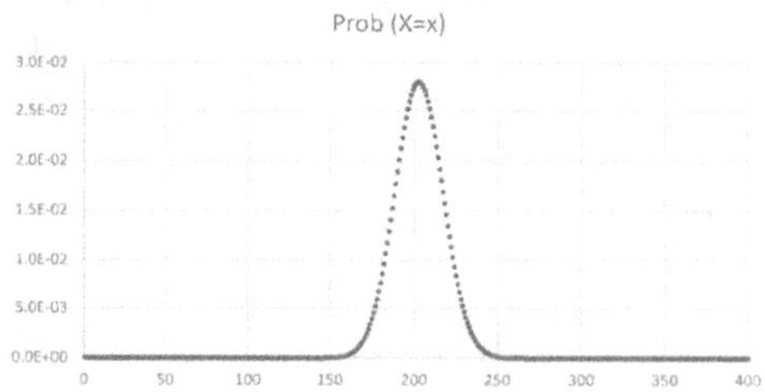

(Continues)

The expectation value or mean (μ) and standard deviation (σ) of the distribution are found by the following formulas:

$$\mu = np \tag{3.3}$$

$$\sigma = \sqrt{np(1-p)} \tag{3.4}$$

Hence $\mu = 0.00203 \times 100{,}000 = 203.5$ and $\sigma = 14.3$.

Whereas our sample is 100,000, the graph is shown up to 400 only as the probabilities beyond ~250 are negligible.

The above distribution is centred on the mean, ~203 and shows this as the most likely outcome. Below X~170 and above X~240 the probability is negligible.

To calculate (A), the probability to include in our table of likelihood, we need the cumulative probability of all the events plotted in the graph, up to and including 203. This is the sum of Equation 1 from $x = 0$ to $x = 203$, shown as a 'cumulative distribution function'.

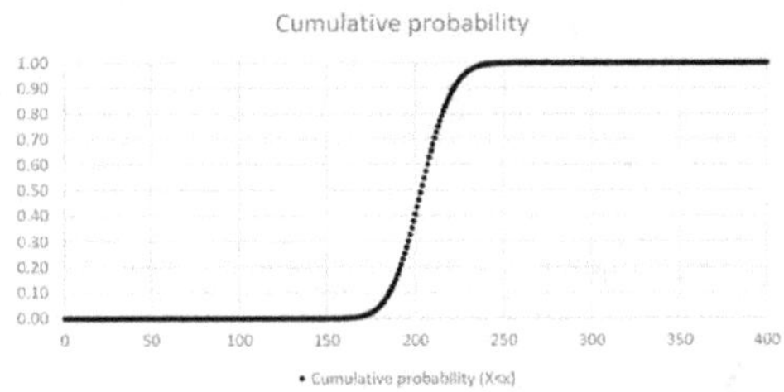

The graph plots the cumulative probabilities, showing trials from $x = 0$ to 400. From the calculations for this:

- the probability that up to and including 203 cars are stolen by relay attack is the sum of probabilities between 0 and 203, or sum of $P(x \leq 203) = 49\%$;
- the probability that more than 250 cars are stolen is 1 − (sum of $P(x \geq 250) = 1 - 99.9\% = 0.1\%$.
- the probability that between 203 and 250 cars are stolen is the sum of $P(203<x<250) = 99.9 - 49 = 50.9\%$.

In summary, if the manufacturer has equivalent remote locking systems security to other manufacturers, they might expect a 48% probability that up to 203 of their cars are stolen by relay attacks but a very low probability that (a) more than 250 cars and (b) less than 170 cars are stolen by this method. Note that this is a feature of the binomial distribution chosen and the small probability of cars being stolen. Other distributions could be used, and examples applied to cybersecurity are given in [4].

(Continues)

Box 3.1 (Continued)

The 'success rate' for the other types of attack is much lower, 4.5×10^{-5}, and therefore the binomial distribution has an expectation value of 5, with a very small probability that up to 203 cars are stolen. We may therefore need to adjust our estimate of impact for the other types of attack. If they were the only types of attack, it is unlikely that a loss of five cars would have significant impact on branding or require recall.

Another consideration for our example is whether the manufacturer decides to set a design target for improving its security with respect to the relay attack. If we increase the security by a nominal 20%, the change in the above graphs is as below, suggesting a very low probability of losing more than the industry average, a potential strong marketing point to sell more cars and an overall reduction in the impact of losing up to 200 cars from the relay attack.

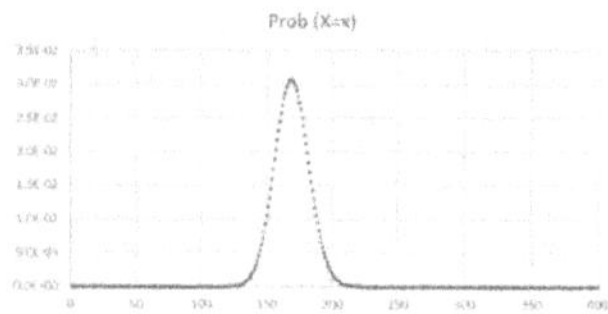

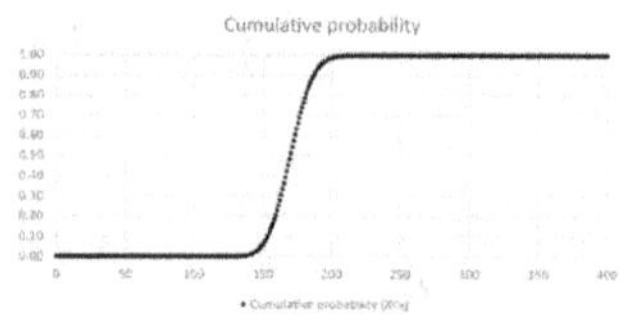

Should the design of controls reduce this risk, then other threat scenarios may become more important.

A further point to note is that Table 3.9 gives the quantified likelihood of each threat scenario as the 'expectation' of the numbers of cars stolen. The expectation value is the same as the mean μ. The cumulative probability of up to 'X = Expectation Value' cars being stolen should be 50%, but there is some rounding to include whole numbers of cars. Thus, the quantitative likelihood may be interpreted as 'a 49% probability that up to 203 cars will be stolen by relay attacks'.

3.3.3.3 Determining risk

The actual risk for each threat scenario is a combination of the impact of the scenario and the likelihood of it occurring. To do this for the qualitative approach, we use a qualitative risk matrix, Table 3.10, which is typical of many in use.

Table 3.10 Risk matrix

		Impact			
		Insignificant	Minor	Moderate	Major
Likelihood	Rare or unlikely	Low	Low	Medium	Medium
	Likely	Low	Medium	Medium	High
	Highly likely	Medium	Medium	High	High
	Very likely	Medium	High	High	High

For threat scenario 3 (relay attack), and from the owner's perspective, the impact is 'Moderate' (see Table 3.3) and the likelihood is 'Likely' (see Table 3.9). Looking these values up from the risk matrix gives a 'Medium' risk.

For the quantitative approach, we multiply impact and likelihood, which is €20k $\times$ 2.0 $\times$ 10^{-3} = €40. From the manufacturers' perspective, also for the relay attack, the respective figures are a 'High' risk or €31M $\times$ 49% = €15.2M. The risks from the manufacturers' perspective are shown in Table 3.11.

For the other two threat scenarios, we find it necessary to make an adjustment of the previously estimated impact, as the binomial theorem (Box 3.1) tells us that the expectation value for cars stolen is 5 rather than 203. Looking at the previous estimates in Table 3.4, it is unlikely that there will be a recall or branding impact that affects sales from just the loss of five cars. The impact may be negligible, but we can estimate a worst case of repair costs to each vehicle of no more than €1,000.

Note that we could have avoided making the preceding adjustment by analysing the impacts more closely in the first place. However, there is no problem in iterating the risk assessment as we work through it and better understand the issues. Allied to this point is the distinct benefit of having a diverse group of people involved in the risk assessment.

3.3.3.4 Risk evaluation

Following the risk analysis, we evaluate how to deal with the risks, using the criteria discussed in Section 3.1.5, that is, to tolerate/accept a risk, treat/mitigate it, transfer it to another party or terminate/avoid it. Most organisations will have a limited budget and resources, so there is also a need to prioritise risks that will be treated.

An important part of risk evaluation is also to document the decision-making process to justify the decisions made, which may be returned to in future iterations of the risk assessment.

Table 3.11 Risks from manufacturer's perspective

Scenario	Approach	Impact	Likelihood	Risk
1.Theft of key fob	Qualitative	Major	Likely	High
	Quantitative (expectation value)	€5k	53% (5)	€2.5k
2.Interception of wireless signal	Qualitative	Major	Very likely	High
	Quantitative (expectation value)	€5k	53% (5)	€2.5k
3.Relay of wireless transmission	Qualitative	Major	Very likely	High
	Quantitative (expectation value)	€31M	49% (203)	€15.2M

From Table 3.11, Scenario 3 is the only scenario that presents any significant degree of risk for the manufacturer. Hence, we may decide to treat the risk of relay attack but tolerate the risks of 'theft of key fob' and 'interception of wireless signal'. There may not be an argument to terminate the risk, as the RKS is seen as a standard item in new vehicles. It may be possible to transfer risks under a product recall insurance. However, this would only partially transfer the risk and the manufacturer would probably still need to meet an industry standard for cybersecurity.

A further point to consider is how the risks might grow over time, so some sensitivity analysis may be needed, with assumptions around the capability and methods of attackers in future. Should relay attacks become better mitigated against, attackers may adopt more sophisticated methods of interception, hence the potential for R&D to be launched or continued for Scenario 2.

Risk appetite/tolerance

As a guide to the risk evaluation, it is highly desirable that the organisation defines its 'risk tolerance' or 'risk appetite'. Risk appetite/tolerance is the maximum individual and combined risk to operations and the safety of people that an organization is willing to tolerate. Defining risk appetite is a matter for an organisation's board. If the board do not do this they will be accepting the default, which will be an undefined level of risk.

In the transport sector, decision-making around safety improvements is generally linked to target accident rates. For example, the global aviation accident rate is between 10^{-6} and 10^{-7}, so an airline might set its own tolerable level of risk to be 10 times higher, say 10^{-8}, equivalent to an accident every 100 years, which is a pragmatic way of setting a target of zero accidents.

The target level of safety would equally apply to safety in the transport sector, and it would not make sense to define a separate target in respect of cybersecurity. Hence, an additional risk appetite could be defined with respect to key performance areas, such as capacity, operational continuity, as well as related financial losses.

3.3.4 *Risk treatment*

Those risks to be treated can be assessed against their priority and the budget available. If there is insufficient budget to treat all risks, then a re-evaluation may be required as to whether to tolerate, transfer or terminate the risks or to extend the budget.

As with threats and vulnerabilities, there exist catalogues of controls that can be applied, such as NIST SP 800–53 [62], which also has an accompanying spreadsheet with nearly 1,200 controls listed and cross referenced to related controls. Cross referencing controls helps identify a defence in depth or breadth strategy, which is discussed in Chapter 6. While the sheer number of controls may appear daunting, there are natural groupings of them, listed as families in NIST SP 800–53 and including:

- access controls
- awareness and training
- identification and authentication

Controls can also be mapped to threats and vulnerabilities, initially by experienced security practitioners and then a knowledge base can be built up over time. Controls are essentially standing between a threat and the vulnerability of an asset, as illustrated in Figure 3.6.

In choosing controls, they must relate to a threat or a vulnerability. For example, the risk may be articulated as a specific software vulnerability and the control is a 'patch' or update to the software. Or the threat may be an SQL injection attack, where the required control is to sanitize the data that a user can input into a form. It is generally considered that a control is not needed if there is a threat without a corresponding vulnerability. If security risks are articulated as both a set of threats and a set of vulnerabilities, the controls must address the intersection of that set only, as shown schematically in Figure 3.7.

Continuing our worked example, we decided to treat the risk of a relay attack, Scenario 3, and tolerate the other two risks. As a reminder, the relay attack works by rebroadcasting the signal from car's security unit to the key-fob signal, which would

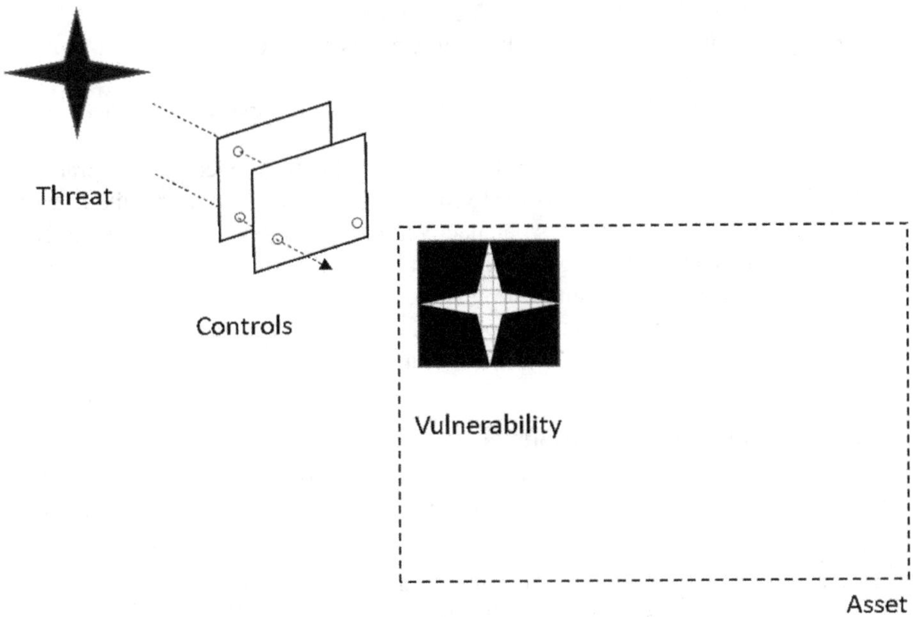

Figure 3.6 Role of controls

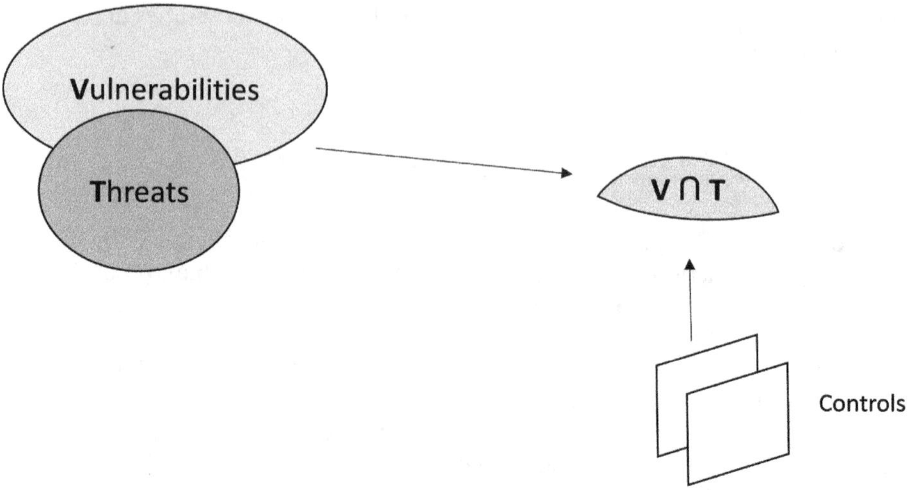

Figure 3.7 Controls address threats and vulnerabilities

otherwise be out of range, and vice versa. This results in the car unit responding as if the key was in range, by unlocking the car and allowing the engine to start. Possible controls are therefore:

- Protecting the key from receiving the relay transmission by placing it within a metallic foil bag or metallic screening (Faraday cage).‡‡
- Adding security controls to the key fob such as a switch to activate it or even fingerprint recognition [63].
- The addition of mechanical barriers (locked gate, garage and steering wheel lock).

These and further ideas for controls are discussed in Reference [64]. In practice, they would require further analysis for feasibility, particularly given the likely costs involved. From the manufacturer's perspective, it would undermine their product to suggest customers use a steering wheel lock, although this may be good advice for the owner [65]. Perhaps more interesting is the action taken by some manufacturers in adding a motion sensor to the key fob, so that the keys enter a sleep mode when stationary and do not then respond to the car unit's transmission.

3.3.4.1 Implementing controls

Clearly, the risk is not treated until the controls are actually applied. From the authors' experience with 'design-in' security risk assessments, administrative controls may

‡‡A Faraday cage blocks wavelengths larger than the gaps in the cage. RKS systems operate at around 434 MHz, which has wavelength of 70 cm making a Faraday cage pouch easily effective. A metallic box is effective due to the 'skin effect', which limits the penetration of electromagnetic waves in conductors. A 434 MHz signal would have a skin depth of ~3 μm in copper and could not penetrate even a thin layered box.

be defined a long time before the system is actually developed, giving the impression that risks are being addressed, but raising the prospect that the controls are not then applied in future.

3.3.4.2 Control specifications

Once controls have been defined in outline, they will need to go through system engineering processes, including specifying requirements, design, development etc. Controls may be additions to software and included in the system's software development life cycle, or external 'add-ons' or simply administrative. Controls are discussed in more depth in Chapter 6.

3.3.5 Communicating and consulting

The risk communication and consultation stages of the risk management process ensure sharing of information between the decision-maker and other stakeholders at key intervals. A basic form of risk communication is the risk register, which records the risks and mitigations. However, given that risk is an abstract concept, not all decision-makers and other influential parties will be familiar with it. Where risk is purely qualitative, the register may comprise a list of high medium and low risks, which are difficult to distinguish and prioritise. In this case, an approach may be to set a qualitative level of risk tolerability to 'all risks must be reduced to low'. However, when quantifying risks, the distinction between them is clearer, and it is also possible to compare quantified cybersecurity risk against other enterprise risks.

3.3.6 Risk monitoring and review

As already mentioned, risk management is a continuous process, requiring organisations to keep abreast of new threats and vulnerabilities, update their risk assessments and add new controls. There are challenges to maintaining an overall and up-to-date risk picture, particularly in ensuring that changes to systems, and the rise of new threats and vulnerabilities, are quickly captured and acted upon. Whereas risk management frameworks and standards include the requirement for regular updates, a broader organisational framework is likely to be needed for this, in the same way as adopted for safety through Safety Management Systems. For security, this can be provided through a 'Security Management System', which is addressed in the next chapter.

3.3.7 System level risk management

In this walk-through, we have described a small but important part of transport security to focus on the approach to risk management. For many operators, the scale of the exercise will be much larger as they seek to secure their whole operation. The risks will also be higher, also taking account of the potential safety implications of cyber attacks. This requires a top-down approach, analysing the critical systems of the operations and ensuring that these are prioritised to minimise risk.

Figure 3.8 shows the part of an airport 'system of systems' and Reference [66] describes an approach to scope and identify critical systems. In practice, the high level systems view of an airport could comprise 10 times the number of systems with several segmented networks. The challenge for risk management is to capture the high level systems and then work downwards, building the risk picture over time while hardening systems as new vulnerabilities are discovered. A further challenge is that while some systems may be critical, there can be no complacency in hardening less critical systems as they may provide a pathway into the critical systems. These aspects are expanded on in Chapter 7.

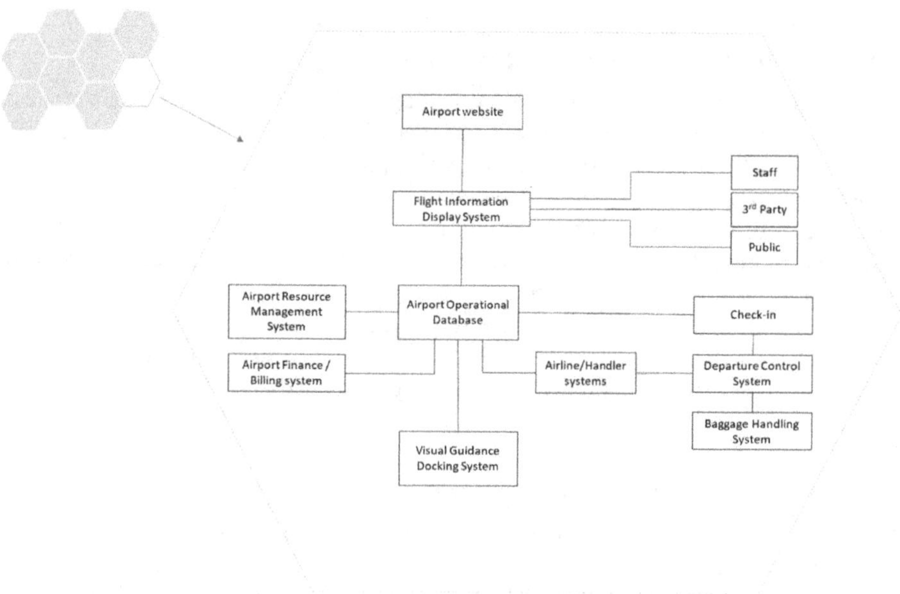

Figure 3.8 Part of an airport system of systems

Table 3.12 Risk evaluation decisions

Risk	Tolerate	Treat	Transfer	Terminate
1. Theft of key fob	Yes, risk is very low	Largely out of manufacturer's control	N/A	No, RKS established as a standard feature
2. Interception of wireless signal	Yes, risk is very low	Launch research as possible future higher risk	Recall insurance	No
3. Relay of wireless transmission	No, risk is significant	Yes, apply controls	Recall insurance	No

3.4 Conclusion

Risk management is fundamental to effective cybersecurity. It enables an organisation to understand its current and target 'cybersecurity posture' by evaluating its tangible and intangible assets, the risks to them and the controls needed. In the transport sector, cybersecurity risk assessment has generally been high level, compared to the detailed technical approaches taken with safety risk. This means that controls are often applied sensibly, but with only a loose connection to risk, based on the expertise of security teams. This loose connection risks missing the bigger picture in understanding risks and opportunities for different system architectures and business models that protect the goals of the organisation.

The challenge to improve cyber risk assessment in the transport sector also extends to adopting quantified risk approaches. Pure qualitative risk approaches are increasingly seen as insufficient. In this chapter, we have performed a walk-through of a risk assessment in both qualitative and quantitative terms. The comparison shows that there are distinct advantages in quantitative approaches. They provide more granularity in the results, which supports risk prioritisation and cost-benefit analysis of risk treatments. Qualitative methods, however, are quicker to do and easier to understand, even if the results make it hard to prioritise risks other than between low, medium and high risks.

Risk management is a major component of cybersecurity management, which is addressed in the next chapter.

References

[1] ISO/IEC 27005:2011(E). *Providing guidelines for information security risk management.* Geneva, Switzerland: International Organization for Standardization;

[2] Viscelli T.R., Beasley M.S., Hermanson D.R. 'Research insights about risk governance: implications from a review of ERM research'. *SAGE Open.* 2016;6(4):1–17.

[3] Hubbard D.W. *The failure of risk management: why it's broken and how to fix it.* 2nd edn. New Jersey: John Wiley & Sons; 2009.

[4] Hubbard D.W., Seiersen R. *How to measure anything in cybersecurity risk.* 1. Hoboken, NJ: Wiley; 2016.

[5] Kutsch E., Browning T.R., Hall M. 'Bridging the risk gap: the failure of risk management in information systems projects'. *Research-Technology Management.* 2014;57(2):26–32.

[6] Kaplan S., Garrick B.J. 'On the quantitative definition of risk'. *Risk Analysis.* 1981;1(1):11–27.

[7] Zio E. 'The future of risk assessment'. *Reliability Engineering & System Safety.* 2018;177(1):176–90.

[8] Redmill F. 'Risk analysis – a subjective process'. *Engineering Management Journal.* 2002;12(2):91–6.

[9] Knight F.H. *Risk, uncertainty and profit*. Boston & New York: Houghton Mifflin; 1921.

[10] Langner R., Pederson P. *Bound to Fail: Why Cyber Security Risk Cannot Simply be "managed" Away*. Brookings; 2013.

[11] Kosub T. 'Components and challenges of integrated cyber risk management'. *Zeitschrift für die gesamte Versicherungswissenschaft*. 2015;104(5):615–34.

[12] National Institute of Standards and Technology. Computer Security Resource Center Glossary. Available from https://csrc.nist.gov/glossary.

[13] ISACA. *Effective IT Governance at Your Fingertips* [online]. Available from https://www.isaca.org/resources/cobit.

[14] Dubois E., Heymans P., Mayer N., Matulevičius R. 'A systematic approach to define the domain of information system security risk management'. *Intentional Perspectives on Information Systems Engineering*. Berlin, Heidelberg: Springer; 2010. pp. 289–306.

[15] Ross R. Guide for conducting risk assessments (NIST SP-800-30rev1). Gaithersburg, MD: National Institute of Standards and Technology; 2012.

[16] Sestorp I., Lehto A. *CPDLC in practice: a dissection of the controller pilot data link communication security*. Linköping, Sweden: Linköping University; 2019.

[17] Cook A.J., Tanner G. European airline delay cost reference values. Brussels: EUROCONTROL Performance Review Unit; 2015. Available from https://www.eurocontrol.int/publication/european-airline-delay-cost-reference-values.

[18] EUROCONTROL. Network manager outage on 3 April 2018 – summary brief. Brussels: EUROCONTROL; June 2018. Available from https://www.eurocontrol.int/publication/network-manager-outage-3-april-2018.

[19] Moore T., Dynes S., Chang F.R. *Identifying how firms manage cybersecurity investment. Southern Methodist University* [online]. 2015. Available from http://blog. smu. edu/research/files/2015/10/SMU-IBM. pdf [Accessed 14 Dec 2015].

[20] EUROCAE. 'EUROCAE ED201'. *Aeronautical Information System Security (AISS) Framework Guidance*. 2015.

[21] Faris S., Medromi H., El Hasnaoui S., Iguer H., Sayouti A. 'Toward an Effective Information Security Risk Management of Universities' Information Systems Using Multi Agent Systems, Itil, Iso 27002,Iso 27005'. *International Journal of Advanced Computer Science and Applications*. 2014;5(6).

[22] ISO 27005 information security risk assessment standard. Available from https://www.itgovernance.co.uk/iso27005.

[23] National Institute of Standards and Technology. *NIST risk management framework RMF*. Available from https://csrc.nist.gov/projects/risk-management/about-rmf

[24] Joint Task Force Working Group. Security and Privacy Controls for Information Systems and Organizations. Gaithersburg, Maryland, United States: National Institute of Standards and Technology; 2020.

[25] Stouffer K., Lightman S., Pillitteri V., Abrams M., Hahn A. 'NIST special publication 800-82, revision 2: guide to industrial control systems (ICS) security'. *National Institute of Standards and Technology*. 2014.

[26] Chapple M., Stewart J.M., Gibson D. *(ISC) 2 CISSP certified information systems security professional official study guide*. Indiana, USA: John Wiley & Sons; 2018.

[27] Itil Open Guide [online]. Available from https://www.itlibrary.org.

[28] OCTAVE - Related Assets. Available from https://resources.sei.cmu.edu/library/asset-view.cfm?assetID=309051.

[29] National Technical Authority for Information Assurance. Analysis of information risk management methodologies. Available from https://www.gov.uk/guidance/analysis-of-information-risk-management-methodologies#octave-allegro.

[30] Wangen G. 'Information security risk assessment: a method comparison'. *Computer*. 2017;50(4):52–61.

[31] Comparing IT Risk Assessment and Analysis Methods Transcript [online]. Available from https://resources.sei.cmu.edu/asset_files/Podcast/2014_016_100_85630.pdf.

[32] Windelberg M. 'Objectives for managing cyber supply chain risk'. *International Journal of Critical Infrastructure Protection*. 2016;12:4–11.

[33] 'Regulation (EU) 2016/679 of the European Parliament and of the Council of 27 April 2016'. *On the Protection of Natural Persons with Regard to the Processing of Personal Data and on the Free Movement of Such Data, and Repealing Directive 95/46/EC (General Data Protection Regulation)*.

[34] *Information Commissioner's Office. 2020. What Responsibilities And Liabilities Do Controllers Have When Using A Processor?* [online]. Available from https://ico.org.uk/for-organisations/guide-to-data-protection/guide-to-the-general-data-protection-regulation-gdpr/contracts-and-liabilities-between-controllers-and-processors-multi/responsibilities-and-liabilities-for-controllers-using-a-processor/ [Accessed 26 Aug 2020].

[35] Deloitte. Using blockchain to drive supply chain transparency [online]. Available from https://www2.deloitte.com/us/en/pages/operations/articles/blockchain-supply-chain-innovation.html [Accessed 26 August 2020].

[36] NIST. 'Guide for conducting risk assessments'. *NIST Special Publication (NIST SP) - 800-30 Rev 1*.

[37] Freud J., Jones J. *Measuring and managing information risk: a fair approach*. Oxford, UK: Elsevier; 2015.

[38] Ponemon Institute *The Billion Euro lost laptop problem, benchmark study of European organizations*. Michigan, USA: Ponemon Institute; 2011.

[39] Information Commissioner's Office. *Monetary penalty notice 16 October 2020. Case reference COM0783542 British airways PLC*. Wilmslow, UK: Information Commissioner's Office; 2020. Available from https://ico.org.uk/media/action-weve-taken/mpns/2618421/ba-penalty-20201016.pdf

[40] BBC. British Airways fined £20m over data breach. 2020. Available from https://www.bbc.com/news/technology-54568784.

[41] European Parliament, Council of the European Union. *Regulation (EU) 2016/679 of the European Parliament and of the Council of 27 April 2016 on the protection of natural persons with regard to the processing of personal data and on the free movement of such data, and repealing Directive 95/46/*

EC (General data protection regulation) (text with EEA relevance). Brussels, Belgium: European Commission; 2016. Available from https://eur-lex.europa.eu/legal-content/EN/TXT/?uri=CELEX:32016R0679

[42] Gordon L.A., Loeb M.P. 'The economics of information security investment'. *ACM Transactions on Information and System Security*. 2002;5(4):438–57.

[43] von Solms R., van Niekerk J. 'From information security to cyber security'. *Computers & Security*. 2013;38(6):97–102.

[44] Hird J., Hawley M., Machin C. Air traffic management security research in SESAR. 2016 11th International Conference on Availability, Reliability and Security (ARES); Aug 2016. pp. 486–92.

[45] ISO/PAS 22399:2007. *Societal security – guideline for incident preparedness and operational continuity management*. Geneva, Switzerland: International Organization for Standardization.

[46] SecRAM 2.0 Security Risk Assessment methodology for SESAR 2020. Edition: 02.00.00. Sep 2017. Source [online]. Available from https://wwwsesarjueu/sites/default/files/documents/transversal/SESAR%202020%20-%20Security%20Reference%20Material%20Guidancepdf [Accessed 30 January 2021].

[47] Gotz K., Hawley M., Machin C., Hird J. 'CTRL_S'-A security tool for SESAR's design-in security approach'. *2016 11th International Conference on Availability, Reliability and Security (ARES)*; IEEE, 2016 Aug 1. pp. 499–502.

[48] SecRAM 2.0 Security Risk Assessment methodology for SESAR 2020. Brussels: SESAR Joint Undertaking; 2017. Available from https://www.sesarju.eu/sites/default/files/documents/transversal/SESAR%202020%20-%20Security%20Reference%20Material%20Guidance.pdf.

[49] SecRAM 2.0 Security Risk Assessment methodology for SESAR 2020 [online]. Edition: 02.00.00. 25 Sep 2017. Source. Available from https://www.sesarju.eu/sites/default/files/documents/transversal/SESAR%202020%20-%20Security%20Reference%20Material%20Guidance.pdf [Accessed 30 January 2021].

[50] Handson Security. Jam and Replay Attacks on Vehicular Keyless Entry Systems [online]. Available from https://s34s0ngithubio/2019/07/18/Jam-and-Replay-Attacks-on-Vehicular-Keyless-Entry-Systems/ [Accessed 1 February 2021].

[51] Ibrahim O.A., Hussain A.M., Oligeri G., Di Pietro R. 'In Security and Safety Interplay of Intelligent Software Systems'. *Key is in the Air: Hacking Remote Keyless Entry Systems*. Springer, Cham; Sep 2018. pp. 125–32.

[52] NIST. Mobile Threat Catalogue. Available from https://pages.nist.gov/mobile-threat-catalogue/ [Accessed 15 Feb 2021].

[53] MITRE. Common Vulnerabilities and Exposures. Available from https://cve.mitre.org/ [Accessed 15 Feb 2021].

[54] Eurostat. Fewer cars stolen in most Member States. Available from https://ec.europa.eu/eurostat/web/products-eurostat-news/-/DDN-20191104-1#:~:text=

Police%20in%20the%20EU%20recorded,in%20most%20EU%20Member%20States [Accessed 1 Feb 2021].

[55] Eurostat. Passenger cars in the EU. Available from https://ec.europa.eu/eurostat/statistics-explained/index.php?title=Passenger_cars_in_the_EU&oldid=485664#Overview [Accessed 1 February 2021].

[56] Acturial Post. Keyless car theft at all time high in 2019. Available from https://www.actuarialpost.co.uk/article/keyless-car-theft-at-all-time-high-in-2019-17893.htm [Accessed 1 February 2021].

[57] Honest John. Mercedes-Benz Crime Statistics. Available from https://www.honestjohn.co.uk/crime/mercedes-benz [Accessed 1 February 2021].

[58] Bayesian network & Causal AI software. Available from https://www.bayesserver.com/.

[59] 'For an easy introduction see: Scott Hartshorn'. *Probability With The Binomial Distribution And Pascal's Triangle. ASIN.*

[60] Fenton N., Neil M. *Risk assessment and decision analysis with Bayesian networks*. 2nd edn. Florida, USA: Chapman & Hall/CRC; 2019. pp. 637–61.

[61] Comparitech. What is a relay attack (with examples) and how can you prevent them? Available from https://www.comparitech.com/blog/information-security/what-is-relay-attack/ [Accessed 31 January 2021].

[62] 'NIST special publication 800-53. security and privacy controls for information systems and organizations. revision'. 2020.

[63] Joo K., Choi W., Lee D.H. 'Hold the door! fingerprinting your CAR key to prevent keyless entry CAR theft. arXiv preprint arXiv:2003.13251'. 2020.

[64] Francillon A., Danev B., Capkun S. Relay attacks on passive keyless entry and start systems in modern cars. *Proceedings of the Network and Distributed System Security Symposium (NDSS)*; 2011.

[65] Zurich Insurance U.K. How to tackle the growing threat of keyless car theft. Available from https://www.zurich.co.uk/news-and-insight/how-to-tackle-the-growing-threat-of-keyless-car-theft [Accessed 15 February 2021].

[66] UK CAA. *Cyber security critical systems scoping guidance*. Crawley, UK: UK Civil Aviation Authority;

Chapter 4
Security management systems

Rainer Koelle and Martin Hawley

4.1 Introduction

The preceding chapter on risk management provides a good argument for considering security as 'good management practice'. There is no shortage of slogans such as 'built-in security' or 'security is as weak as its weakest link' to highlight the holistic nature of security thinking. Nonetheless, organisations reportedly struggle to address security appropriately.

This book and, in particular, this chapter, is rooted in security management. If there is one message from this book about security from a management perspective, it is that today's security leadership has evolved to taking a more holistic approach to security [1]. Security is no longer an exclusive matter of avoiding and banning 'bad people and bad things' in the physical domain [2]. Accordingly, security is no longer just a question of technical security measures it involves the whole organisation and includes every member of staff.

Air transportation changed forever from 11 September 2001, which has had knock-on effects on other transport modes. It also changed the way how we look at and deal with security [3]. Security is no longer limited to the domain of the military or police forces and security industry. It is a vital part of today's organisational portfolio of activities. The dynamics of security require that security be managed day in, day out. As vulnerabilities develop and emerge, technical and organisational compliance erode, risk perceptions change and influence decision-making. It is important to understand the interplay between security management and how security plays into limiting and reducing the potential impact of threats and vulnerabilities on an organisation's objectives.

Across all industries, there has been an increasing use of management systems. Textbooks regularly point at quality management as one of the first widely used management systems. This initial development was strongly driven by emerging regulatory requirements for a management system. Accordingly, the initial implementation of management systems was primarily driven by compliance requirements as a key motivator.

Over time, organisations have been faced with the need to establish several management systems (quality, safety, environmental management etc.), and have been driven to action by regulatory requirements or influenced by modern management philosophies. Today, attempts are ongoing to align these management systems and seek synergies amongst them, to reduce the resources required to maintain multiple systems, and also to reduce the burden on staff reporting and process-related work. Today's coexistence of a variety of management systems has triggered a wider discussion of the alignment and implementation of integrated management systems (IMS) [4, 5].

When speaking about security, traditionally there is room to apply a wider resilience type of thinking. Security should not be thought of as a purely preventive discipline. Related disciplines such as business continuity, contingency planning, disaster recovery and crisis management, to name a few, form a spectrum of activities that ensure the organisation's capabilities to prepare for, prevent, respond to and recover from incidents that endanger the achievement of business objectives. Accordingly, none of the various disciplines should be thought of as standing alone. They form part of an interwoven set of activities, as shown in Figure 4.1. The conceptualisation of management systems for these disciplines goes beyond the classical expression reflected by the term. The notion of continual improvement entails ongoing planning, evaluation and change implementation, and through this improvement mechanism broader approaches and practices are established over time.

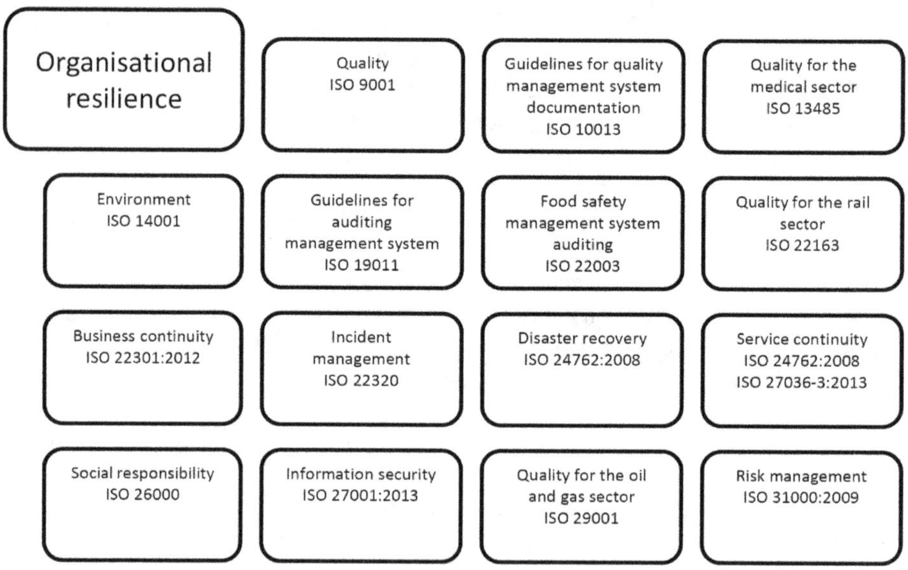

Figure 4.1 Organisational Resilience at the convergence of disciplines
The International Standardisation Organisation provides an overview of their standards at: https://www.iso.org/management-system-standards-list.html

The advent and dependency of organisations on their digital information and systems has created a larger attack surface. In recent years, the observed rise in cyberattacks has accelerated the integration of cybersecurity within operational and enterprise risk management. It has also accelerated the development of associated disciplines and the deployment of resources such as business continuity and crisis management. All of these now form a spectrum within a larger concept of '360-degree risk management' and the conceptualisation of a resilient organisation.

Ross *et al.* [6] set out the challenges of security in the 21st century, and an extract is provided in Box 4.1. This chapter aims to address these challenges and place them in the context of organisational management systems. The chapter will first establish a good common understanding of what we mean by security, and in particular organisational security. We then address security and resilience of operations as a critical success factor in today's business context, which supports the need to implement or adapt existing management systems to account for them. Next, the chapter highlights an emerging requirement: collaborative security management, as the challenges of securing operations go beyond a single organisation and require collaborative effort with regional/national organisations, in both the public and private sector. This is increasingly likely to be the case for multimodal transport and mobility as a service operation. Equipped with these building blocks, the chapter discusses the implementation aspects of security management systems and provides a high-level overview of digital support tools. Finally, the chapter addresses some contemporary issues of organisational resilience management.

Box 4.1 The challenges of security in the 21st century [6]

To begin to address the challenges of the 21st century, we must

- understand the modern threat space (i.e., adversary capabilities and intentions revealed by the targeting actions of those adversaries);
- identify stakeholder assets and protection needs and provide protection commensurate with the criticality of those assets and needs, and the consequences of asset loss;
- increase the understanding of the growing complexity of systems – to more effectively reason about, manage and address the uncertainty associated with that complexity;
- integrate security requirements, functions and services into the mainstream management and technical processes within the life cycle processes of systems; and
- build trustworthy secure systems capable of protecting stakeholder assets.

4.2 Security and operational continuity – organisational resilience

A unifying characteristic of all undertakings is the execution of the work – that is, the operations. Given its nature or complexity, and in particular its criticality, there is a vested interest in mitigating risks to the outcome of the operations. Although everybody agrees that 'the show must go on', today's risk management practices still seem to deal with security primarily as an afterthought or compliance requirement. This section introduces organisational resilience as an enabler and a decisive factor in the success of the organisation. By understanding why and how security is linked to an organisation's operations, managers will be able to navigate the landscape and confusion caused by the proliferation of security terms and disciplines, and place them properly within the business model of their organisation.

4.2.1 Security

The roots of a changing security world can be linked to the 1990s, with fundamental changes to the geopolitical world: the reunification of Germany, new and refined objectives of NATO, the collapse of the USSR and the rise of China. In parallel, the world witnessed the emergence of ever higher levels of computerised production and operational systems with increasing interconnectivity.

These developments led to today's security landscape, where the borderline between large-scale terrorism* [7] and local activism, state actors, terrorists and crime is blurring. The resulting challenges of this landscape are part of what the scholar Lawrence Freedman calls 'the transformation of strategic affairs' [8]. The classical view that security is an issue to be dealt with by state authorities, traditional police, counterterrorism and military forces on a national state level is no longer applicable. Societal and organisational aspects of security include all organisations. In this sense, private organisations become security actors and need to address as well as manage security challenges within their remits [3].

Security, in general, refers to any of the various measures with a view to safeguarding elements at risk (assets) against a broad range of threats – for example, theft, fire, accidents, espionage, explosion and stand-off attack. Looking at an organisation as a sociotechnical system, security is tightly coupled to the overall operations of an organisation and its other business processes (e.g., safety, quality, reliability) and must not be ignored or relegated to incidental concerns. Moreover, the resilience of such a system should be a key management concern.

*Although there exists no globally accepted definition of terrorism, it can be understood as the use or threat of violence to further a political cause [1].

4.2.2 Organisational security

To further conceptualise the organisation as a sociotechnical system, we introduce the business engineering model. This model supports the discussion of the management approach within an organisation (e.g., security organisation, security culture). The model conceptualises an organisation as comprising several levels [9, 10]:

- Strategic – by its very definition, the strategic level refers to the strategic management of an organisation. This entails the identification, implementation and refinement of the long-term goals and objectives of the organisation.
- Process – while the strategic level answers the *what*, this layer addresses the *how*. The process layer revolves around the major business processes – that is, operations and supporting business processes (e.g. logistics, human resources). It also contains the management processes to monitor and assess success, and potentially replan the operations.
- System – this level comprises, in the widest sense, technological systems for both management processes and operations.

It follows from this conceptualisation that organisational security is identified as an element of the organisation's strategy. Based on this, an appropriate approach is chosen, implemented, and continuously monitored.

Another key point of the model is that the organisation is embedded in a culture frame. Organisational culture is a widely researched construct. Culture, in general, influences how processes are managed and performed, how knowledge is shared and communicated and how technologies and interventions are perceived. From a security perspective, there is no shortage of examples of how culture, both external to the organisation (e.g., national, regional or local cultural norms) and internally within the organisation, can influence security. It is important to understand that within a sociotechnical system, humans and their interactions play a pivotal role when it comes to the success of policies, adherence to procedures and levels of cooperation. Within global and international settings, and also in diverse societies and workforces, the mix and variety of cultural behaviours and norms and how these influence the organisational culture can vary widely. Effectively addressing such differences in culturally complex or international contexts can be challenging, as totally different or additional interventions may be needed to remedy issues.

Conceptually, organisational security is about managing security across an organisation. Security is increasingly viewed as an integral part of business risk management and an important distinguishing aspect. In the end, the risks to the operation of the business are the real justification for investments in security. Hence, there has been a move from exclusively regulatory/legislative compliance-driven implementation to the active establishment of an organisational security system.

This poses the question *why do we need secure organisations?* The aforementioned approach to viewing security as an integral part of the overall organisational risk picture is an inward view. From a societal perspective, organisations have an important role to play when it comes to the provision of services or goods. There is a societal dimension to trust in the delivery of such services and goods, and also trust that organisations will not harm the public or customer through their enterprise. In that respect, the public can place trust in secure organisations.

4.2.3 Resilience spectrum – beyond defending the fortress

The introduction of this chapter touched upon the move towards 360-degree risk management and the resilient organisation. Many organisations are realising that the traditional approaches to the protection of staff, infrastructure and information span manifold activities. Resilience thinking enables an organisation to absorb and respond to disruptions, including adapting to an ever-changing context, an operating environment with growing complexities and potential threat scenarios and risks.

The imperative for resilience is high. The paramount goal is to ensure continuity of essential processes or services in the face of, literally, all significant threat and hazard scenarios.

The concept of resilience has gained a high level of attention. The term itself is used in various contexts and disciplines. It is used, amongst other things, at personal, group/community and organisational level, in a variety of engineering and scientific domains, and disciplines such as safety and security. The widespread use of the term has also led to differing definitions and meanings. In this chapter, we take resilience as a fundamental property of a system, which is able to respond to, and recover from, impacts that do not exceed its operational limits [11–13]. Such a resilience capability is of course a desirable property of a sociotechnical system. In that respect, we see resilience management as a holistic approach to managing the prevention, preparedness, response and recovery capabilities of an organisation.

Resilience is also often used as an umbrella term to describe the properties of a system (or organisation) to address its risks and respond to and recover from disruptions to normal operations [14]. A comprehensive approach to security risk management should therefore address measures applicable at different points in time of a potential security incident. One example for this approach is the Prevention–Preparedness–Response–Recovery model [15] where:

- Prevention concerns the operational risk to a high-value service. The aim is to implement measures for critical assets to ideally prevent the realisation of the risk or at least mitigate it as far as possible.
- Preparedness ensures availability of resources or implements additional measures to account for changes in the threat and risk landscape where the preventive measures will only be useful to a limited level.
- Response sustains a high-value service if risk is realised by implementing measures suitable to ensure the continuation of operations and services without major service degradations.

- Recovery effectively addresses the consequences to the organisation if risk is realised and returns the organisation to a 'normal' state.
- Resilience management is to monitor, assess and optimise the achievement of these objectives, to maximise the effectiveness with reasonable effort and costs.

Understanding security as part of the operational risk spectrum of an organisation and the establishment of proper security management practices will inevitably lead to a wider and more comprehensive appreciation of resilience. In this respect, security is more than 'building walls' to prevent the previously mentioned 'bad people and bad things' from entering an organisation.

4.2.4 Critical infrastructure thinking

Since the turn of the 21st century, states have been reviewing their security posture, and began early implementation of a large number of local, regional and national initiatives. For example, the European Commission established a programme for the protection of (European) critical infrastructure in 2006 [16].

Around the world, the national or regional transportation systems were identified as part of the critical infrastructure, next to the power networks and communication networks. The unifying criterion is the role such an infrastructure plays with a view to maintain societal functions, and the impact of the loss or partial unavailability of such infrastructure, on the national or regional security and well-being of the population. The latter represent the overarching goals of states, the protection of citizens and integrity of the territory, and assurance of the functioning of society.

The interesting aspect is that the critical infrastructure thinking revolved around the reduction of vulnerabilities, thus, putting the *critical assets* as the focus of concern.[†] In the light of September 11 attacks, and also large-scale power outages, critical infrastructure protection programmes recognised the threat of terrorism and introduced an all-hazards approach. This more holistic approach is useful as it breaks the traditional risk-based approach and does not limit itself to specific threat scenarios and attack spaces. Hence, 'acts of god', or natural events, were considered equally to 'man-made' intentional attacks. This thinking also paved the way for a more resilience-oriented thinking, as the understanding of criticality also included a strong interest in re-establishing services and functions.

The resilience notion also introduced another important aspect – the interdependencies between sectors and organisations. This widened the view on securing assets and infrastructures, as parties have a role to play to ensure the mitigation of risks with other parties involved in the infrastructure and beyond the specific sector. Within air traffic management (ATM), this led to the introduction of a second dimension of security termed 'collaborative support'. Collaborative (security)

[†]Criticality to be understood here as the relative importance or necessity in achieving or contributing to the functioning of the organisation, internal and external services.

support is understood as 'the provision of (support) services to other authorities/ organisations dealing with aviation security' [17].

In the light of critical infrastructure protection programmes, various organisations linked to societal functions and infrastructures, and both public and private, undertook vulnerability and risk assessments on a large scale. Depending on the domain, the criteria for the identification of critical assets were established through regulations, typically based on the experiences from other industries. This spurred a clearer approach to the categorisation of assets that is also helpful in risk management.

The concept of criticality is now inherent in today's operational risk management practices. A resilience approach will also uncover interdependencies between organisations and/or their resources and services. Securing an organisation's operation inevitably requires consideration of these interdependencies, which may even become decisive risk factors.

4.3 Management systems

Not all management happens in the form of organisational structures. In the end, management can apply to single persons or small groups.

One of the basic premises of management theory is that all organisations (e.g. commercial enterprises, public services) are built on a mission describing 'why the organisation exists', for example, its purpose, vision and 'where the organisation will go'. As part of strategic planning, the mission is typically translated into strategic goals, which are then further broken down to operational goals and objectives. Accordingly, the goals and objectives of staff, and also their level of participation, are essential for achieving the mission and high performance. Only through such an organisational framework will an organisation be able to establish or contribute to the vision over time.

It follows that the day-to-day operations of an organisation, as well as mid- to long-term planning and adaptation, require management activities to be carried out, both within departments and across the organisation. Depending on the size of the organisation, structures are in place to support this type of management activity. The level of complexity and maturity of an organisation has therefore led to 'management systems' being established. Conceptually, a management system is a structured and documented set of interdependent policies, practices, processes and procedures used by the organisation to plan and execute activities that are required to achieve the organisation's objectives [18]. This definition requires such a framework to promote continual improvement following the Deming cycle of 'Plan–Do–Check–Act' [19].

Based on this discussion, we can summarise the basic tenets and benefits of a management system approach for an organisation, as shown in Table 4.1.

Accordingly, a management system is a structured approach as to how an organisation 'manages the inter-related parts of its business in order to achieve its objectives' [20]. Depending on the management paradigm established in the organisation, several topics may be managed through dedicated management systems: product

Table 4.1 Benefits of management system approach

Benefit	Characteristics
Compliance	Assurance of legal compliance with regulation/legislation, international standards. Management systems provide a recognised means of achieving the appropriate technical and organisational measures.
Reputation	Higher credibility through the thorough application of standards and best practices, including associated certification.
Risk management	Operational risk management is a key component of all management systems. Structured and reproducible analysis of operational risks with a systematic approach to risk treatment.
Cost-efficiency	Reduced and targeted costs, as the focus is on the operational risk, ensuring the implementation of necessary and adequate controls for risk mitigation.

or service quality, workforce and workspace safety, environmental protection and sustainability, and security.

It is noteworthy that the International Standards Organisation (ISO) introduces the notion of domain- and sector-specific standards. In addition to ISO, there exists a variety of communities of interest, both formal or semi-formal, that have developed guidance and standards. For example, within the field of information security, further standards and reference material are produced by the National Institute of Standards and Technology [21] and published in the form of so-called *Special Publications*. Further details on standards can be found in Chapter 2.

4.4 Aspects of security management system implementation

4.4.1 Introduction

Managing security requires that the goals of security management be forged from, and aligned with, the high-level objectives of the organisation [22]. There is a tendency to consider security as a technology-driven activity; however, it is a business or organisational problem that must be framed and solved in the context of the organisation's strategic objectives. The evolution of a risk-based paradigm for security has made it clear that a secure organisation does not result from securing its technical infrastructure and production base alone. A security approach that is mission-centric – that is, based on strategic objectives – strives to secure the organisation's critical assets and processes regardless of where they reside [23] (Figure 4.2).

The security management approach provides a framework addressing the wider societal and internal organisational view on security. For example, within the field of transport systems, the ultimate goal is to ensure the provision of transportation services. This applies to public transportation systems such as local mass transportation (e.g., bus services) and international travel (e.g., maritime or air). This wider view also includes supporting services. Air traffic control is an essential service to

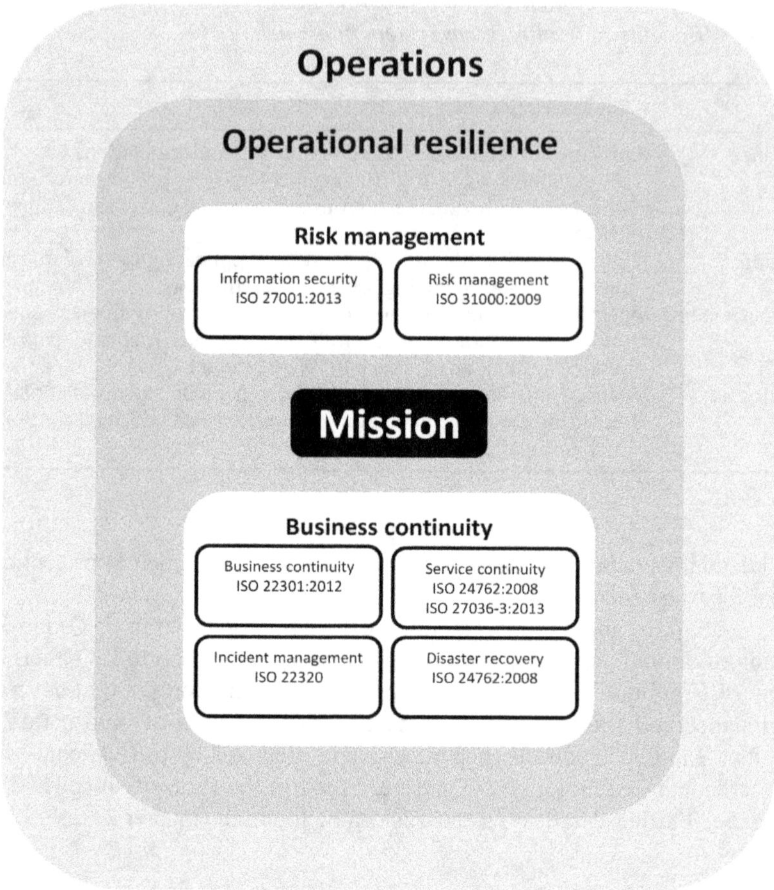

Figure 4.2 Organisational mission protected by operational resilience [23]

ensure the orderly flow of air traffic. In this respect, is the security of air traffic control organisations, and ensuring the availability of air traffic services, wider than just securing the organisation?

Furthermore, management systems introduce the notion of continual improvement. Based on the principal ideas of the Deming Plan–Do–Check–Act cycle, there must be a continual review of the organisational practices, rules and procedures. This recurring activity ensures the ability to meet the organisational top-level requirements, as well as ensuring that improvement activities are structurally identified, planned and implemented. This is in contrast to more reactive approaches requiring remedial action, which are often 'tough lessons learnt'.

Within air transportation, the management system practice is well developed in the field of safety, where it is an integral part of all activities turning a system into a manual that is 'lived' day-to-day, interweaving roles, tasks and organisational culture and practices.

4.4.2 Implementing a security management system

In Section 4.3, we described a management system as a framework of policies, procedures, guidelines and associated resources and activities, jointly managed throughout an organisation. With respect to security, there are three key elements:

1. Security organisation
2. Security processes/procedures
3. Security culture, and awareness

Here, the security organisation refers to the people in the organisation with their defined roles and responsibilities, which are to maintain and achieve the security objectives of the organisation.

Different guides for the implementation of security management systems exist, and a good generalisation of the approach is found in Reference [24], which identifies the following key steps:

- secure executive support and set the objectives
- define the scope of the system
- evaluate assets and analyse the risk
- define the security management system
- train and build competencies for the roles
- system maintenance and monitoring
- certification/audit
- system maintenance and continual improvement

These steps should ensure the proper implementation of the security management system, ultimately evidenced with the achieved certification or positive audit of the framework. However, the steps per se will not guarantee that the security system will be 'lived' within the organisation. An issue for security practitioners is that security is largely seen as a cost on the business, with no clear return on investment or understanding, until after an incident.

The long-term success of a security management system lies in both its framework and day-to-day execution. Throughout the organisation, management and staff need to be made aware of security management, so that security becomes part of the organisational culture. This includes acceptance of critical feedback and truthful follow-up of reported weaknesses or identified opportunities for improvement. In the authors' experience, this is a mammoth task and requires open and transparent communication within the organisation.

Continual improvement is an essential aspect of a living management system. In today's business-driven world, this can also include substantial costs: investment in training and implementing additional security controls. In particular, continual improvement revolves around the review, strengthening and expansion of current practices. It can be challenging to make the case for security at the executive level. But it also offers a lever for management to empower the organisation to develop and grow its own security posture. Accordingly, the very first step in the list above is essential – the executive support and buy-in.

By way of example, we present the security management system defined within the context of European air traffic management, which became a requirement under the Single European Sky Common Requirements for providers of air navigation services [25]. This requirement began as the intention to harmonise the different national approaches to security management in ATM. Hence EUROCONTROL, in close collaboration with its stakeholders, developed a framework for the establishment of an organisational security management system [26], as shown in Figure 4.3.

POLICY	
Element 1	Policy
Security Risk Assessment and Planning	
Element 2	Security risk assessment
Element 3	Legal, statutory, regulatory and other security requirements
Element 4	Security management objectives
Element 5	Security management targets
Element 6	Security management programmes
Implementation and Operation	
Element 7	Structure, authority and responsibility for security management
Element 8	Competence, training and awareness
Element 9	Communication
Element 10	Documentation and document control
Element 11	Operational control
Element 12	Emergency preparedness, response and security recovery
Checking and Corrective Action	
Element 13	Security performance measurement and monitoring
Element 14	System evaluation
Element 15	Security related failures, incidents, non-conformances and corrective and preventive action
Element 16	Control of records
Element 17	Audit
Management Review	
Element 18	Review and continual improvement

Figure 4.3 EUROCONTROL security management system framework

The framework should not be seen solely as an instrument and checklist, which can create a false sense of managing security properly. It is essential to understand that the success of any management system is dependent on the fostering of an associated culture within the organisation. The framework alone is not enough.

4.4.3 Human factors

The human factor is recognised as a major concern in organisational security and has been termed by Bayern as an 'evergreen problem' [27]. In particular, any security system, no matter how well designed and implemented, relies on people. The Kaspersky Lab 2018 report noted that almost half of all organisations face critical security issues related to staff errors [27]. It is worrying that the human factor plays such a big role in today's security landscape. Apparently, while we can implement appropriate technical security solutions, we seem to fail miserably with the human element.

This highlights the need for training and the use of the right tools. However, there is a deeper requirement for establishing security know-how and situational awareness with staff.

Situational awareness is the 'perception of the elements in the environment within the volume of time and space, the comprehension of their meaning, and the projection of their status in the near future' [28]. Situational awareness informs time-critical decision-making, and ultimately human action.

There are many factors that affect security know-how and associated situational awareness. For example, compliance with security measures can be impacted by the higher priority of work given to staff by managers. Conceptually, security procedures may require additional time that needs to be found in addition to the time that staff has to do their normal work. Focusing solely on the organisation's tangible production can therefore result in security being perceived as a burden and time thief, and ultimately ignored.

For security situational awareness, there are two key considerations: security risk perception and incident situational awareness.

4.4.3.1 Security risk perception

Research in decision-making in time-critical domains (e.g., police, firefighter, military, medicine) suggests that certain patterns of behaviour and action are based on the similarity with previously learned patterns [28]. This type of situational awareness, however, and dealing with the risk situation at hand, is not well understood in normal work environments.

Various factors influence risk perception, such as gender, one's health and age. It is still an open area of research how these factors work together when it comes to organisational security. However, the model of Behnke identifies explanatory factors that can be taken into account when forming an organisational culture and raising the general awareness of staff. This is useful to address cognitive biases in both directions: extreme risk awareness and a false sense of security. The former can lead to an inability to perform tasks, while the latter may expose the organisation to security breaches. The 2018 Ponemon Institute benchmark report [29] on the costs of data breaches, attributes 27% of the root causes of breaches to human error, such as poorly followed procedures.

4.4.3.2 Incident situational awareness

Incident situational awareness requires in situ compliance with procedures and processes, including reasoned application of principles to new situations. The core aim is to provide a framework and set of principles for the individual to arrive at a decision, and start the implementation of actions based on it, even in the absence of direct communication with line managers or commanding officers. Broadening these concepts to security functions in general, and also extending into the normal training of staff, will be a critical success factor for maintaining security. Staff training for incident situational awareness is focused on follow-up decision-making based on the assessment of the situation.

4.4.4 Organisational and security culture

The previous section highlighted the drivers that can lead to *human error*, which can be traced back to poor situational awareness. This can be compounded by a lack of personal motivation or the understanding of procedures. Well-known examples include passwords being written on Post-it® notes and allowing visitors to be unaccompanied in restricted areas. Hence knowledge, defined in Ref. [30] as 'cognition and the capacity for action and decision-making', is also an enabler of situational awareness.

Recognising that individuals work within a context of organisational goals, policies and procedures, what then is the impact of culture? Smolnik and Hornung [10] refer to reference [3] to show the interplay needed within knowledge management to create content, build competence and encourage collaboration. Culture plays a significant role in supporting these actions [30].

Organisational culture in general refers to concepts such as trust, norms and standards, unwritten rules, symbols or artefacts. The ultimate goal is that these concepts are shared by the members of an organisation and provide orientation in a complex world. Organisational culture is, to a large extent, an implicit phenomenon and thus difficult to observe and open to interpretation [30]. It is the result of a learning process and is handed on to new members of the organisation in a process of socialisation [31].

For security management, a security-oriented culture can be developed to influence the behaviour of members of the organisation in general and, more specifically for security, their willingness to share knowledge [32]. A supportive organisational culture is considered one of the most important success factors for faster organisational learning [33] or the implementation of a knowledge management initiative [34]. Culture positively affects knowledge creation and especially knowledge sharing, even across subcultures, such as those of executives, engineers and operators [35]. A supportive organisational culture has been conceptualised as a resource [21] reflecting the character of social relations within the organisation, known as organisational social capital [36]. However, the concept is only vaguely defined and it remains largely uncertain if, how and to what extent organisational culture can be assessed and influenced in a systematic way [37–39].

4.4.5 Reinventing the wheel – what can we learn from safety?

The aviation industry is built around the concept of safety and organisational learning, with a view to create one of the safest operational systems and including one of the safest working environments. Safety has a rich tradition in aviation and has resulted in a steady stream of activities and research. Throughout the years, the concept and approach to (operational) safety has been continuously developed.

In seeking to share information between staff members and across organisations more formally, new barriers become apparent. Although the safety domain indicates the potential for improved knowledge through formal sharing of information, it has not been without its problems. For instance, the absence of a 'just culture' environment in many states has led to a reduced knowledge of incidents and their severity. 'Just culture' refers to a culture where staff are not punished for 'honest mistakes', thereby creating an open reporting environment. Where states have implemented just culture frameworks, there has been a noticeable improvement in incident reporting rates.

Security on the other hand is less developed and acknowledging the fact that since the beginning of the 21st century, little organisational developments have taken place on a large scale, and there is also limited progress. It may often seem that security is spinning in circles, with constant reinvention but limited effect. As mentioned above, confusion in terminology from different risk disciplines can lead to a lack of clarity. Over time, the number of risk management-related disciplines and standards has outgrown the organisational capabilities and resources to properly implement the associated management systems. This points to integrated or combined risk management approaches.

A good example is security-related research and development. While the topics shift and new sets of measures are identified, the classical approach is to refine and augment existing asset classification and risk assessment methodologies. Depending on a particular call for research, 'novel' measures for the proliferation of disciplines like 'crisis management', 'cyber defence', arise, but are proliferating different approaches rather than fundamentally advancing the management science. The result is that few organisations have implemented comprehensive resilience or security management systems and measures.

Why is this? Despite its strategic relevance, organisational security is competing with other strategic objectives. From an operational perspective, resilience is an enabler for most organisations. Accordingly, security has to compete for resources under heavy scrutiny. The concept of 'local rationality' is often used to describe human decision-making based on the in situ local knowledge. This can explain why strategic decisions are also driven by predominantly compliance thinking. At the same time, understanding, adherence to and application of security measures by staff may be limited as exposure to certain risks within the organisation or related context has not taken place. In hindsight, rules might be perceived as too complex and simply not applied or broken.

One of the key questions is therefore, how can we use organisational learning to achieve a similar success as that of the safety approach in aviation?

4.4.6 Technological support for security management

This chapter provides for a broad understanding of security management. When discussing technological support for security management, it is also essential to separate tools for

- managing the (security) management system
- risk management
- security functions

As mentioned throughout this chapter, there exists a variety of security (or integrated) management standards. For virtually all of these standards, tools have been developed to support the implementation and maintenance of the management system. This comprises checklists as well as policy and procedure templates that can help organisations in preparing and implementing the management system.

Organisations have an interest in the investigation and proactive management of incidents before these escalate and impact brand or reputation. Accordingly, a set of security risk management tools have emerged over the last few years, such as Reference [40]. These tools support the security organisation to identify, track and follow-up on identified security risks in the organisation. On top of this, community networks such as Computer Emergency Response Teams (CERTs) provide additional risk identification and prevention-oriented input for in-house security risk management. A noteworthy proposal for inter-organisational risk and control information exchange was developed under the umbrella of SESAR. Information from security risk management as well as identified minimum security risk and selected security controls were collected in the form of a database. Such information can be made available to other experts dealing with a similar system.

Security functions beyond the traditional guarding of premises or systems are emerging in many organisations. Earlier we mentioned the dedicated (cyber) security operations centres implemented in air traffic control in Europe, and many other organisations have or are investing in these. This chapter suggests building on this approach by orienting initial 24/7 CERT or crisis cells towards wider inter-organisational security operations, based on a more collaborative security management approach, as discussed in the following section.

4.5 Collaborative security management

4.5.1 Introduction

Security management systems tend to be unique to a single organisation, reflecting a standard framework but tailored to the specific organisational requirements. However, there is great potential for organisations to share information, which can go beyond the usual sharing of threat intelligence and analyses.

It is also argued [41] that it is virtually impossible for any organisation to prepare and respond to cyber incidents without collaboration. In the following paragraphs, we distinguish between cooperation, which may be defined as a transactional approach based on 'quid pro quo' information sharing, and collaboration, where there is a strong imperative to truly work together for mutual benefit. We note that the European Union Agency for Network and Information Security (ENISA) has described collaboration as 'a common objective of every European national cyber security strategy'.

At the state level, government agencies and authorities traditionally have the means and instruments to exchange security-related information between partnering state organisations. These networks often originated from bilateral agreements or multilateral organisations such as NATO, the International Criminal Police Organization (commonly known as Interpol) and the European Union Agency for Law Enforcement Cooperation (better known as Europol). There is a limit, however, as to what can be achieved through formal sharing mechanisms. Hence, driven by various communities of interest, networks of collaborating partners have evolved.

For example, within the field of information security, the CERT concept sparked off incident- and prevention-oriented information exchange between a wide network of organisations.

Over the last 20 years, 'Information Sharing and Analysis Centres' (ISACs) have emerged as the predominant cooperative vehicle for organisations to share information and analyses on threats, incidents and other topics. Such information sharing through membership groups is a centralised model, where an ISAC gathers and analyses threat intelligence, as well as developing its own products for the benefit of its members [42]. ISACs have typically been set up within industry sectors, and mostly within a single State, although there now exist multinational ISACs.

Even with ISACs, there are continued calls for more information sharing, so in this section, we consider the wider possibilities for information sharing, leading to potential collaboration within the framework of organisations' security management systems. We also discuss the concept of 'collaborative support', which was developed in ATM as part of the European Union's SESAR programme [43].

Within the air traffic control sector, the SESAR development phase also promoted information exchange within the system engineering domain, by proposing a database-driven concept to collect security risk assessment information and associated recommended controls [44].

4.5.2 Information Sharing and Analysis Centres

Information Sharing and Analysis Centres (ISACs) have become the de facto model for information sharing, although many organisations have established less formal bodies, such as the Lithuanian 'Forum for information exchange' [45] or the 'NATO-EUROCONTROL ATM Security Coordination Group' (NEASCOG) [46].

ISACs comprise mostly private sector organisations with public sector (agency) support and are funded by a membership fee. There may also be cooperation with academic institutions. An essential component is that there is a facilitator to coordinate the exchange of information between members and software platforms for day-to-day exchange of technical threat intelligence.

The ENISA regards the whole concept of ISACs as being 'connected with sharing information and analysis concerning cybersecurity incidents'. ENISA has recently reviewed European ISACs and has identified a number of issues important to their future success [45]. These include creating the appropriate governance, establishing mutual trust and investing in analysis capability. ISACs may gather and analyse more strategic information, such as industry security trends, but the extent that this is done depends on the level of funding for the ISAC.

ISACs have been defined around national, sectorial and international themes. For example, the European Centre for Cyber Security in Aviation (ECCSA) has recently been established as a joint initiative by the European Union Aviation Safety Agency and the EU's Computer Emergency Response Team (CERT-EU) for EU institutions, agencies and bodies [47]. Membership of ECCSA is open to organisations relevant to civil aviation safety and security, and the focus is on protecting the 'critical elements of the system such as aircraft, navigation and surveillance systems, datalinks, airports, etc.' [48]. CERT-EU will provide cybersecurity tools and share threat intelligence and analysis with ECCSA. More informal groups are also created where there are common interests, for example, a dependency on a specific technology such as industrial control systems [49].

4.5.3 Information exchange

The information exchanged through ISACs concerns a variety of topics [45, 49], as shown in Table 4.2.

While the content of the information exchange addresses the security mission, the actual implementation and means of information exchange can still be hampered by silos between organisations and government agencies, with the potential for miscommunication or communication break downs.

Considerable value can be gained by sharing such information. Members may be informed by new threats, activities of threat actors to develop their own mitigations or borrow from others. Any member who shares the details of an incident enables others to avoid the same attack [50].

Organisations can also re-evaluate their risk assessments as new threats and vulnerabilities are exposed. Sharing of best practices, such as in backup, recovery

Table 4.2 Examples of information exchange in ISACs

Type of information	Examples
Incident	Detailed description of an actual attack, whether successful or not, including what is known about the attackers, the tactics used, the impact of the attack, etc.
Vulnerabilities	Specific vulnerabilities discovered in hardware/software, including networks and associated devices, procedures etc.
Threats and threat actors	Details of malicious software, compromised IP addresses/URLs etc. discovered in an organisation's network or files. May be communicated as indicators of compromise. Threat actors may include information about specific groups.
Mitigations	Software patches, antivirus updates, incident recovery.
Best practices	Processes, procedures, metrics used by organisations to manage cybersecurity.
Trends	Analysis of threats and threat actors to reveal trends relevant to the members.

and conducting exercises, enables organisations to better handle real events [42]. It is also noted that many cyberattacks can only be detected by gathering and correlating evidences obtained at different locations [42].

The collection of information from numerous organisations can also give a wider context, such as whether an organisation has been uniquely targeted. Further benefit derives from the diversity of members, who bring different insights into threat intelligence from the different security tool backgrounds of their experts [51]. There is also a capacity-building benefit through supply chains, as hackers may attack an organisation's less secure supply partners as a way into the target firm [50]. From an economic perspective, Gal-Or and Ghose [52] conclude that there are strong economic incentives for firms to 'indulge in such security breach information sharing'.

Although the information shared through ISACs clearly adds value, a number of issues have been raised:

- ISACs tend to exclude professional services and vendor organisations, being distrustful of their commercial motives [45, 48].
- There is a concern that information shared with other organisations could be passed on without prior agreement [51].
- Timeliness of exchange: a key concern of sharing information is in its timeliness. Haas *et al.* [42] note that damage from a cyberattack may occur and propagate in a very short timescale (seconds, minutes and hours) while the discovery and containment of a threat takes days, weeks or even months.
- The volume of technical threat intelligence now provided through automated methods is well beyond the capabilities of human analysts. The case study of ACTRA [41] refers to automated feeds with 30,000+ events per month.

- Hernandez-Ardieta [41] argue that effective policies for near real-time information sharing must rely on both formal models to estimate the value of the information shared and trust/reputation models that consider the dynamics of the sharing community.

Hence, although there are benefits in information sharing within member groups, the exclusion of vendors, professional services firms and connections with other groups tends to limit their effectiveness and very likely lead to duplication.

4.5.4 Information sharing methods

There are different methods of information sharing, from face-to-face exchanges to intelligence shared through technology platforms such as 'MISP' [53]. The detailed information shared through platforms such as MISP tends to be very technical and potentially time-limited, such as the Internet addresses used by attackers. The treatment of such detailed threat intelligence is discussed in Chapter 7.

Informal information sharing, such as through conversations between members, is valuable, particularly concerning more strategic threat intelligence. Meeting to exchange and agree best practices from peers should have a positive effect. However, a concern is that information exchange can be hindered by the reluctance of some members to reveal gaps in their knowledge. Members join ostensibly as 'experts' and may not want to admit to shortfalls in their knowledge. This could result in discussions limited to easier topics, for example, methodological issues, lack of resources and regulations. Informal information also suffers from being mostly tacit knowledge, which could mean that as memories fade or staff transfer, the corporate body of knowledge diminishes [54].

4.5.5 Developing a collaborative approach

A fundamental principle for collaboration is trust. Mutual sharing of information can create positive feedback, so that as one organisation shares, others are more likely to share and invest in security-enhancing technologies [52]. However, security managers, practitioners and other professionals may struggle with collaboration with competing organisations.

Corporate concerns may be around what information is shared, particularly if shared vulnerabilities might be leaked and result in commercial impacts. Farral [50] takes a game theory approach to cybersecurity information sharing. The results show that when the potential damage cost is low, organisations may be better off not sharing, compared with the risk of sharing their vulnerabilities. In Wagner *et al.* [52], IT executives revealed more concern with the impact of online security problems on consumer confidence and trust in e-business than the actual financial losses of physical infrastructure.

There is also a concern that some organisations limit their own information, relying on shared information from others. This is also known as free-riding.

Furthermore, there are concerns that a competing firm or a third party might use the shared information against the reporting firm by anonymously reporting it to the public. Wagner *et al.* [52] report that even the US CERT passed vulnerability information to third parties and potential competitors of the reporting firm.

Collaborative security management can only build on shared values, principles and transparency. A way to establish a common ground is also to develop a deeper understanding of the system and service context in which different organisations operate.

Steps in this direction have been built into EUROCAE's ED-201 'Aeronautical Information System Security (AISS) Framework Guidance' [55]. ED-201 provides guidance for different aviation organisations to cooperate on aeronautical information systems security. The aim is to reduce risks to the safety of flight and significant disruptions to operations. A fundamental principle is that organisations share information about security wherever there is shared responsibility for systems and operations. ED-201 assumes that the organisations involved have formal security management systems and risk assessment methods, although these may not follow the same standard. The guidance proposes a comparison of organisations' risk management methods against the ISO 27005 framework. Once comparability of risk assessment is established, organisations are able to identify risks that may be shared by virtue of the interfaces with partner organisations.

The European (more than EU states) Aviation Safety Agency (EASA) and EUROCONTROL have been developing the concept of 'Shared Trans-Organisational Risk Management' (STORM). This is a framework to support sharing of information as foreseen in ED-201. It requires methods to harmonise risk assessment and share appropriate outputs, which map to organisations' functions and the interfaces between them. Across these interfaces are flows of information, and there exist 'interface risks' according to the security controls applied by individual organisations. Sharing information about these risks enables the different organisational partners to ensure that appropriate controls are in place.

4.5.6 Collaborative support

The concept of collaborative support was developed in ATM as part of the European Union's SESAR programme [56]. It was created to guide development of new ATM concepts towards wider systems thinking for security. The intention of collaborative support is to provide services or information from ATM systems to external agencies such as law enforcement, the military, emergency services or incident investigation agencies in the context of acts of unlawful interference on aviation.

Collaborative support is intended to be a pre-planned coordination between ATM and outside agencies. Hence, it concerns incidents such as hijacks, cyberattacks, air policing, entry into restricted airspace and so on. The roles of different parties are predefined and coordinating procedures are developed and rehearsed. The concept was introduced into the SESAR programme's security activities [18] with the aim of encouraging projects to adopt collaborative support requirements into their operational descriptions and system requirements. For example, a new operational service could have a coordination requirement leading to the dispatch of search and rescue aircraft, or it could support the interception of unknown aircraft by the military (air policing) by clearing the surrounding airspace.

To date, the airspace security coordination cell in Germany represents the only existing centre for a permanently staffed collaborative support cell in air transportation in Europe. Staff from the Air Navigation Service Provider (ANSP), national forces and police cooperate in all matters on security of the airspace. In other parts of the industry, similar cooperation agreements would be activated in specific conditions, referred to as 'crisis cells'. With a view to airspace security and the inherent time-criticality for decision-making, such cells cannot support the required level of responsiveness. Hence, there remains a need for better integration of existing security operation centres and information exchange capabilities between different stakeholder organisations.

4.6 Conclusions

This chapter introduced and discussed the organisational approach to managing security. A management system in general – and in particular for security – cannot be isolated from the organisational culture and supporting processes. These may be strengthened by technological tools. However, at the core, a management system is the combination of policies, processes, culture and tools employed by an organisation to ensure it can meet its objectives. In other words, a management system is how an organisation gets things done. Inherently, a management system is targeted to protect the business operation of an organisation and manage associated risks.

This view on management systems is independent of whether we speak about government or public organisations, non-profit foundations or commercial enterprises. With regard to security, the context or role of these organisations may, however, differ significantly.

In discussing security management systems in this chapter, we took a broad approach to operational risk. Depending on the business sector, including the respective regulatory requirements, security may be managed in combination with, or complementary to, other domains (e.g. business continuity, critical infrastructure protection). It was not the intention of this chapter to establish 'One Ring to rule them all' [57]. The main aim was to demonstrate that security – and other management disciplines – forms a spectrum where the definition and delineation of one topic versus another is more a question of style and discipline. Similar to a ring, the key takeaway is that there is no end or beginning. It is the management task to conceive a system through policies, rules, processes and organisational culture, which ensures the continual delivery of the organisation's objectives, including ensuring organisational security and collaborative support.

This chapter has also conceptualised security in two dimensions: (i) self-protection and resilience of the organisation and (ii) collaborative support across organisations. This view is fuelled by considerations of the ever-increasing interdependencies of our production systems and processes, the link between service provision and service consumption and the higher level of Internet working within the cyber domain.

Concepts like critical infrastructures or service providers and the view on an organisational role for national and regional security provide for a wider discussion of security beyond individual organisations. The future security functions of an organisation will include operational elements for the whole life cycle of security that are tightly bound with the processes and functions of other organisations. Security in that respect will be a community task rather than one's own responsibility. In this sense, we see security management as becoming partly a collaborative endeavour, requiring the development of greater trust and deeper sharing of information. The benefits from collaborative cybersecurity management are likely to be extensive, not least in the leveraging of resources and combined effectiveness of multiple security teams, often duplicating efforts as they respond to a 'fire hose' of threat intelligence. Greater collaboration does, however, need organisations to manage commercial concerns, which was part of the rational for EUROCAE ED-201 proposing 'external agreements' between partners [55].

Another important aspect in embracing a resilience approach to security is the move towards system thinking in terms of continual improvement and not just factual framework implementation. One essential stepping stone in this is the move from pure compliance to a lived and empowered organisation when it comes to security. Addressing security as one aspect of the wider organisational risk picture in the form of a more integrated approach to management systems is a logical consequence. In the end, security should not be seen as an additional discipline, but as an integral part of ensuring the goals of the organisation.

A key enabler for enhancing security in all its aspects is the development of an open and transparent culture within an organisation. This includes the principle of trust between all parties. Within the organisation, between senior management and the security organisation, and between the security processes and functions and the staff. 'Building security in' is a common term used in system engineering; however, organisations have to tackle the challenge of embedding security in their organisations, and free it from the perception of being a burden or costly practice. While this is essential within the organisation, it is also important for wider security collaboration outside the organisation.

Security management is therefore making a security management system 'lived' by the organisation through an open, trusted, and transparent culture.

References

[1] Ritchie D. '2012 Security leadership issue: The security leader – Beyond guns, guards, and gates'. *Security*. 2012.

[2] Koelle R., Markarian G., Tarter A. 'Aviation security engineering: a holistic approach'. *Artech House*. 2011.

[3] Koelle R. 'Aviation/ATM Security: An Introduction to Resilience'. *Proceedings International Summer School of Aviation Psychologist (ISAP)*; 2007.

[4] Barbosa L.C., de Oliveira O.J., Santos G. 'Proposition for the alignment of the integrated management system (quality, environmental and safety) with the business strategy'. *International Journal for Quality Research.* 2018;12(4).

[5] SAI Global *Integrated Management Systems, a Framework for Business Success, Whitepaper*. 2015.

[6] Ross R., McEvilley M., Carrier Oren J. *Systems Security Engineering - Considerations for a Multidisciplinary Approach in the Engineering of Trustworthy Secure Systems*. NIST; 2014.

[7] Roser M., Nagdy M., Ritchie H. *Terrorism – Our World in Data'. Global Change Data Lab*. University of Oxford; 2018.

[8] Freedman L. *The transformation of strategic Affairs*. Routledge; 2017.

[9] Riempp G. *Integrierte Wissensmanagement-Systeme: Architektur und praktische Anwendung*. Springer-Verlag; 2012.

[10] Smolnik S., Hornung O. 'Knowledge management, course unit 1: foundations'. *University of Hagen, Faculty of Business Administration; Economics*. 2016.

[11] Gluchshenko O., Foerster P. 'Performance based approach to investigate resilience and robustness of an ATM system'. *Tenth USA/Europe Air Traffic Management Research and Development Seminar*; Chicago; 2013.

[12] Martin S., Deffuant G., Calabrese J.M. 'Defining Resilience Mathematically: From Attractors to Viability'. *Viability and Resilience of Complex Systems*. Springer; 2011. pp. 15–36.

[13] Nieuborg A., Koelle R., Gluchshenko O., Foerster P. 'Towards Evaluating Air Navigation Performance for Estimation of Its Resilience'. *2014 Integrated Communications, Navigation and Surveillance Conference (ICNS) Conference Proceedings*; 2014. pp. L4–1.

[14] Tiernan A., Drennan L., Nalau J., Onyango E., Morrissey L., Mackey B. 'A review of themes in disaster resilience literature and international practice since 2012'. *Policy Design and Practice*. 2019;2(1):53–74.

[15] Caralli R.A., Allen J.H., Cutis P.D., White D.W., Young L.R. 'Cert resilience management model, version 1.0'. 2010.

[16] European Commission. 'Communication from the Commission on a European programme for critical infrastructure protection'. *European Union*. 2006.

[17] Koelle R., Hawley M. SESAR Security 2020: How to Embed and Assure Security in System-of-Systems Engineering? *2012 Integrated Communications, Navigation and Surveillance Conference*; 2012. pp. E8–1–E8–11.

[18] 'ISO 9001:2015 – Quality Management System – Requirements'. *International Organisation for Standardisation*. 2015.

[19] Deming E.W. *Out of the crisis. Cambridge, MA: center for advanced engineering study*. Massachusetts Institute of Technology; 1986.

[20] International Standardisation Organisation. 'Management system standards'. *ISO*. 2018.

[21] NIST [online]. Available from https://www.nist.gov/.

[22] Caralli R.A., Wilson W.R. *The Challenges of Security Management*. Software Engineering Institute; 2004.

[23] Caralli R.A., Allen J.H., Cutis P.D., White D.W., Young L.R. 'Cert Resilience Management Model, version 1.0. Improving Operational Resilience Processes'. *Software Engineering Institute*. 2010.

[24] How to implement an Information Security Management System [online]. Available from https://ins2outs.com/implement-information-security-management-system/.

[25] Regulation C.I. 'Commission implementing regulation (EU) 2017/373 of 1 March 2017 laying down common requirements for providers of air traffic management/air navigation services and other air traffic management network functions and their oversight, repealing regulation (EC) NO 482/2008, implementing regulations (EU) NO 1034/2011, (EU) NO 1035/2011 and (EU) 2016/1377 and amending regulation (EU) NO 677/2011'. 2017.

[26] EUROCONTROL. *Security management handbook – A Framework* [online]. 2008. Available from https://www.eurocontrol.int/publication/eurocontrol-security-management-handbook.

[27] Schwab W., Poujol M. The state of industrial cybersecurity 2018. *Trend Study Kaspersky Reports*; 2018.

[28] Endsley M.R., Garland D.J. *Situation Awareness Analysis and Measurement*. Boca Raton, FL: CRC Press; 2000.

[29] Ponemon Institute LLC. *2018 Cost of a Data Breach Study: Global Overview: Benchmark Research Sponsored by IBM Security Independently Conducted by Ponemon Institute LLC*. IBM; 2018.

[30] Schein E.H. 'On dialogue, culture, and organizational learning'. *Organizational Dynamics*. 1993;22(2):40–51.

[31] Schreyögg G. 'Zur Logik Der Strategischen Unternehmensführung'. *Management revue*. 1992;3(H. 3):199–212.

[32] Hofstede G., Neuijen B., Ohayv D.D., Sanders G. 'Measuring organizational cultures: a qualitative and quantitative study across twenty cases'. *Administrative Science Quarterly*. 1990;35(2):286–316.

[33] Schein E.H. 'Legitimating clinical research in the study of organizational culture'. *Journal of Counseling & Development*. 1993;71(6):703–8.

[34] Davenport T.H., De Long D.W., Beers M.C. 'Successful knowledge management projects'. *MIT Sloan Management Review*. 1998;39(2):43.

[35] Schein E.H. 'Three cultures of management: the key to organizational learning'. *Sloan Management Review*. 1996;38(1):9–20.

[36] Leana C.R., Van Buren H.J. 'Organizational social capital and employment practices'. *Academy of Management Review*. 1999;24(3):538–55.

[37] Freiling J., Fichtner H. 'Organizational culture as the glue between people and organization: a Competence-based view on learning and competence building'. *German Journal of Human Resource Management*. 2010;24(2): 152–72.

[38] Nieswandt M. 'Changing organisational culture: a review of the literature'. *Fast Cultural Change*. 2015:5–60.

[39] Drumm H.J. 'Probleme der Erfassung und Messung von Unternehmungskultur' in Dülfer E. (ed.). *Organisationskultur*. Stuttgart: Poeschel; 1988.

[40] Investigate and Resolve Incidents [online]. Available from https://www.amag.com/incidentmanagement [Accessed 27 Jul 2021].

[41] Hernandez-Ardieta J.L., Tapiador J.E., Suarez-Tangil G. 'Information sharing models for cooperative cyber defence'. *2013 5th International Conference on Cyber Conflict (CYCON 2013)*; 2013. pp. 1–28.

[42] Haass J.C., Ahn G.J., Grimmelmann F. 'Actra: A case study for threat information sharing'. *Proceedings of the 2nd ACM Workshop on Information Sharing and Collaborative Security*; 2015. pp. 23–6.

[43] 'EUROCONTROL. Security Management Handbook – A Framework'. *Edition 1.0*. 2008.

[44] Gotz K., Hawley M., Machin C., Hird J. 'A security tool for SESAR's design-in security approach'. *2016 11th International Conference on Availability, Reliability and Security (ARES)*; 2016. pp. 499–502.

[45] E.N.I.S.A. 'Information sharing and analysis centres (ISACs) cooperative models, ISBN 978-92-9204-239-4'. 2017.

[46] NATO and EUROCONTROL strengthen collaboration to master current and future challenges for civil and military aviation in European airspace [online]. Available from https://www.eurocontrol.int/press-release/nato-and-eurocontrol-strengthen-collaboration-master-current-and-future-challenges [Accessed 27 Jul 2021].

[47] EU Computer Emergency Response Team (CERT-EU) [online]. Available from www.cert.europa.eu.

[48] EASA cooperate with CERT-EU on cybersecurity [online]. Available from www.easa.europa.eu/newsroom-and-events/press-releases/easa-cooperate-cert-eu-cybersecurity.

[49] Goodwin C., Nicholas J.P., Bryant J., *et al.* 'A framework for cybersecurity information sharing and risk reduction. Microsoft'. 2015.

[50] Kamhoua C., Martin A., Tosh D.K., Kwiat K.A., Heitzenrater C., Sengupta S. 'Cyber-threats information sharing in cloud computing: A game theoretic approach'. *In 2015 IEEE 2nd International Conference on Cyber Security and Cloud Computing*; 2015. pp. 382–9.

[51] Farral T. 'The definitive guide to sharing threat intelligence'. *Anomali*. 2018.

[52] Gal-Or E., Ghose A. 'The economic consequences of sharing security information'. *Economics of information security*. Boston, MA: Springer; 2004. pp. 95–104.

[53] Wagner C., Dulaunoy A., Wagener G., Iklody A. 'MISP: The Design and Implementation of a Collaborative Threat Intelligence Sharing Platform'. *In Proceedings of the 2016 ACM on Workshop on Information Sharing and Collaborative Security, ACM*; 2016. pp. 49–56.

[54] Hawley M., Howard P., Koelle R., Saxton P. 'Collaborative security management: Developing ideas in security management for air traffic control'. *In 2013 International Conference on Availability, Reliability and Security*; 2013. pp. 802–6.
[55] 'EUROCAE Document ED-201 – Aeronautical Information System Security (AISS) Framework Guidance'. 2015.
[56] SESAR Joint Undertaking. Available from https://www.sesarju.eu.
[57] Tolkien J.R.R. 'The fellowship of the ring'. *Harper Collins*.

Chapter 5
Security and safety

Chris Machin[1], James Hunt[2], and Chris Johnson[3]

5.1 Introduction

There has been a historic partition between safety and security risk management, which could be seen as an oversimplification as the two are becoming increasingly blurred. This is particularly the case with the increased use of automation and autonomous systems in the transportation sector. Furthermore, there is a misguided assumption that by performing a safety risk assessment, security risk is also addressed.

The distinction between accidental and deliberate causes of an unwanted event is becoming harder to distinguish, especially when nation state offensive teams or cybercriminals have developed cyberweapons that mimic more routine software or hardware bugs.

The premise of this chapter is that security and safety are inextricably linked, and organisations with safety risk management practices should begin to introduce more formalised security risk management. A key question is how these approaches to risk management differ and whether, and how, they might be integrated.

In this chapter, we address this issue, first by describing safety risk management and 'safety cases', which are used for the introduction of new systems or significant system changes and to existing operational systems. We then describe an approach to a 'security case'. The remainder of the chapter considers the extent to which safety and security can be brought together and ways of doing this.

The examples, processes and statements in this chapter are drawn from the aviation industry. Similar organisational structure and processes are defined for other transport sectors, such as the International Maritime Organisation or the European Railways Agency.

[1]General Manager, Digital Acceptance at Discover Financial Services, USA
[2]Director, PwC, UK
[3]Engineering and Physical Science, Queen's University Belfast, UK

5.1.1 Safety management and assurance

Safety is concerned with protecting people from death or injury and avoiding any degradation in operations that may be caused by an unintentional (i.e., accidental) event. Safety is a cumulative property of all the components of the system interacting with a particular environment or context rather than a property of individual components. For example, software by itself is neither safe nor unsafe; the same applies to hardware, procedures, operations and humans. Safety is the sum of all the components of the system operating in a particular environment, which make it either safe or unsafe.

From a transport perspective, work on safety is done to ensure that the transport system is

- safe when working normally and
- safe when a system component, or components, fails then the risk of harm is within a given tolerable boundary.

Safety *assurance* involves demonstrating that operations are safe, and the consequences of failure do not degrade the safety performance below an agreed level. Failure in this respect may mean component failure, system design flaws, unsafe interactions among components and unsafe interactions with other elements of the transport system. Safety assurance also involves maintaining acceptable levels of safety throughout the system life cycle, from concept to decommissioning.

As readers will know, the transport sector is highly regulated and approvals are often required, for example, to operate a train service, introduce a new aircraft, change a procedure or introduce new equipment in an operational environment. The approvals are given by a Safety Regulator, often in the form of certificates or permits to operate, and these functions can be devolved to certification or 'inspection' bodies, which are authorised by states or multinational entities, such as the European Commission. Approval to operate requires a set of conditions to be met, one of which is to show that the system will be safe enough to operate. The certification decision – that is, the granting of the written assurance or 'certificate' – is based on the inspection report, possibly complemented by other information sources. Certification is always done by a third party, and the certificate demonstrates that the operator complies with certain standards. These may or may not be the same agencies that regulate the cybersecurity of a transport system.

At the time of writing, approval to operate is generally carried out with little regard to the impact of security. However, if an organisation's systems are not secure, can they be safe?

5.1.2 Differences in risk management approaches

Both safety and security management lead to a set of mitigations (or controls) to manage the consequences of a failure or attack. There are some differences, however:

- Safety risk management approaches risk by analysing unintentional acts resulting in unintentional events – the actors, designers and developers have no intent on causing harm. These are typically approached by assessing how often a *failure* of a system component (person, hardware, software, operational procedures, etc.) will result in the undesired consequence. The assessment is followed by specifying or 'designing in' mitigations to reduce the risk of failure causing the event.

- Security risk management approaches risk by analysing the potential for malicious actors to do harm. These actors are generally assumed to have a deliberate intent to attack system components. To protect the system components, security management sees the implementation of 'countermeasures' or 'controls', which either reduce the likelihood of an attack being successful (pre-event control) or reduce the impacts of the attack if successful (post-event control). It is important to stress that security threats are typically more dynamic than safety hazards; security threats come from a range of adversaries innovating new forms of attack at a rate that is far different from the more gradual changes in operating environment that might force a revision to safety risk assessments.

We can say that safety and security risk management processes have the same objectives, to manage avoidable incidents by addressing risk and implementing mitigations to reduce the risk. They similarly achieve this by:

- setting risk performance targets
- finding out what could go wrong
- finding the causes
- mitigating the causes (e.g. reduce the likelihood of the cause happening or reducing the impact)
- providing evidence that the mitigations reduce the risk
- monitoring the performance of the system to ensure that the mitigations continue to meet their performance targets

5.2 Safety management

What is striking about the comparison between safety and security is the extensive legal and management frameworks that have been established for safety, but not yet security. Security is seen as a separate activity, with only loose links to safety. However, this is being addressed in government agencies, academic and industrial circles leading to a more integrated approach to safety and security.

Safety management [1] is a key component of safety regulatory requirements and is implemented through safety management system (SMS), described in a Safety Management Manual [2]. This chapter discusses safety management in the context of aviation, but the approaches in other transport sectors are similar. A SMS defines a systematic approach to managing safety, including the necessary organisational structures, accountabilities, policies and procedures.

At the international level in aviation, a framework for aviation SMSs has been agreed through the International Civil Aviation Organisation (ICAO). ICAO standards require the implementation of a SMS by the following organisations:

- aircraft operators
- aircraft maintenance organisations
- air navigation services providers
- airport operators
- training organisations
- aircraft manufacturers

The effect is that the suppliers, operators, installers, maintenance staff or generally anyone who interacts with these organisations or are part of the supply chain are all responsible for safety.

The ICAO safety management system [3] framework consists of four components:

1. safety policy and objectives
2. safety risk management
3. safety assurance
4. safety promotion

In the following paragraphs, we focus on safety risk management.

5.3 Safety risk management

Safety risk management addresses the transport system when working normally, and in the case of a failure of a system component or components. There are two approaches to safety that work together. The 'success approach' aims to protect the inbuilt safety features of an operation. The 'failure approach' aims to evaluate safety when a failure of a component occurs and to ensure it does not introduce additional safety risk to the user.

5.3.1 Safety management without failure

Safety management under normal operations leads to requirements that state what should happen when operating without failure. These safety requirements are typically stated in terms of the quality of service, which can be measured: latency, integrity, accuracy and availability of system elements (equipment, humans and procedures). Some specific examples of safety performance requirements are:

- that aircraft altitude shall be delivered to an air traffic controller within 10 s of its measurement on board an aircraft
- that all driver messages shall be delivered within 5 s
- that the accuracy of speed measurements shall be with ±2 km/h

- that an emergency stop message shall be acknowledged within 3 s
- the accuracy of the altitude estimate shall be within 100 feet

5.3.2 *Safety management in failure conditions*

Safety risk management for failure conditions is concerned with identifying and mitigating hazards to an acceptable or tolerable level. The process followed is to develop a set of safety objectives, specifying what should not happen, identify the potential causes of failure (e.g., in terms of equipment failure, human and procedural errors) and then place safety requirements on those causes to mitigate the risk. Safety objectives and requirements can either be absolute (a number) or qualitative (a relative statement of performance).

Failure risk management uses the concept of tolerable risk, also referred to as risk appetite. This is simply a statement of how much risk an organisation is willing to accept around the occurrence of an event, such as caused by unintentional component failures. The aviation sector typically defines risk as the combination of the overall probability, or frequency, of occurrence of a harmful effect induced by a hazard and the severity of that effect [4]. As discussed in Chapter 3, risk can be expressed in qualitative terms (high, medium and low), but in safety, it is often expressed as the probability of a fatal accident. A target level can then be set for this risk, which is known as a target level of safety (TLS). For example, a large airline might set a TLS of 10^{-8} per flying hour, equivalent to one accident in a hundred years, with the equivalent for the industry being 10^{-6}–10^{-7} per flying hour. In European air traffic management, the TLS has been defined as 'the maximum tolerable probability of ATM directly contributing to an accident of a commercial air transport aircraft shall not be greater than 1.55×10^{-8} per flight hour' [5].

Aviation risk appetite is derived from past events and focused on unintentional acts; that is, it does not include security incidents. It is also expressed at the system level, comprising the failure rates of multiple components. Failure rates are difficult to express for some system components, particularly for software and human interaction.

The traditional activities in a safety failure risk management process are:

- A hazard assessment (using techniques such as a functional or operational hazard analysis) to produce safety objectives stating the frequency or probability a hazard can occur and the overall system safety risk of the operational service remains acceptable;
- Developing a set of safety requirements and assurance levels for the system elements to limit the frequency of the hazards or their impacts. Such requirements are sometimes referred to as mitigation means, of which there are two categories: 'internal mitigation', which is the responsibility of the system developer, and 'external mitigation', which refers to requirements placed on other systems.
- Implementation and monitoring of the safety requirements.

The first stage of a safety assessment will develop a set of 'safety objectives'. These objectives state a tolerable rate of occurrence for which a particular failure event, irrespective of the cause of the failure, can occur. The safety objectives are qualitative or quantitative statements that define the maximum frequency or probability at which a given hazard can be expected to occur. Examples are:

- the accidental undetected corruption of a display in the control centre shall not occur more that 10^{-4} per operational hour;
- the accidental undetected loss of train coupling shall not occur more than 'frequent';
- the accidental creation of a false event tree analysis shall not occur more than 'infrequent';
- the accidental loss of the communications between the driver and the centre controller shall not occur more than 'improbable'.

The second stage of the failure assessment is to identify the system components that could accidentally cause the hazard, and then place safety requirements on the rate of component failure. Requirements are placed on the system component such that if the causes do occur the safety objective is still achieved. Safety requirements are defined in terms of:

- The failure mode, for example, loss (full or partial) or corruption of data;
- Asset, for example, arrival schedule, engine, clearance message or flight identity;
- Failure rates for hardware and component failure expressed in failures per flight hour or operational hour;
- Assurance levels (for all components) capturing the level of rigor to be applied during the design process and operation of the component. Assurance levels are relative and may take the form of 'one to *x*', or low, medium and high. The higher the criticality of the software, procedural or human component, then more rigour in the definition and valuation of the component is required.

Quantifying hardware failure is based on the mean time between failure and mean time to repair or recover [6], which attempts to capture how often a hardware component will fail based on historical observations and how long will it take to replace or recover the function. However, quantifying software, human or procedural failure rates is essentially not possible.

Modelling the system architecture and failures are often used to show the propagation from component failure to unwanted incident. Modelling can be used to represent the combined impact of all system components (e.g. procedures, people, hardware, software, weather) with some modelling techniques better suited to model particular component failures. There are a variety of modelling techniques [7] and those in common use include the following:

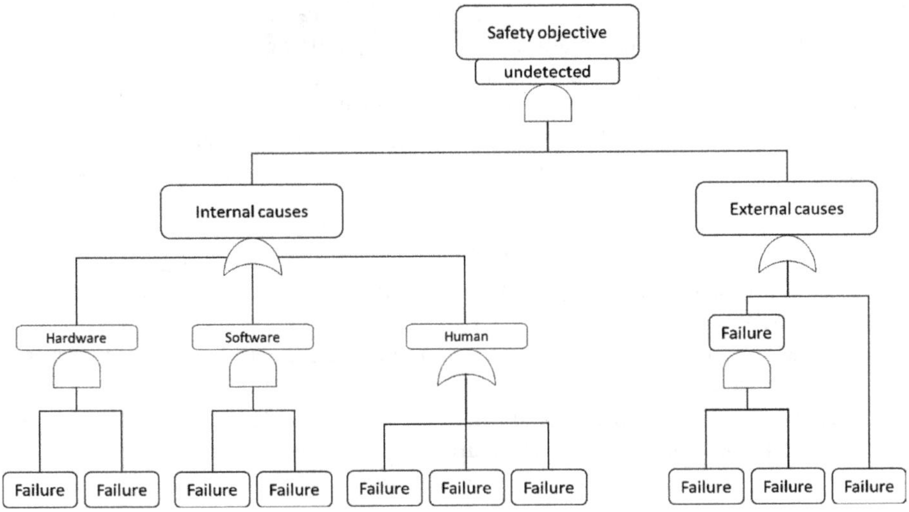

Figure 5.1 *Safety modelling*

- fault tree analysis, which is a top down technique
- failure mode and effects analysis, which is an inductive, bottom-up analysis
- event tree analysis
- human error analysis
- operator error analysis
- bow tie modelling

Figure 5.1 provides a generic example of a fault tree. It shows a top level safety objective which should be achieved and which is broken to internal (to the system) causes and external causes. Internal causes may include a hardware failure, software error or procedural error. External cause may include weather, flooding, bird strike or animals on the runway, which are typically outside the control of the system developers. Each of these may only cause the system failure if they occur independently or in combination with other failures.

The third stage of safety management is to ensure that the safety requirements are implemented when the design is released, brought into service and operational. Safety requirements are also needed for monitoring of safety performance during the operational life of the system and ensuring that the system will remain safe when decommissioned.

5.4 Safety assurance

Safety assurance comprises the planned and systematic actions necessary to be confident that a product, service, system or organisation achieves acceptable or tolerable safety. Assurance is shown by the applicant requesting permission to operate, for example, service provider or manufacturer, to the regulator.

The amount of assurance is related to the criticality of the component in Figure 5.2 The more critical the component then the more the assurance, that is, testing, which is needed to demonstrate to the regulator that the component is safe.

Safety assurance is gathered during the implementation of the system. Evidence typically includes

- validation, with more validation depending on the criticality of the component;
- that the team participating in the safety management activities and risk assessment are credible and competent to do so;
- that all phases of the operational life of the service, including installation, operations and decommissioning are addressed.

A number of software standards help in providing safety assurance, including ISO/IEC 12207, ED109, ED-12B/DO 178B, IEC 61508 and CMMI.

Safety assurance may be provided in a 'safety case', which contains claims made, safety arguments and evidence of safety activities such as design, testing and integration.

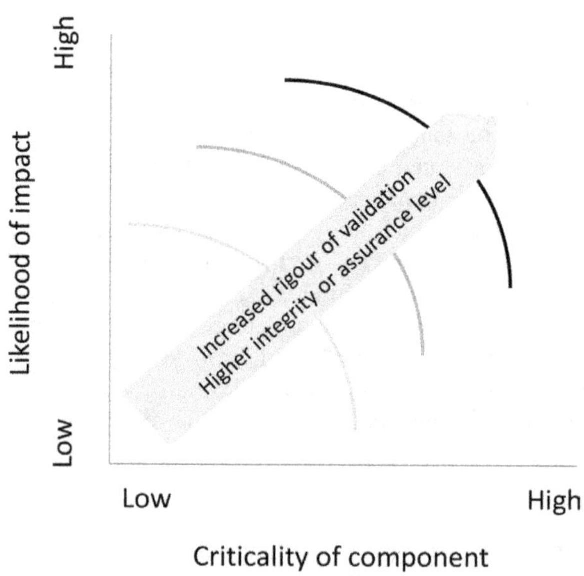

Figure 5.2 Assurance required related to criticality of the components

5.5 The safety case

A safety case is a communication tool between the applicant seeking permission to operate and the regulator for the introduction of a new system or operation. The safety case comprises a statement (or a set of statements) that asserts that the service or system concerned is and will continue to be safe. In more detail, Office for Nuclear Regulation [8] defines a safety case as

> a logical and hierarchical set of documents that describes risk in terms of the hazards presented by the facility, site and the modes of operation, including potential faults and accidents, and those reasonably practicable measures that need to be implemented to prevent or minimise harm. It takes account of experience from the past, is written in the present, and sets expectations and guidance for the processes that should operate in the future if the hazards are to be controlled successfully. The safety case clearly sets out the trail from safety claims through arguments to evidence.

The concept of a safety case was introduced following the Piper Alpha accident in the UK North Sea in 1988, which led to 167 deaths. The United Kingdom subsequently introduced a new safety regime [9], which required a 'safety case'. This was intended to demonstrate that a company has identified all hazards posed by an operation, adequately assessed the risks and reduced them to be 'as low as reasonably practicable' and has in place adequate SMSs to ensure that risks are controlled. This was adopted by the European Union through the 'Seveso' Directive [10], which extends to include a new general duty 'to take all measures necessary to prevent major accidents and limit their consequences to persons and their environment'. Seveso also adopted the concept of 'as low as reasonably practical' [11] and introduced a legal framework for a continuous improvement cycle of prevention, preparedness and response to major accidents.

The development of a safety case is not an alternative to implementing a SMS but is a means of structuring and documenting the results of the safety management process for a system, including a risk assessment. Safety cases may be made for an ongoing operation or a new function or operation. The work on the safety case is done during the project that delivers the new operation or operational change. Developing a safety case is a cooperative exercise between safety experts and those involved in the delivery of the system (operational, technical, managers) and the systems users.

The content of the safety case is derived from the knowledge of those who are designing and using the systems. Safety experts provide the structure and guidance on developing the safety case. They may provide input to stimulate the discussions on threats to the system from their knowledge of the threat landscape, but the consequences and controls (mitigations) must be driven and endorsed by those involved in the system delivery, and its users.

5.5.1 Safety case structure

The process for producing safety cases takes into account the needs of those who will use the safety case to ensure safe operations. It is essential that the safety case documentation is clear and logically structured so that the information is easily accessible to those who need to use it. This includes designers, operations and maintenance staff, technical personnel and managers who are accountable for safety [8].

There is no predefined or agreed format of a safety case, its only purpose is a presentation by the applicant to the regulator. It should, however, contain

- the stated safety goal of the system (e.g. statements about risk)
- how this is achieved (claims, argumentation)
- evidence that the safety goal is achieved (e.g. through the safety assurance process)
- evidence that the assurance is credible

Goal structured notation [12] is a popular means of illustrating a safety case, although it is not mandatory. Figure 5.3 shows an example of the safety argument. The safety case makes a claim about the safety performance of a project and its resulting operation, and the contents of the safety case provide evidence that the claim is valid. The top-level claim is decomposed into a set of sub-claims (arguments) expressed as a simple predicate (true or false), which, when proven, shows that the top-level claim is true. It is the purpose of the safety management process to provide evidence that each argument is true, therefore proving the top-level claim is valid and meets its safety performance criteria.

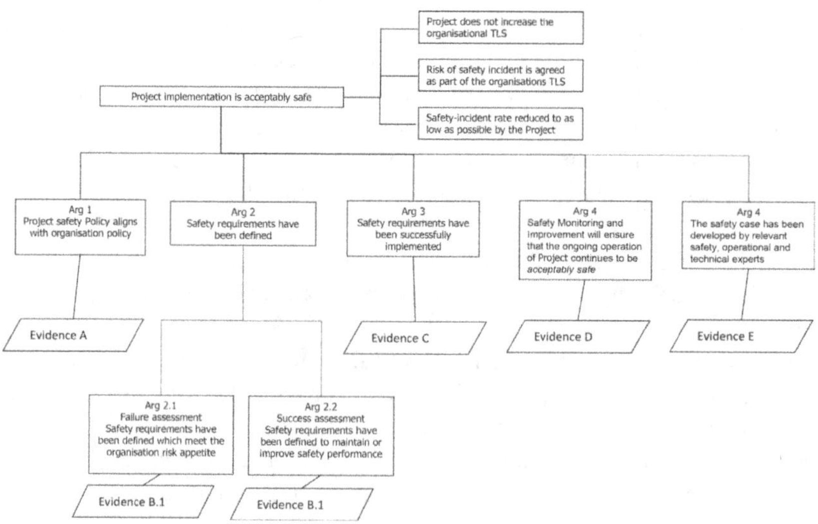

Figure 5.3 Example safety argument

The safety claim for the above project is that it will not increase the safety risk to the organisation, that it has an agreed contribution to the TLS and that it helps in reducing safety incidents. The decomposition of the claim into lower-level arguments is structured based on the SMS process and provides the essential links between the claim and the wealth of evidence needed to show that the claim is valid. Note that Argument 2 develops the success and failure approaches discussed earlier.

5.6 The security case

5.6.1 Introduction

The authors have previously defined a security case within the European air traffic management 'Single European Sky ATM Research' (ESAR) programme [13]. The concept of a security case was introduced early on in SESAR [14] as part of the documentation for SESAR's 'Solutions'. A SESAR Solution comprises the operational and technological requirements and specifications needed to introduce a new air traffic management innovation [15]. Part of the contents of the Solution documentation was a security risk assessment and a security assurance case, adopting the principles of a safety case. The idea was to support the eventual implementation, deployment and operation of SESAR Solutions through their lifetime, starting in the research and development phase. The production of security cases was limited in the programme timescale, but valuable insight was gained.

As for a safety case, the purpose of the security case is to document and communicate evidence that a system or one of its components manages security risk within the accepted risk appetite. There is as yet no regulatory requirement for a security case, but it has the same benefits of a safety case in structuring and presenting evidence to the regulator. In the following paragraphs, we describe our views on what a security case should look like and the process to develop it, building on the approach to safety cases.

The starting point is that the security case demonstrates that an organisation

- understands the scope of the system and the threats to that system
- has defined objectives for the security of that system
- has developed or implemented controls to ensure the risk is managed
- has managed security risk throughout the design, implementation and operations of a system, which could be part of a transport system

5.6.2 Structure of a security case

The structure (sometimes called a security 'pattern') of the security case is based on a top-level claim that the system will be secure (with definitions of what secure means for the users) and arguments about why that claim is true (based on the security management system). Each argument should have a unique claim relating to the top claim and sufficient evidence that the combined sub-claims achieve the top claim. If one of the claims is proven untrue then security has not been adequately addressed, potentially resulting in system redesign, rescoping or rejection.

Arguments are developed and assurance presented to those in control (e.g., business leaders, generals, chief executive officers, politicians, general public), and the regulatory authorities typically act on behalf of the state to ensure that they fall within 'acceptable guidelines' for the citizens of the state or organisation. The case is then presented to a decision-maker to allow an assessment of the merits and trade-offs between the processed solutions. There is no standard for a security case, and each case is developed on a 'case-by-case' basis. There is also no guidance onto how much detail a case should comprise, it may be a few pages through to thousands of pages, as long as it demonstrates to the regulator that the new or updated system is and will continue to be secure.

The security case is more than risk management, encompassing the wider concept of security management (see Chapter 4). It includes not only risk assessment but also more widely how the security function will perform in a wider context of supporting other aspects of the transport or critical infrastructure.

A representative security case structure is presented in Figure 5.4. This is simplified and each argument may have many sub-arguments and associated evidence. The point is not to overcomplicate the argument however, and structure the case so it can be readily understood.

The following paragraphs explain the different elements of the security case.

5.6.3 Security claim

A claim should be easily understood. It has no predefined structure as long as it can be answered either *yes* or *no*. It should be phrased in a manner that users understand and may contain references to threats, countermeasures and threat agents; indeed, whatever the system developer is concerned about, the claim may be phrased as.

5.6.4 Argument 1: security policy

The security policy describes the holistic security aspects of the project throughout the complete life cycle. This includes the organisation, procedures and practices used to bring the project into operational use and the technical solutions (software, hardware) that enable the project.

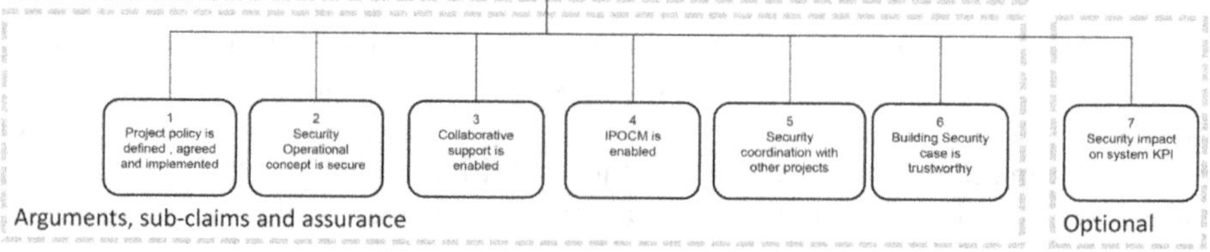

Figure 5.4 Generic security case

The security policy for the project should comply with the organisation's policy and should cover all aspects of security governance, including:

- personnel management, from recruitment (e.g. screening, identification and criminal checks), training, deployment and roles in the organisation and the eventual termination processes (e.g. return of access control or removal of access rights and company material);
- third-party management;
- compliance both regulatory;
- security processes and expectations on the project.

5.6.5 Argument 2: security concept

The security concept describes how security will be delivered for the new system (general principles) and the scope of the security case, including the interfaces that are internal and external to the system covered by the security case. It will also describe:

- the stakeholders involved or impacted;
- the objective of security in the event of an attack (prevention, detection, containment and recovery);
- how the project supports security functions of outside systems;
- the limitations and constraints imposed by other key performance indicators (KPIs).

The security concept is difficult to validate. As a minimum, the concept shall be agreed by all stakeholders at the project outset and reviewed/refined as the project progresses.

5.6.6 Argument 3: collaborative support

Collaborative support was described in Chapter 4 and considers that transport operations may have a security role in support of other systems and stakeholders in the event of a security incident, such as emergency services. If collaborative support is relevant to the operations, the security argument would claim that the requirements of collaborative support functions were met, including their performance attributes (timeliness and accuracy of information) and then evidence their correct implementation.

5.6.7 Argument 4: incident preparedness and operational continuity management

The incident preparedness and operational continuity management sub-claim concerns risk management and should aim to demonstrate that the prevention, detection, containment and recovery requirements have been correctly defined, validated and implemented in the intended operational environment. The evidence should show how, in the event of an incident, the system will not compromise the safety of passengers and will have sufficient resilience to recover the service quickly.

5.6.8 Argument 5: security interaction with outside systems

The interoperability with external systems is important, as transport systems interact with other systems in the provision and receipt of information that may be critical to their operations. Airports are a good example of this, exchanging information between airlines, air traffic control, handlers, caterers, government agencies and so on. Indeed, a cyberthreat to one could be a threat to all, potentially propagating and damaging systems widely.

Hence, Argument 5 examines the bigger picture – how the system interacts, influences and impacts the security of the wider transport system. The security case should provide evidence that interoperability requirements have been defined, accepted, implemented and validated. Making this argument ensures that the new system has considered not only its impact on other systems but that these other systems do not impact the security of the new system. Where requirements are put on external stakeholders, it is important to evidence how these will be maintained, such as through formal arrangements.

5.6.9 Argument 6: credibility of the security case

This argument should demonstrate that the security case has been developed by those qualified to do so. Unlike the other arguments, the assurance is qualitative and subjective. The assurance will examine if:

- experienced security experts were involved
- operational and technical experts from the transportation sector were involved
- peer review was performed
- best practices and standards were applied in developing and validating the requirements
- the evidence has been approved by the project manager
- all regulatory requirements are met

5.6.10 Argument 7: relationship between security and other KPIs

This optional argument considers the impact of security on other KPIs, such as safety or other economic indicators. The concern is that security controls might impair the performance of other key performance areas, which may need to be accepted as a trade-off or require further design. Examples are where security controls may delay passenger processing, impinge on privacy, impact operational performance, add additional workload to staff and so on. The argument explores what the impacts could be and the remedial measures to reduce them and accepts any remaining reduction in organisational performance.

5.7 Linking cybersecurity with safety

5.7.1 Introduction

To return to our earlier question, if an organisation's systems are not secure, can they be safe? A passenger might not be interested whether an incident is caused intentionally or accidentally, so there is perhaps an artificial barrier that needs to be crossed to link safety and security. Today, safety cases and, if done, security cases are largely developed in isolation. There is an argument for both, however, to become part of a larger coordinated approach, either as a security-informed safety case, part of a business case or a more holistic 'resilience case'.

For large projects, business cases have evolved to include a wide variety of feasibility issues, developing from the economic (cost–benefit analysis) to environmental impact, safety impacts, political arguments and so on. The set of combined arguments are typically high level in the business case and then progressively developed and validated throughout the project to ensure that the benefits of the project are ultimately delivered. An initial security case could well fit into such an approach, but may not drive a full integration of security and safety assurance.

A more promising approach is the concept of a resilience case. In this section, we explore the challenges in developing a broader resilience case.

5.7.1.1 Resilience

Resilience is a concept closely linked to incident management and business continuity [16], and it has multiple definitions depending on the areas of application, ranging from the physical recovery of objects to the capabilities of organisations to recover. Resilience refers to the capability of a system to accommodate variable and unexpected conditions without catastrophic failure. This is an important goal of security, which aims at safeguarding against threats and the exploitation of vulnerabilities.

Resilience is a function of an organisation's or system's management of threats and vulnerabilities and its ability to adapt in complex, dynamic and interconnected environments. The key characteristic is the ability to adapt prior to, during and after disruptions, so that the organisation can sustain operations and service provision under both expected and unexpected conditions. To develop a resilient system, security should ideally be 'designed-in' from concept, through design, operational deployment and eventually decommissioning.

5.7.1.2 Challenges in creating a resilience case

The aim of integrating the security and safety assurance processes is so that the safety consequences of cyber vulnerabilities are both understood and the risks are proportionately and appropriately mitigated. This creates a number of challenges and areas to work on in the future:

- transport industry awareness of cybersecurity
- a common cyber and safety taxonomy

- the extent that cyber and safety should be integrated
- methodologies for integrating safety and security processes
- operational challenges in linking cybersecurity with safety
- professional development – skills, qualifications and the profession generally
- regulation and guidance

Further details on each of these recommendations are set out below.

5.7.2 Improve transport industry awareness and supporting guidance

Organisations need to be more aware of the vulnerabilities in components and other products provided by their supply chain and what impact these vulnerabilities might have on safe operation. It is therefore a fundamental principle [17] to ensure that staff has appropriate awareness, knowledge and skills [18]. Operators of essential services may be able to draw on safety culture to build up the organisation's cybersecurity culture.

5.7.3 Agree a common cyber and safety taxonomy

When aligning cybersecurity and safety processes, a key initial activity will need to be agreeing to a common taxonomy, or where it is not possible to align, agree how to manage this.

5.7.4 The extent that cyber and safety should be integrated

It is possible to align cybersecurity with safety, but for full consideration of the safety consequences of a cybersecurity vulnerability, integration of safety and security risk management processes should take place. There are therefore various degrees to which cybersecurity and safety might be integrated. Suggested degrees of integration are set out in Table 5.1.

Table 5.1 Levels of safety–security integration

Level
Level 0: cyber and safety assessments performed in isolation.
Level 1: cyber processes hooked within safety augmentation. For example, initial safety assessment questions confirming cyber compliance.
Level 2: alignment of impact-scoring criteria and matrices, for use by both cyber and safety assessments.
Level 3: vulnerabilities identified through a cyber assessment linked with safety consequences.
Level 4: vulnerabilities identified through a cyber assessment are interconnected at every stage of the safety risk assessment process, thus demonstrating that security has been considered at each safety decision point.
Level 5: a singular, converged cyber–safety assessment process.

Clearly, Level 0 is not desirable, but it is not clear whether Level 5 is either visible, pragmatic or proportionate.

5.7.5 Methodologies for integrating safety and security processes

In recent years, various methodologies for integrating safety and security assurance processes have been proposed. Some of these processes are set out in this section.

5.7.5.1 Cyber-physical systems safety and security alignment approach

Sabaliauskaite and Mathur proposed a Cyber-Physical Systems safety and security alignment approach, which would perhaps achieve a Level 4 integration. The alignment is achieved by merging safety and security life cycle phases and developing a unified model, failure–attack–countermeasure graph [19], as shown in Figure 5.5.

5.7.5.2 HAZOP-based security analysis

Use cases are widely used for functional requirements elicitation. However, security non-functional requirements are often neglected in this requirements analysis process. As systems become increasingly complex, current means of analysis will probably prove ineffective. In the safety domain, a variety of effective analysis techniques have emerged over many years. Since the safety and security domains

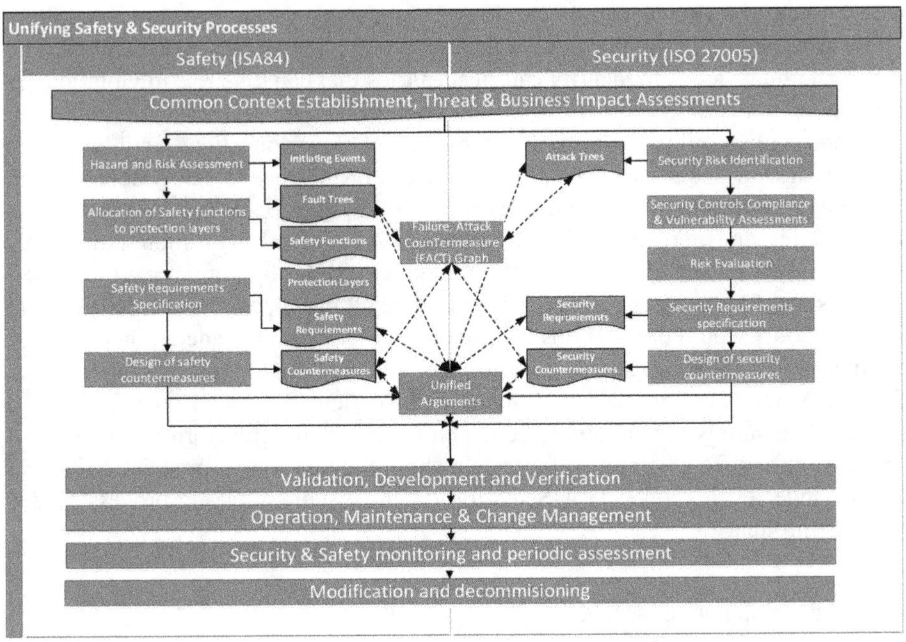

Figure 5.5 Failure–attack–countermeasure graph [19]. With kind permission © Springer International Publishing Switzerland 2015.

share many similarities, various authors have suggested that safety techniques might usefully find application in security. A hazard-based security analysis could apply HAZOP (HAZard and OPerability) methods to 'use cases' to systematically analyse potential security issues at the start of system development [20].

5.7.5.3 System-theoretic process analysis applied to security

Applying lessons learned from nearly three decades of research in safety engineering for complex systems, William Young and Nancy Leveson have proposed a more powerful hazard analysis technique, System-Theoretic Process Analysis (STPA) [21]. An extension of STPA, 'STPA for Security' (STPA-Sec), has been developed for securing software-intensive systems against intentional disruptions. This method reframes the problem as one of strategy, which involves shifting the majority of security analysis away from guarding against attacks (tactics) and more towards the broader sociotechnical vulnerabilities that allow disruptions to propagate throughout the system (strategy). Adversary activity is a common source of these disruptions, but it is not the only source. Trusted insiders can also take action to disrupt the operations of systems. The key question facing today's security analysts is how to control vulnerabilities, not how to avoid threats.

'STPA-Sec' shares the same four process steps applied within the system risk management process of STPA, although the results and detailed procedures may be different. The first step is establishing the systems engineering foundation for the security analysis. Then the control actions that threaten system security is identified. These control actions are used to create security requirements and constraints. The fourth and final step is to identify causal scenarios that can give rise to violations of the security constraints.

Some vulnerabilities may only be evident if the connections or interactions between the various subsystems are examined. For instance, a safety constraint in a train door controller may require that the door never be opened unless the train is at a station or an emergency exists. If a terrorist seeking to kill or injure individuals through a cyberattack is able to attack the door controller by mimicking the 'emergency' state, then the controller's logic might send the 'open door' command and the train doors would open. If this command was sent with a loaded train operating at full speed, it is easy to see how loss of life or damage to the system could occur.

STPA-Sec also approaches scenario construction in a much more structured manner than simply assembling experts and having them brainstorm scenarios that could go wrong from scratch. After establishing the necessary appreciation for the system under evaluation, STPA-Sec's top-down, systems thinking process guides analysts through:

- the development of scenarios
- interaction failures (e.g. feedback delays, conflicting control actions)
- the combination of component *and* interaction failures capable of producing the generated scenarios

This provides a deep insight into technical and non-technical aspects of the system, so that security analysts are then better prepared to select and apply the most appropriate controls. Unlike other approaches, security analysts using STPA-Sec are not forced to depend on their creativity to generate the full list of scenarios from scratch. Rather, STPA-Sec helps illuminate loss scenarios in ever-increasing detail, while allowing analysts to maintain their perspective on the larger system. In informal evaluations of STPA-Sec by security analysts and operations personnel, participants were surprised that using it helped them to consider threat scenarios that they had not thought of previously.

STPA-Sec does not provide answers about what specific countermeasures should be taken. Identifying protection mechanisms is and remains the realm of the security specialists. What STPA-Sec does provide is a potentially useful tool for identifying those scenarios that should be the focus of cybersecurity efforts to secure specific systems. Additionally, STPA-Sec provides traceability between the scenarios and the losses.

The United Kingdom's National Cyber Security Centre has considered using STPA alongside causal analysis based on system theory [22]. By using both frameworks, it is hoped that the following two questions will be answered in the event of a safety event that has been caused by a cyber incident:

1. How do we find an inadequate control in a design or in operations?
2. How do we find an inadequate control that may have caused an accident?

5.7.6 Operational challenges in linking cybersecurity with safety

For safety and security assurance processes to be integrated, there are many challenges that need to be considered and overcome. Some of these challenges are set out below:

- **Conflicting approaches to managing operational systems**. The approach to managing the cybersecurity of an operational system differs to managing safety. From a safety perspective, robust testing takes place prior to operations, then the system is declared to be in a known good state and a 'hands-off' approach is applied. Safety of the system is only assured if the configuration state has not been changed. The risk of inaction is not assessed, and malicious intent is not considered. In contrast, from a cybersecurity perspective, *continuous monitoring* is necessary. This 'hands-on' approach is necessary to ensure that the cybersecurity of the system is continuously monitored and defended in the ever-evolving threat environment. This includes assessing the risk of inaction and malicious intent.
- **Fail-safe versus fail-secure**. Safety and security have opposite fail conditions. An example would be to consider the consequences of implementation or not implementing a patch to address a zero-day vulnerability in a safety-critical system. From a cybersecurity perspective, every moment the patch is not implemented renders the system exposed to exploitation. However, from a safety

perspective, the system may need to stay operational to support a safety-critical process and therefore deny the immediate opportunity to turn the system offline and update, resulting in the deferment of patching. This decision to defer would need to be made by a suitably trained, empowered and authorised individual and captured in some form of cyber-safety risk balance case.

- **Safety in 'air-gapped' systems undermines cyber policies.** The 'air gap' between the device and any network improves the cybersecurity of safety-critical applications because it limits the opportunities for remote attacks. However, the 'air gap' also limits the effectiveness of intrusion detection systems, discussed in Chapter 7. In an air-gapped system, there is no easy way for system administrators to automatically update intrusion detection systems with malware signatures. Operators must manually install any updates on each isolated device across the plant. This leads to a paradox. It is hard for any attacker to compromise a stand-alone device, such as a programmable logic controller unless it is hooked to another device – for example, to install a patch or update the intrusion detection system. The more often these updates occur, the greater the risk of cross-contamination. Systems managers of safety-critical systems, therefore, often deliberately ignore conventional cybersecurity advice, preferring to leave isolated devices unpatched. Other problems limit the application of whitelisting in air-gapped systems. Without network access, it may be weeks or months before an engineer can examine the logs in sufficient detail to note an infection on a remote device.

5.7.7 Professional development

In the previous section, decisions to defer patching for safety reasons would need to be made by appropriately trained, empowered and authorised individuals. These individuals will need to understand not only the cyber consequences of not patching but also the safety consequences should an attacker impact the integrity or availability of the system. In Figure 5.6, we illustrate how the professionalism of safety engineering is well established and that of cybersecurity is maturing. The overlap in the middle, however, is less defined (Figure 5.6). The unknown quantity is the

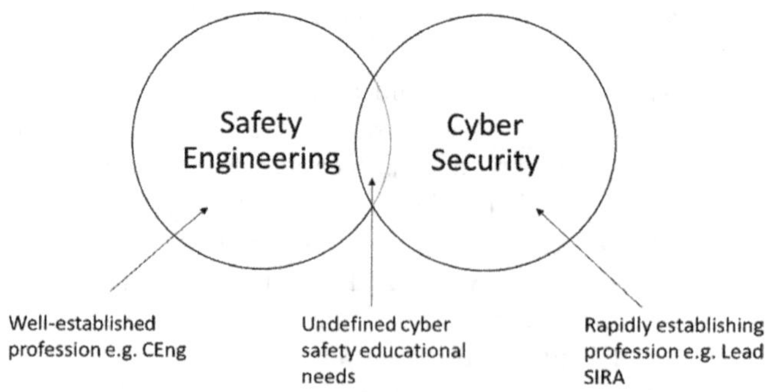

Figure 5.6 Overlap of safety and cyber disciplines

transfer of expertise: what additional cyber training would a safety engineer require to operate in the overlap? Conversely, what safety engineering training would a cybersecurity professional require to operate in the overlap?

5.7.8 Regulation and guidance

Neither government nor private industry has moved at the rate required to maintain our defences against the growing threats from cybercriminals and terrorist states. Public–private partnerships have failed to deliver regulatory guidance and appropriate audit mechanisms [22]. It is widely recognised that more rigorous and enforceable mechanisms are required to ensure that cybersecurity and its impact on safety is considered throughout the development life cycle.

The European Network and Information Systems Directive and the United Kingdom's National Cyber Security Centre's Cyber Assurance Framework are the first step towards establishing cybersecurity regulation in the context of the transport industry. It sets out the need to consider safety consequences of cybersecurity vulnerabilities as an outcome; however, it does not prescribe how. This would be down to the organisation to leverage internal and industry-leading practice in order to achieve the outcome. Regulators will need to understand what sector-specific prescriptive guidance is missing that will support companies to become compliant with the Cyber Assurance Framework or its international equivalents, and therefore the Network and Information Systems Directive (in Europe).

A challenge to identifying this gap in guidance is the limited understanding of this discipline within the regulatory bodies. In addition to upskilling the industry, the regulators too need to have access to the necessary training. In many ways, it is more important for the regulators to complete any necessary training, as it will be their responsibility to issue licences to operate. The decision to issue these licences will be based upon consideration of the safety and security assurance cases, in order to be satisfied that all risks have been appropriately and proportionately managed.

Manufacturing companies may introduce safety-critical systems today that presently are not subject to any form of regulation. To gain customer confidence in the security and safety of the product, initiatives such as kite marks are being explored. This provides a rating of the assurance placed around the product and might be used by the manufacturer as a differentiator in the market.

Ensuring these mechanisms are implemented and enforced will require a 'whole of industry' response, with some mechanisms being enshrined in regulation and others to be implemented through industry practices. To avoid the risks associated with overregulation (e.g. inflexibility with changing priorities), a balance needs to be struck between government oversight through regulation and industry freedom to innovate.

5.8 Conclusions

Safety and security cases provide high-level support for the development of critical systems. They present an overview of the arguments and evidence, which can demonstrate that any application is acceptably safe within a particular context of use. This chapter has described approaches for, and the challenges of, linking security with safety to provide high-level support for the secure and therefore safe development of critical systems. The intention of linking security into safety assurance processes is to provide an integrated, risk-based approach to help assess the resilience of safety cases to security threats.

- Security professionals need to understand the safety consequences of cyber vulnerabilities identified and ensure residual risks after mitigations can be accepted from a safety perspective. Security professionals must also ensure they are engaging with safety engineers at every stage of the development process.
- Regulations are maturing and evolving; however, they are not keeping pace with technical advancements. Safety cases continue to be signed off today that do not consider security, which we believe is unacceptable and putting lives in danger.
- To be confident that systems are both secure and safe, there is a need to ensure that security is considered not only within future safety cases but also within existing safety cases too.

References

[1] Safety Management [online]. Available from https://www.icao.int/safety/SafetyManagement, https://www.caa.co.uk/Safety-initiatives-and-resources/, https://www.faa.gov/about/initiatives/sms/ [Accessed Apr 2019].

[2] ICAO. *Doc 9859 Safety Management Manual*. 4th Edition; 2018.

[3] ICAO. Safety Management Implementation [online]. 2020. Available from https://www.unitingaviation.com/publications/safetymanagementimplementation/content/#/ [Accessed Jan 2020].

[4] EUROCONTROL. 'EUROCONTROL safety regulatory requirement: risk assessment and mitigation in ATM. ESARR 4'. 2001.

[5] Quantitative Safety Levels [online]. Available from https://skybrary.aero/articles/quantitative-safety-levels [Accessed July 2021].

[6] Health and Safety [Executive]. Failure Rate and Event Data for use within Risk Assessments [online]. Available from https://www.hse.gov.uk/landuse-planning/failure-rates.pdf [Accessed 24 Jan 2020].

[7] Everdij M. *Review of techniques to support the EATMP safety assessment methodology*. EEC Note No. 01/04 *EUROCONTROL Experimental Centre*. I; 2014.

[8] Office for Nuclear Regulation. *Safety Assessment Principles for Nuclear Facilities* [online]. 2014. Available from http://www.onr.org.uk/saps/saps2014.pdf.

[9] The Offshore Installations (Safety Case) Regulations 1992 [online]. 1992. Available from https://www.legislation.gov.uk/uksi/1992/2885/made.

[10] 'Directive 2012/18/EU of the European Parliament and of the Council of 4 July 2012 on the control of major-accident hazards involving dangerous substances, amending and subsequently repealing Council directive 96/82/EC'. 2012.

[11] Health and Safety Executive. ALARP "at a glance" [online]. Available from https://www.hse.gov.uk/managing/theory/index.htm [Accessed Jun 2019].

[12] Kelly T.I., Weaver R.O. 'The goal structuring notation–a safety argument notation'. *Proceedings of the Dependable Systems and Networks 2004 Workshop on Assurance Cases*; York; 2004. p. 6.

[13] SESAR [online]. Available from https://www.sesarju.eu/ [Accessed 20 Jul 2021].

[14] Hird J., Hawley M., Machin C. 'Air traffic management security research in SESAR'. *2016 11th International Conference on Availability, Reliability and Security (ARES)*; Salzburg, Austria, 2016. pp. 486–92.

[15] SESAR Solutions Catalogue [online]. Available from https://www.sesarju.eu/ [Accessed 20 Jul 2021].

[16] International Organization for Standardization. *ISO/PAS 22399:2007(en) Societal Security – Guideline for Incident Preparedness and Operational Continuity Management*. Geneva: ISO; 2007.

[17] 'Directive (EU) 2016/1148 of the European Parliament and of the Council of 6 July 2016 concerning measures for a high common level of security of network and information systems across the Union'. 2016.

[18] National Cyber Security Centre. NCSC CAF Guidance: B.6 Staff Awareness and Training [online]. Available from https://www.ncsc.gov.uk/collection/nis-directive?curPage=/collection/nis-directive/nis-objective-b/b6-staff-awareness-and-training [Accessed Jan 2020].

[19] Sabaliauskaite G., Mathur A.P. 'Aligning cyber-physical system safety and security' in Cardin M.A., Krob D., Lui P., Tan Y., Wood K. (eds.). *Complex Systems Design & Management Asia*. Cham: Springer; 2015.

[20] Srivatanakul T., Clark J.A., Polack F. 'Effective security requirements analysis: Hazop and use cases'. *International Conference on Information Security*; Berlin, Germany, Springer, Sep; 2004. pp. 416–27.

[21] Young W.I., Leveson N.A. 'Systems thinking for safety and security'. *Proceedings of the 29th Annual Computer Security Applications Conference (ACSAC '13), ACM*; New York, NY; 2013. pp. 1–8.

[22] National Cyber Security Centre. Safety v Security: Challenges and Applications in the Cyber Security Era. *Safety v Security Workshop – An NCSC Perspective* [online]. 2019. Available from https://cybersouthamptonacuk/sites/cybersouthamptonacuk/files/Safety%20V%20Security_NCSCpdf.

Chapter 6
Prevention security controls

Martin Hawley[1], Andrew Watson, Lucy Caldwell, Stuart Southern, Chris Machin, and Pearl Noble-Mallock

6.1 Introduction

The preceding chapters have focused on the context and management frameworks for decision–making in cyber security, for example, the regulatory requirements, how to assess risks, how to manage security and approaches to managing safety and security. In this chapter we take a look at the types of technical controls used to prevent attacks to give readers an understanding of how they work. While we have focused on technical controls, this is not to dismiss the importance of administrative controls. The controls discussed are some of the UK National Cyber Security Centre's (NCSC) 'Cyber Essentials' controls: firewalls, anti-malware, patch management and passwords [1]. We also discuss some of the underlying security methods of encryption, Internet security and email security.

Our focus in this chapter is pre-event or 'preventative' controls, which reduce the likelihood of an attack being successful, as compared to post-event controls which limit the impact of a successful attack. Preventative controls may also be thought of as perimeter controls, as they aim to prevent attackers from intruding on the perimeter security of an operation or organisation. However, it is increasingly difficult to define what the perimeter of an operation is, as operations are distributed between various premises and cloud computing. This has led to thinking around concepts such as 'zero trust' and the Cloud Security Alliance's 'software-defined perimeter' (SDP) [2]. Zero trust broadly means that there is no default level of trust for any device connected to a network. An SDP requires that network devices are identified and authenticated before connection, as opposed to connecting first and then authenticating [3].

In Chapter 7, we delve into a category[*] of control known as 'detection', which assists in the discovery of cyberattacks that have already bypassed the prevention controls and are somewhere in the organisation's systems and devices. The

[1]Winsland Ltd, UK

[*]There are other types of control, such as 'corrective', which help to restore systems to normal after an incident. (ISC)^2 also defines deterrent, compensating, recovery and directive controls.

distinction between prevention and detection has been helpful for organising these chapters, but in practice there is a lot of similarities, and no precedence is implied of perimeter controls over detection controls.

The controls for security are analogous to safety, covering physical, administrative and technical controls. Readers will be very familiar with physical controls, which facilitate authorised entry into premises. For example, physical controls may prohibit access to a device on a network endpoint†, such as a locked room. Administrative controls may be less familiar, so we present some examples in Box 6.1. In this chapter, we focus on technical controls but note that security comes from a combination of physical, administrative and technical controls.

Box 6.1 Administrative control considerations

Technological controls alone cannot defend the organisation. Further consideration should also be given, amongst other things, to

- Processes, such as secure processing of data.
- Policies and procedures, for example, outlining organisational expectations in terms of security management, from technical cryptography policies to Acceptable Use Policies and segregation of duties. Overarching policies should be agreed upon at the board level and cascaded throughout an organisation. Detailed policies can then be set by senior managers.
- Personnel screening to ensure that personnel involved in the provision of a service possess appropriate security vetting. However, this cannot guarantee against the insider threat, as demonstrated by Edward Snowden when he revealed the extent of the NSA's surveillance [4].
- Security awareness training to ensure that all staff (including contractors and sub-contractors) are fully aware of security policies, standards and requirements. Monitoring of who has received what training is important to establish how well risks are being addressed.
- Third-party assurance and due diligence processes should be established; security should be considered throughout the procurement process.
- Change management within organisations should include security input throughout, to design security into all projects.
- Governance structures.
- Business continuity and disaster recovery plans.

†'Endpoint' refers to the networked device but may also refer to a process on a device identified by its IP address and port number.

We aim to explain the principles of how controls work in a user-friendly manner. The aim is to demystify technical security controls and provide insights for those with an interest in managing security. This means we have simplified our explanations and drawn short of full coverage of each control. Further details can be found from the chapter's references.

To determine which controls to apply is a combination of prioritisation and expert judgement, which is why risk assessment, discussed in Chapter 3, is such an important activity. Managers need to be assured that their operation is as secure as reasonably practicable and that any failure is such that the operation falls into a safe state. Controls should be applied to counter specific types of threats, mitigating their likelihood and impact, and need to be updated as threats emerge and evolve, as discussed in Chapter 7. Identification of threats and risk management is therefore a continuous process, which underlies the importance of security management systems discussed in Chapter 4.

It is also important to be able to navigate the wide variety of technical controls on the market and recognise that there are no 'silver bullets'. The selection of technical controls also depends on the infrastructure being protected, system architectures and, of course, budgets.

6.1.1 Post-event controls

The chapter does not cover post-event controls, which reduce the impact of an attack, and hence the resilience of an operation. Having a stand-by system, such as a power supply or other duplicate equipment, can help maintain operational continuity. The extent that such post-event controls should be deployed depends on the risk of losing an operation for a certain period of time. Duplicating systems can be costly and careful design is needed to avoid common modes of failure of the system.

6.1.2 Defence in depth and breadth

Most organisations will already apply a set of standard security controls, which range from administrative and procedural to highly technical solutions. These controls will ideally be defined from risk assessments, as well as regulatory requirements, national standards and generally accepted best practices. These controls will be applied in layers, a concept is known as 'defence in depth', which sees the combined application of, for example, firewalls, network segmentation, two-factor authentication and so on, alongside physical security and administrative controls (rules and procedures to be followed).

Defence in depth in security has parallels with safety defences and it is useful to think of an analogy to Reason's human error model of safety barriers [5]. This conceptual model, known as the 'Swiss Cheese' model, shows the trajectory of an accident through different defensive layers (Figure 6.1). Each layer has some holes in it, hence the 'Swiss Cheese' metaphor, which means that there can exist an accident trajectory through several layers. Prevention of an accident therefore depends on the combined probability of failure for all of the defensive layers. However, the

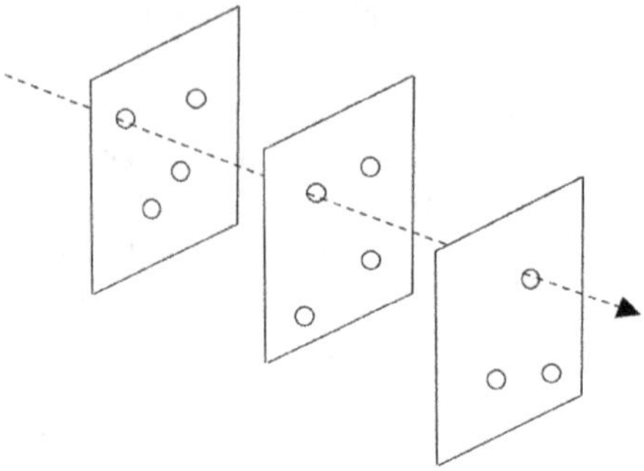

Figure 6.1 *Swiss Cheese Model (after J. Reason)*

Swiss Cheese analogy only goes so far; a determined attacker will deliberately aim to align the holes and penetrate the whole system.

To illustrate these concepts in the context of cyber security, we present an example with three defensive layers, each with a probability of failure of 5% (or effectiveness of 95%), the combined probability of failure will be $0.05^3 = 0.0125\%$ or effectiveness of 98.75%. For example, protecting data stored on a laptop with a firewall, anti-virus (AV) and full drive encryption. These controls may all fail individually, due to, for example, a misconfigured firewall, a new 'zero-day' exploit unknown to the AV software, or a failure of Bitlocker[‡] [6] due to a Windows upgrade.

Similar to 'defence in depth' is the concept of 'defence in breadth'. Defence in breadth distinguishes the defensive layers as comprising different technologies. Cleghorn [7] describes defence in breadth in the sense that 'attacks missed by one technology are caught by another'; for example, several layers of firewalls from the same manufacturer are a 'defence in depth', but from different manufacturers they are a 'defence in breadth'.

A question for the layered 'defence in depth' model is whether the controls are actually working as layers, i.e. 'in series', against a particular threat [8]. From a systems perspective, the combined effect of the controls is not likely to be the sum of individual effectiveness but may be considered as part in series and part in parallel (Figure 6.2). Control effectiveness is not defined here but alludes to the probability of a control or set of controls successfully stopping attacks [9]. These series or parallel models only go so far as a basis for systems modelling but may be a useful mental model.

6.1.3 Organisation of the chapter

Figure 6.3 maps the topics in the chapter. A fundamental issue for cyber security is poorly written software (although there are many other issues) so we start the

[‡]BitLocker Security Feature Bypass Vulnerability, see https://nvd.nist.gov/vuln/detail/CVE-2018-8566.

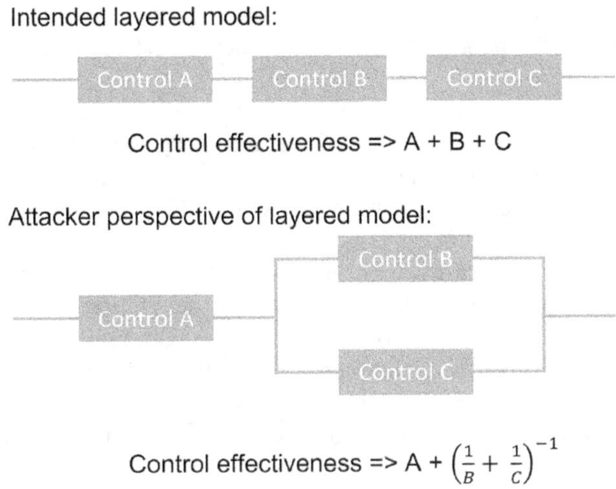

Figure 6.2 Series/parallel systems perspective of controls

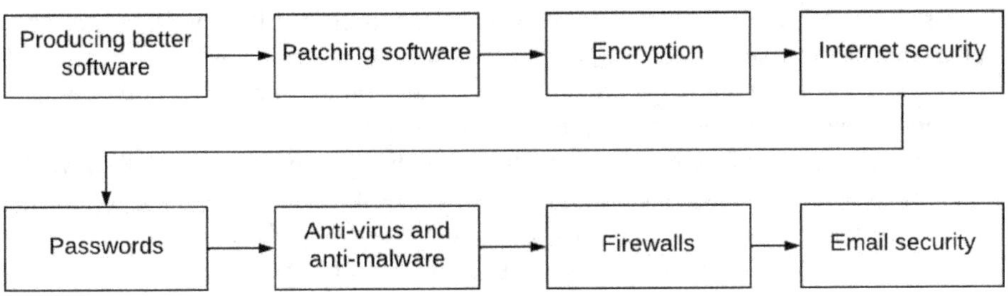

Figure 6.3 Navigating the chapter

chapter with this topic. Well-written software is still subject to vulnerabilities, but basic mistakes in coding are like an open barn door for attackers. All of the controls discussed should be found in some form on a home computer system.

6.2 Designing in security through better software

6.2.1 Introduction

It is now hard to come across any device that does not operate without software and, with the 'Internet of Things' (IoT), even more devices are relying on software that is connected to the Internet.

Given this, good software development is critical, and this is well put by Ed Amoroso [10]: 'The most valuable contribution government can make to cyber security involves providing incentives for software makers to create more correct code'. Amoroso proposed that governments introduce incentives, such as the government procures software with a requirement for 'no exploitable bugs', and a push towards simplification, as complexity is a driver of bugs. This was in 2006 and it is arguable that there are still poor software development practices and ever-increasing complexity.

In spite of increasing numbers of vulnerabilities reported on the Common Vulnerabilities and Exposures (CVE) programme [11], there are more and more software products, with each one producing 2.3 vulnerabilities on average [12]. However, there are also concerns that fewer vulnerabilities are being captured, with some criticism of the CVE process [13].

We do not cover secure software development practices in this chapter, but note that these have been slow to be taken up, with many developers mostly focusing on functional requirements. This is particularly the case with agile methodologies, an issue discussed by de Vicente Mohino *et al.* [14], who propose a Secure Software Development Lifecycle (S-SDLC) to work with agile methodologies. For this section, we focus on why we need such secure development practices in the first place.

6.2.2 A hardware primer

Because software allows electronic devices to function in different ways, we first look at the hardware on which it operates. Any hardware device using software will have some form of memory where the software and other data reside, a processing unit and a means of entering and retrieving data (input and output). For the average computer (laptop/desktop), we refer to the central processing unit (CPU) which performs calculations, random access memory (RAM) which is used to temporarily store the results of those calculations, persistent storage such as a hard disk drive (HDD), and connections to networks and users via input/output (IO) devices.

At the most basic level, computers work with binary numbers (zeros and ones) and process these numbers through logic gates that emulate Boolean algebra. For example, operations such as 'AND' or 'OR'. Sets of logic gates are created to provide different functions and in modern CPUs there are billions of gates on a single integrated circuit. These gates process instructions in nanoseconds, which means computers can process billions of instructions per second, delivering the processing power that supports the software functions we are familiar with today. Note the 'billions of times per second', which we will return to as a source of bugs/vulnerabilities later in the section. Box 6.2 provides a potted history of modern computing and for further introductory reading we recommend Charles Petzold's 'Code' [16] and Jon Stokes' 'Inside the Machine' [15].

Whilst computers have advanced considerably over the last 50 years, it should also be noted that the basic architecture of a general-purpose computing device has not changed much in this period. One notable change has been the amount of memory available to the CPU, and computing architectures have evolved to be able to address this larger memory, most recently evolving from 32 to 64 bit computing. The higher the number of bits the more data that can be accessed in memory. This evolution has led to increased complexity in software and created a set of legacy interoperability problems, which is also a source of vulnerabilities [17].

Box 6.2 A potted history of modern computing

The history of modern computing could be said to have begun in 1936 with Alan Turing's conceptualisation of the 'Universal Turing Machine', which pioneered the idea of the stored programme, what we think of today as software. In 1941, Konrad Zuse created the Z3, which was the first program-controlled computer and then in 1942 Zuse created the first high-level language known as Plankalkül. By 1946, John von Neumann had created the 'von Neuman architecture', and this is the basis of most general-purpose computing devices available today. In 1948, the first piece of software was successfully executed on the 'Manchester Baby' computer at the University of Manchester, UK. In 1960, COBOL 60 was created on behalf of the US Department of Defence. In the early days, software was developed for specific CPU architectures, but this was superseded by a development of the Instruction Set Architecture to become an interface layer between the CPU architecture and software, which then enabled backward compatibility of software as new faster CPUs were developed [15]. The pace of computing developments rapidly accelerated throughout the following decades, with more and more advanced architectures (the Intel 8008 CPU in 1972 and 8080 CPU in 1974).

In 1976, the Intel 8086 CPU was introduced becoming the forerunner of all mainstream desktop/server CPUs to the present day. The first shift from the x86 architecture began with the ARM CPU created in 1980 but did not gain traction until ~2000, accompanied by the proliferation of mobile/battery-powered devices. The ARM CPU uses a different instruction set to x86. It was only in ~2015 that ARM chips started to appear in data centres and in 2018 ARM announced its move to producing laptop/desktop CPUs.

6.2.3 Introducing software

Software is a set of computer instructions, known as algorithms or programs, which operate on data. The purpose of the software is to assist in completing a real-world task. The distinction between instructions and data is important and in the software engineering industry, this is referred to as the 'code segment' and the 'data segment'. This distinction is the root of most fundamental software vulnerabilities. Before exploring this, we quickly review different types of software.

- **Microcode.** All software is converted into CPU microcode instructions through a 'microcode interpreter'. This translates microcode instructions into processor actions, such as to add or multiply two numbers and store the result in memory.
- **BIOS.** The 'Basic Input Output System' of a computer, which is responsible for transferring data from disk/network into memory and then into the CPU and back. BIOS is stored on a separate chip to the CPU and is therefore a type of 'firmware'.

- **Firmware**. This is software stored on the physical integrated circuits ('chips') of the machine in which it resides. Examples include the software in a microwave clock/timer, TV remotes and a car's central locking key. Firmware is written for a specific purpose and device, and tends to have relatively small memory requirements (from kilobytes to megabytes). It is created in the same way as any other software but tends to be harder to update, requiring specific procedures during which the device may be inaccessible.
- **Operating system**. This is the software that enables a device to perform the plethora of functions that make it useful. Operating systems provide memory and device management and usually an input–output mechanism. These can be command line based (such as Linux or 'DOS') or Graphical User Interface (GUI) based such as MS Windows or MacOS. Without an operating system, desktop/laptops would be unable to function.
- **Operating system 'Kernel'**. It is worth noting that an operating system is often split into different parts with the core part being known as the Kernel, which controls and orchestrates the system and performs security checks. If the Kernel can be hijacked maliciously then the hijacker has complete control of the operating system.
- **Device drivers**. A device driver is a special type of software which the operating system kernel uses to communicate with other hardware. Common examples are network interface card drivers, graphics card drivers, keyboard and mouse drivers and printer drivers. These drivers are written by the manufacturers of these devices rather than the developer of the operating system. So, HP would develop the printer driver and AMD (Advanced Micro Devices) would develop the graphics driver for Microsoft Windows. Most computer crashes are caused by faulty device drivers, including the infamous Windows 'Blue Screen Of Death'.
- **Operating system application programming interfaces (APIs)**. Operating system APIs are a type of API by which application software accesses the underlying hardware resources (such as RAM and the HDD for file access).
- **Application software**. This is the software which most people are familiar with, such as word processing, spreadsheets and Internet browsers. Virus scanners and firewalls are often developed as application software.

Software is typically structured into a 'stack' comprising a set of subsystems and components. Users interact with application software, which is at the top of the stack, and relies on additional software below it in the stack. The quality and reliability of the software at one layer depends on that of the layer below it. Application software will not be reliable if the operating system on which it runs is unreliable. Likewise an application will not be secure if the operating system on which it runs is insecure.

Many vulnerabilities in software exist not because of the software itself but because the operating system on which it is installed is vulnerable. Most operating systems are over 20 years old, and many were designed in an era where compatibility was paramount and security an afterthought§. An example security issue is that there is no way to verify that an API call came from the operating system rather than other software.

6.2.4 Creating software

Commercial software is typically the result of a collaboration between a large team of people working to deliver a specific product or goal. These people have different specialisms and different levels of understanding of the product/goal, different personal experiences and different levels of engagement and commitment. Teams of programmers develop software according to different methods or practices that are formal, such as the 'Waterfall' model, or informal such as 'Agile'. The software is generally written by hand with the support of automation tools.

Programming languages have been characterised according to different generations:

- First-generation languages were 'machine code', where the software was written in a binary format exactly as the CPU would use it.
- Second-generation, or 'assembly' languages, improved on the first-generation by being more user friendly for the programmer but requiring interpretation into machine code. Second-generation software required the programmer to be an active participant in low-level hardware actions such as memory allocation and address location calculations. These actions are performed automatically in third-generation languages, which are known as 'high level' languages.
- Third-generation languages are distinguished by a high level of abstraction from the details of the computer hardware on which they operate. This makes them suitable for general purpose use and highly efficient in the creation of complex software. They can also interoperate with second- and fourth-generation languages for specific tasks. High-level languages assist with the generation of 'portable' code, whereby large portions of the software application are not specific to the underlying hardware and can be ported to an alternative hardware environment with minimal editing and re-compilation, for example, C++, C#, Ada and Pascal.
- Fourth-generation languages are also known as domain specific. SQL is a data storage and manipulation domain-specific language that can instruct a computer to store, retrieve, sort and delete data in a database.

Software is generally written as 'source code' by programmers and then 'compiled' to be machine readable. This process, done by a 'compiler', converts the source code into binary software (machine readable 0 s and 1 s). So-called 'binaries' are created for a specific hardware platform and sometimes a specific CPU.

§This was a time when computers were isolated in offices and connectivity between machines was through firewalls and modems with speeds of less than 128 Kbps. This legacy is hard to shake because changes to operating systems would impact large numbers of applications and embedded systems.

They are also specific to whether the machine is '32 bit' or '64 bit'. Note also that the binary code cannot be used to get back to the source code with 100% accuracy. The compilation process often catches errors and problems that were missed by the programmers, which can result in more reliable software than 'scripted' languages.

By comparison to traditional or 'system' programming languages, 'scripting' languages have been developed to connect different software components together [18]. One feature that makes this possible is that scripting languages, such as Perl [19], do not strongly use 'typing' of data. For example, they do not declare in advance if a variable is an integer or a floating-point number. They are also written and executed as plain text files rather than compiled. Hence, the plain text is 'interpreted' at run time, which means that after the programmer saves the file there are no checks on the code until the computer tries to execute it, at which point any errors (including syntax errors) are found. Because the software is based on text files, they can be copied from one machine to another regardless of CPU, as long as an interpreter is available on the given platform. A general security vulnerability with scripted languages is that the software is easy for an attacker to analyse for flaws and may potentially modify and re-upload it.

A hybrid approach of scripted/compiled languages is known as 'Just in Time' (JIT) Compiled or 'Byte Code'. This is somewhere between source code and machine code and is hardware independent as it runs through an interpreter or 'JITer'. This type of software has many advantages over scripted and compiled languages, for example, Microsoft.NET and Oracle JAVA.

6.2.5 How does software become vulnerable?

6.2.5.1 What is a vulnerability?

The Open Web Application Security Project [20] defines a vulnerability as: '....a hole or a weakness in the application, which can be a design flaw or an implementation bug that allows an attacker to cause harm to the stakeholders of an application'.

The main factors that make a design flaw or bug exploitable are as follows:

- The connectivity of the system in which it resides, for example, if the system is connected to the Internet.
- The prevalence/popularity of the technology stack on which it resides.
- The amount of time/resource that an attacker has. For example, an undetected attacker remotely working on the specific vulnerability with extensive computing resources.

6.2.5.2 How do bugs arise?

> Edsger Dijkstra [21]: 'If debugging is the process of removing bugs, then programming must be the process of putting them in'.

Bugs arise for a number of reasons, including:

1. Computer architecture, which makes a vast swath of exploits possible by allowing software to share the same memory as the data on which it operates. This allows classes of attacks that involve tricking computers into accepting data that contain nefarious code. The attacker can then get the target machine to execute the code that they provided to it, i.e. attackers are able to disguise their malicious software as genuine input data, upload it into a target machine and have that machine inadvertently execute it. A related point to note is that when one software application allocates memory to store data and subsequently releases it, the contents of that memory remain. A second software application can then allocate the same memory and has access to the data in it.
2. Bugs are created by the developers often unintentionally but sometimes intentionally. This arises because programmers are humans, prone to error, and these errors may be exacerbated by time limitations, invalid specifications, incomplete or inappropriate testing. A developer may intentionally make a short cut that they know creates a vulnerability because they are under time pressure to deliver. They intend to go back later to address it but either forget or never get the opportunity for one reason or another.
3. Modern software requirements can be so complex, in themselves and in combination with the stack on which they operate, that no single person can grasp the whole system at one time. Furthermore, many aspects of the system may need to operate in a non-linear fashion, meaning that not all of the actions of the system can be comprehended, and they are not all in order, further increasing complexity.
4. Software is an abstract construct, so complex software may not easily be described and discussed.
5. Different users have different requirements for the same software system. For example, different users require different levels of access, some people can read data, some people can write data, some data should be write-only and some data should be read-only. This complexity can lead to flaws in the software.
6. Interconnectivity. As mentioned above, complicated systems may not be fully comprehensible to software developers, with emergent properties from processes occurring non-linearly. Connecting one incomprehensible system to another exacerbates this.
7. Lifespan. The older software becomes the more difficult it is to maintain. Skills shortages, CPU advancements, such as in-built delay timings, are now too slow for faster CPUs, and operating system/vendor support diminishes.
8. Hardware speed and complexity, multi-threading and synchronicity. The software development (creation) environment can be imagined as a hotel with 65,536 rooms and the software as a guest using each room in that hotel in turn.
9. Shared versus dedicated systems. The power of modern computing means that often many software applications are installed on the same server. This 'shared environment' creates compatibility issues and software interactions which cannot be predicted in the real world. An analogy is pharmaceutical drugs having

unexpected and unpredictable interactions in patients. For illustration, Software 'A' interferes with Software 'B' even though vendor A is not even aware of vendor B's software.

10. Human factors. Programmers get tired, have unconscious and conscious biases, have pre-conceived ideas and ideologies, have cultural norms, and they may make assumptions and decisions on limited understanding.

6.2.6 Summary

The above issues all have an impact on security. Returning to our introductory point, software development focuses on the primary functions of the software, bugs and all, and in the author's experience there are few organisations outside of safety-critical sectors using secure software development practices. Fergusen and Schneier [22] describe the software development process as a 'try and fix methodology' which can have a devastating impact on security:

> This process of making fairly complex systems and implementing them with a try-and-fix methodology has a devastating effect on security. The central reason is that you cannot easily test for security; security is not a functional aspect of the system. Therefore, security bugs are not detected and fixed during the development process in the same way that functional bugs are.

A final remark from the author's experience is that if turning bugs into vulnerabilities can be described as the process of 'tricking a computer to execute data as if it were code' then a key aim of security is 'to ensure that data is never treated as if they were code'.

6.3 Patch management

Readers will be familiar with continuous updates in software on their home computers. Software updates can include new functionality, bug fixes and security enhancements such as to 'patch' a security vulnerability. Otherwise these vulnerabilities my be exploited leading to devastating attacks on an individual or organisation [23]. The UK National Cyber Security Centre (NCSC) recommends that organisations have a vulnerability management process to prioritise vulnerabilities, identify relevant patches and ensure resources are available to do so to best effect [24].

An example of a vulnerability is a buffer overflow that allows an attacker to supply their own code to the software – this is an example of being 'exploited'. For example, it was discovered that the Post Office Protocol version 3 (POP3) login service on an email server contained a buffer overflow vulnerability in the password field; if a sufficiently long input was made this would exceed the buffer allocated to the password field and cause a crash. However, if an attacker entered a long password string with code to open a 'back door', the system will execute this code and give remote access to the server. This scenario was leveraged in an exploit against the SLmail program [25].

Security vulnerabilities will have been identified from a variety of sources.

- The vulnerability is discovered by a security researcher (also known as a 'white-hat' hacker) and responsibly disclosed to the software vendor that developed the product. Assuming that the vulnerability was not previously discovered by a malicious threat actor at some point (also known as a 'blackhat' hacker), then it is unlikely that attacks would be seen 'in the wild'. This is a *best case* scenario, where patching immediately upon release of the 'patch' gives protection.
- Software vendors may operate 'bug bounty' programmes. Bug bounty programmes are a type of crowdsourcing of effort and expertise from security researchers or 'ethical hackers'. These researchers find and submit vulnerabilities to software companies, typically through intermediary websites and are rewarded for their efforts. Malladi *et al.* [26] report rewards of $75,000 being made for critical vulnerabilities. Google reports a range of rewards from $500 to $150,000 [27].
- A researcher may be one of many to discover the vulnerability but the only one who responsibly discloses it. This means that attacks may have been going on in the wild for some time by threat actors, with the software vendor having no knowledge. During this time, the vendor is oblivious of any security issue and as such has no patch to release, which means no protection to end users.
- Where a blackhat hacker discovers the vulnerability they will keep the exploit for themselves or release it publicly. This is known as a zero day exploit as it gives the vendor 'zero days' to develop a patch.
- A zero day can also include a vulnerability that was responsibly disclosed to the developing company, but they decided not to release a patch within 90 days[¶]; the industry accepted time frame by which the researcher can publicly release details of the vulnerability alongside a 'Proof of Concept' (PoC) zero day exploit code. PoCs are quickly modified or reverse engineered by threat actors and used to leverage the vulnerability. A PoC that one of the authors, Andrew Watson, released (upon discovering and responsibly disclosing CVE-2018-10088 [28]) was quickly modified and integrated into the Satori botnet within a matter of days. It is also the case that the developing company may ignore the responsible disclosure or acknowledge it, but decide not to patch the vulnerability[**].

Patches are delivered to users as programmes which correct the installed application software. From a small user perspective, these patches are straight forward to install, with users being requested to agree to the update or having previously selected 'auto-update'. The general advice is to do these updates immediately to avoid the vulnerability leading to a security compromise.

[¶]The severity of the vulnerability, whether it is externally facing and exploitable etc. would be factors that would influence any decision.

[**]This also applied to CVE-2019-6973, also ethically disclosed by Andrew Watson.

For larger organisations, rolling out a patch is not so simple as the patch may impact the working of other software applications. Dissanayake *et al.* [29] describe patch management in large and complex systems as a 'hugely challenging task that involves different types of stakeholders making several interdependent technological and socio-technical decisions'. IT and engineering departments therefore must take a considered view of how to manage the patch, balancing the risks of the known vulnerability with the possible impact on operations. This may be more complicated for operational technology as to update systems, including industrial control systems, could disrupt what are normally 24/7 operations and therefore requires careful management of the patching process [30].

6.4 Encryption

Encryption of data is a vast topic, born out of the need for secure communications, and there are some excellent introductions in other books, in particular by Singh [31] and Piper and Murphy [32]. Readers may be familiar with encryption in one form or another, whether it is providing a password to encrypt a Microsoft document, password protected WinZip files, or a password protected hard drive. In all of these instances, the data is not readable without first decrypting it.

In theory, only someone in possession of the password is able to decrypt the content. However, it is possible that the encryption software may have certain 'back doors' or security flaws that render any encrypted content accessible to the creator (of the encryption software) and, typically, state security services. Where any flaws are discovered, i.e. in the underlying algorithm or software, users will normally change to other encryption methods.

Aside from such software flaws, the main day-to-day problem of encryption is the strength of the password used, as attackers may attempt to guess it.[††] This can be done by repeated guesses, for example by using 'dictionary' or 'brute force' attacks, which are described later. First, we delve into the encryption process.

In cryptography, the original (unencrypted) data is known as 'plain text', although it may not be exclusively text, but any other data including sound and video. To encrypt the plain text a 'key' is needed, which we might more commonly think of as a 'password', although as we will discuss later, a password often undergoes a transformation to create the actual 'key'. The key is used in the encryption and decryption process to convert the plain text into encrypted text and vice versa. It is normally represented as follows:

C=P(k)	P: Plain text k: Key C: Cypher text

[††]The strength of the encryption is also very important but is not a day-to-day problem for most users.

Using the same key to encrypt and decrypt a message is known as 'symmetric' cryptography; encrypting and decrypting with different keys is known as 'asymmetric' cryptography. Both of these techniques are critically important in information technology today.

The most secure means of encrypting any data is to use what is known as a 'one-time pad', a concept which has been in use for centuries. The length of the key, k, also has to be at least as long as the message, m, i.e. $k \geq m$. This is symmetric cryptography, as the sender and receiver both need to have previously shared the key. It works as follows:

- Random keys are generated and shared between the sender and the receiver.
- On a later occasion, the sender uses one of the keys to encrypt a message and sends it.
- The receiver uses the pre-agreed key to decrypt the message.
- The key is destroyed so that it cannot be used again.

If the message is intercepted, the plain text cannot be recovered without knowing the key, due to the random nature of the key. If the key is not random and is shorter than the message, statistical analyses can be used to 'crack' the code. This is made harder by making the key longer and more random; up to the point where the key is as long as the message and completely random, which creates what is known as 'perfect secrecy' [33].

Some practical issues arise that make it difficult to have a key as long as the message. In modern cryptography, the key length issue is addressed by using the key (or password) to create a longer pseudo-random key so that the output appears random. This will not create perfect secrecy but will be hard to crack without knowing the key. Once a longer pseudo-random key is created, this is applied to the data being encrypted in two main ways: by a 'stream' cypher or a 'block' cypher. In practice, the stream and block cypher methods may be combined.

Historical methods of cryptography were text based, but in the modern era the cypher is created from the binary representation of the plain text. In a stream cypher, the plain text in binary form (0's and 1's) is transformed into cypher text using a binary operation known as an 'exclusive OR' ('XOR'). The plain text is 'XORed' with the 'key stream'. The key stream is a continuing series of data generated by the key so that it is as long as the message being encrypted. Because the key stream is generated from the key, it is not completely random.

The XOR operation is illustrated below, where each digit of the plain text and key stream is compared using the following logic:

- If either digit is a 1, but not both, then the XOR results in 1.
- If both digits are 0 or 1, the result is 0.

XOR is not the only function applied in cryptography. For example, block cyphers use functions that swap binary digits around, similar to shuffling a deck of cards but in a pre-determined way, and also expand the number of bits in a block, for example, from 32 to 48 bits long.

As mentioned above, symmetric cryptography means that the key is the same for encrypting and decrypting the plain text message. This does not cause a problem when encrypting data on a laptop, say, because the user simply uses the same key. Problems arise, however, when people wish to communicate data over a distance, such as with email. Symmetric cryptography requires that users share the key in advance, which complicates and delays the process of message exchange. Keys could be shared in advance, for example, using couriers and with designated keys for certain days, but this would be a considerable problem for modern secure communications.

This problem was solved conceptually in the 1980s by Diffie, Hellman and Merkle and became known as 'Diffie–Hellman–Merkle key exchange' [34]. The creation of this concept was not practical until the work of Rivest, Shamir and Adleman, which became known as RSA or public–private key cryptography [35]. The story of this work is well told in the book by Singh [31]. The approach relies on the factorisation of a large number into two prime numbers. For example, $21 = 7\times3$, where 7 and 3 are prime numbers. For small numbers this is easily done, but for extremely large numbers it becomes very difficult to compute. This approach is now categorised as asymmetric cryptography. The concept of a public and a private key works as follows.

Using RSA encryption or another algorithm, a user establishes a 'key pair', comprising a public and private key. They make available or 'publish' their public key. Anyone with access to this key can then use it to encrypt a message. However, the only person who can decrypt the message is the one with the private key (Figure 6.4).

In modern applications, symmetric and asymmetric cryptography complement each other. Symmetric cryptography is faster to compute than asymmetric cryptography but cannot be used to establish a secure connection. Hence, asymmetric cryptography can be used to establish a secure connection with public–private key exchange, which is then followed by the exchange of a symmetric key.

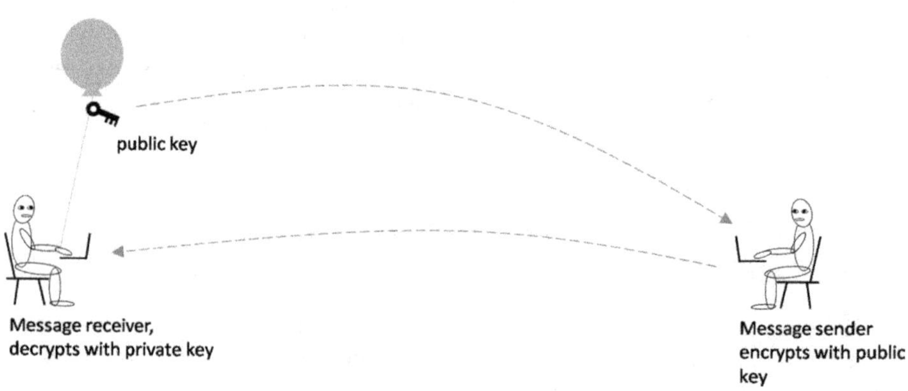

Figure 6.4 Concept of a public and private key

Asymmetric cryptography also provides a method of authentication. If the sender encrypts a message with their private key, it can only be decrypted with the public key. This establishes that the keys are a pair and that while there is no security in this transaction (anyone with the public key can read the message), it authenticates the sender as having access to the key pair.

This technique is used in email security, discussed later. It requires that the sender, encrypting a validation message with their private key, is also linked to the public key by some form of trusted process. This is achieved by registering with a 'Certificate Authority' and storing the certificate in a standardised X.509 format, which includes the public key generated by the certificate authority. The X.509 certificate is then stored in the Domain Name Server (DNS) records of the sender/receiver. While this improves security, there is also a potential vulnerability in that the Certificate Authority itself could be untrustworthy or can be compromised [36].

Authentication and encryption are often used together so that an email message header can be authenticated, and the message body encrypted.

6.5 Internet security

Business networks, and increasingly operational networks, use Internet protocols to transfer data between devices. This also includes some safety relevant applications [37] such as voice communications, also known as 'voice over Internet protocol'. In this section, we give a high-level explanation of how data is transmitted using Internet protocols and some of the security issues.

6.5.1 Transmission of data

Data are transmitted through networks using a variety of different physical links: wireless, ethernet cables, coaxial cables, optical fibre and satellite communications. Data are 'routed' from origin to destination using 'routers' which direct the data traffic according to its destination. Figure 6.5 shows this schematically, where a client (an application such as a web browser on a laptop) connects with a web server to retrieve information, for example, to look up train times.

In Figure 6.5, a connection is established between the client and server application before information is exchanged and, once the user leaves the website, the connection is dropped. When 'inter-networking', now known as the Internet, was first

Figure 6.5 Data transmission over different physical links

implemented, there were no security requirements other than limiting access to the physical network devices, i.e. the cables and routers.

Networks comprise: the computers or devices that are being networked, the routers and the network connections. The computers/devices are generally referred to as 'hosts' [38], but in security parlance they are often referred to as 'end-points'. A router is a switch that receives input data with a source and destination 'address' and routes that data to the appropriate output, [39] to be forwarded to the next part of its route. If the router is connected to the Internet, then it is also known as a 'gateway' or 'IP router'. Note that the addresses referred to here are analogous to postal addresses (House number, Street, Town and Postcode) or telephone numbers (country code, town prefix and local number).

There are a large variety of rules or 'protocols' guiding how data is transmitted. These protocols are mostly defined as layers, which, when metaphorically layered on top of one another, are described as a 'protocol stack'. The Internet protocol stack includes 'TCP/IP' (Transmission Control Protocol/Internet Protocol) and other supporting protocols.

The point of having different protocol layers is to simplify the implementation of networks. For example, the top layer, known as the application layer, does not need to define how data is communicated from origin to destination but rather identifies the data to be communicated and to which server. Below the application layer, the TCP layer concerns itself with ensuring that the data reaches its destination and requests data to be resent if they do not.

TCP/IP defines, amongst other things, how data are split up into 'packets' for transmission, how the data are directed towards its destination and what happens if errors in transmission are encountered. Figure 6.6 shows this schematically. In the first place, a connection is established through a 'handshake' process between the source and destination addresses. Once a connection is established, data are split into packets and sent to the destination. Each packet includes the source and destination address. Periodically, the destination will send an acknowledgement of the data received and any missing data (lost packets) can then be re-transmitted. Further details are given in Box 6.3.

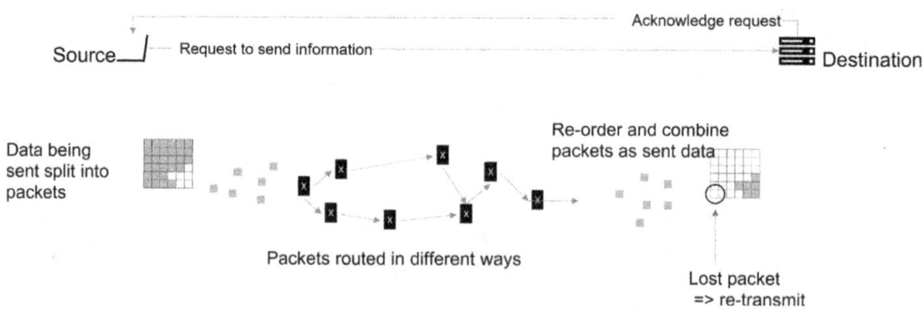

Figure 6.6 TCP/IP packet transmission

Box 6.3 TCP/IP protocols

In this box, we aim to give a sense of how the TCP/IP layers work [38]. There are four layers, each communicating with its nearest neighbour. The network access layer supports the physical communication between computers or devices.

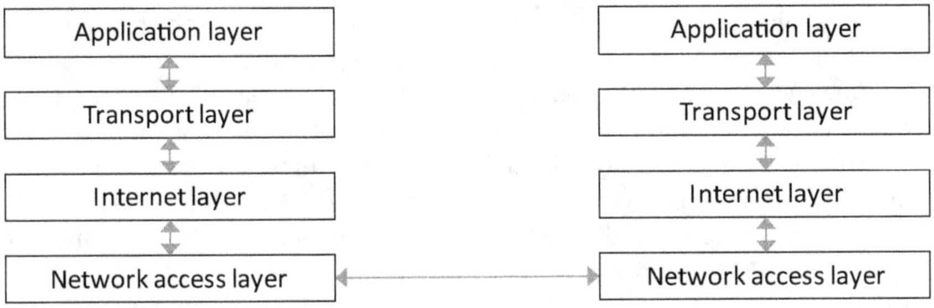

The network access layer comprises the physical layer (wireless/Wi-Fi, ethernet cable, optical fibre, satcom etc.) and the 'link' (or 'datalink') layer:

- The physical layer is the physical instantiation for transmitting data in the form of 'binary digits' or 'bits', which are machine readable. For each different physical link, data are transmitted according to the protocols relevant to that link. Hence, a 'link layer' fits on top of the physical layer. This is analogous to sending data as Morse code (the link layer) over a wired or wireless telegraph system [40]. The link layer requires a system of addressing each device on the link, which is provided by a 'Media Access Control' (MAC) code. The MAC code identifies the physical presence of a device on a network, which is different to the Internet protocol (IP) address.
- The link layer splits data into packets, then coordinates the transmission of packets between different devices/computers on the network. These packets comprise the data itself (the 'payload') boxed between the header and trailer information which supports routing and security. The link layer is also where packet transmissions are coordinated so they do not 'collide' and destroy the data. For example, when data arrive at the same time to a receiver, the bits overlap and become 'garbled'.

For communication to the Internet, the link layer directs packets from a device to the router using the MAC address. The router acts as the 'gateway' to the Internet. Once a packet of data has reached the router it may be transmitted along a different

(Continues)

Box 6.3 (Continued)

link, such as going from a home Wi-Fi datalink to the router then to a coaxial or fibre optic cable provided by a broadband service provider. This next link will have its own means of packaging data and the link layer is responsible for re-packaging the data for the next link for onward transmission. The data therefore travels across a series of datalinks between devices and routers which are collectively known as 'nodes'. Each traversal between two nodes across a particular link is known as a 'hop'. At the end of the hop, the data is unpacked or 'decapsulated' from the frame, either to be encapsulated for a new hop or having reached its destination.

The role of the Internet layer is to determine the direction of the next 'hop' for each packet. This is based on the IP address of the destination. The IP address identifies where the destination device is connected, which is different to its physical or MAC address. As an example, for IP version 4 (IPv4), the IP addresses are of the form 'x.x.x.x', where each 'x' is a number between 0 and 255. For example, if you enter '192.168.0.1' into your search engine when at home it should connect you to your home router. Other devices on your network will have addresses such as 192.168.0.08, i.e. the last part of the address changing between 0 and 255.

The router uses the IP address to direct the packets towards the destination and with each hop between routers the packets are routed closer and closer to the destination. The routing is according to look-up tables in the router's memory. These tables are created by the router when it is first connected and then updated. Various means (protocols) have been established to do this, typically counting the number of hops between one router and another and measuring the response time [41].

The Internet layer is also where IP addresses are assigned to devices by the router with the 'Dynamic Host Configuration Protocol'. The 'dynamic' part reflects that the IP address assigned to your computer (the 'host') is temporary and will be dropped when you log off the network connection.

The transport layer is where connection is first established and the reliability of the transmission is enabled. The application layer, discussed next, sends to and receives data from the transport layer. The transport layer monitors the transmission of data and, if not all is received, re-sends it. This is achieved by each byte of data being numbered in sequence so that the data may be re-assembled in sequence and passed to the application layer intact. The transport layer is also where the rate of transmission from the sender is monitored and adjusted to reduce network overload and loss of data/packets. To achieve this, the sending computer waits for acknowledgements that the sent packets have been received. By measuring how long the acknowledgements take, the sender can adjust the rate of data it sends. The amount of data sent before an acknowledgement is expected is referred to as the 'window size'.

If the connection is heavily loaded with traffic, the acknowledgements may take longer to come, so by decreasing the window size there is less risk of packets

(Continues)

being lost and needing to be resent. Conversely, if acknowledgements arrive quickly, the sender will increase the window size.

The application layer is for user applications, such as a web browser or email client. For example, a web browser has a number of components [42] and therefore software processes operating simultaneously. As well as the visible components such as the user interface and address bar, there are processes to render the content from HTML instructions, execute code and make network calls from the browser to a web server. The 'network' process specifies the data to be sent, and to where, with the detail left to the subsequent layers: transport, Internet and network access layers. This data is addressed to the recipient's IP address and a 'port number', which identifies which process at the recipient's address the data are intended for (e.g. server network process).

Some port numbers are standardised so that one application knows where to address another, such as port 80 for communicating with HTTP (used for web pages) and port 443 for HTTPS (secure connection with web pages). When setting up your email address on a computer you may see that for the IMAP protocol, port 143 is the default for receiving email but port 993 is used when an encrypted link is specified using 'Transport Layer Security' (TLS) / 'Secure Socket Layer' (SSL) security. For sending email using Simple Mail Transfer Protocol (SMTP), the port numbers used are 25 and 465 for secure sending. In most applications, the port number used by the client process does not need to be standardised and 'ephemeral' (temporary) port numbers are used

For further reading the author recommends the following references [43, 44, 45, 46]:

6.5.2 IPSec

Because Internet traffic is not encrypted by default, any device routing the traffic is also able to read it, delete it or change it, affecting the confidentiality or integrity of the data being sent. Hence, for example, if a malicious actor joins a network, they are able to read the data being sent. Most Wi-Fi networks will have a password and encrypt the data being sent on the network, which protects against the data being read by users outside of the network. However, those within the network may still be able to read the data, unless it has previously been encrypted from the originating device. For these reasons, a set of IP security protocols working at the IP layer were developed [47], known as 'IPSec'. IPSec was designed with two connection modes, 'Transport' and 'Tunnel', and two security protocols: 'Authentication Header' (AH) and Encapsulating Security Payload (ESP). Routers can be configured to use either of the connection modes and security protocols.

The security protocols are to ensure the data is authenticated between sender and receiver (AH) or encrypted (ESP). From a security perspective, both authentication and encryption are important; as packets containing encrypted data could be intercepted but not read, but without authentication there is still a risk that data could be tampered with.

Concerning the different modes, ‘Transport’ mode is less secure than ‘Tunnel’ mode, as the security protocol is applied to the IP payload only. In Tunnel mode, the security protocol is applied to the IP header and the payload. This requires a new IP header to be added and the effect is as if data transmitted from one host or router to another are inside a protected ‘tunnel’. The tunnel mode is regarded as more secure than the transport mode, so it is now common usage to combine the ESP security protocol in tunnel mode [48].

6.5.3 Transport layer security

Transport layer security (TLS), as the name suggests, works at the transport layer. This means that data is encrypted between the transport layers of the sender and receiver and confidentiality is maintained between hops at the IP layer. The advantage of encrypting data at this layer is that the encryption can serve different applications, such as web browsers or email. TLS can authenticate the domain of sender and receiver by looking up the certificate that is stored alongside the domain name, although this is not always carried out, even by email providers [49]. Note that TLS was an evolution of Secure Sockets Layer (SSL), with the first [50] TLS specification released in 1999 by the Internet Engineering Task Force (IETF).

6.6 Passwords

An important method for verifying the identity of a user is the identity and authentication question–response process. The combination of a username and password represents a unique token that grants an individual access. The username and password are stored in a password file accessible to the application or system being accessed. For quite some time it has been recognised that storing the password in plain text is a bad idea [51], as an attacker can associate the passwords with relevant usernames and accounts; nevertheless this is still fairly common practice [52]. Furthermore, because users tend to re-use passwords, a breach of one password file can lead to a breach of multiple user applications. Hence, it is now common practice for passwords to be protected through a one-way ‘cryptographic hash’, as shown in Figure 6.7. Hash functions are described in Box 6.4.

An important characteristic of hash functions is that they are ‘one-way’ functions, meaning that the input data (password) cannot (in theory) be recovered from the hash, so storing the hashes of passwords greatly reduces the impact of incidents where user credentials are stolen compared to storing passwords in plain text.

To authenticate a user, the computer or software application will look up the username entered and the stored hash of the password in its records. The password entered is run through the hashing algorithm and compared against the stored hash. If the hash values are identical then the user is granted access.

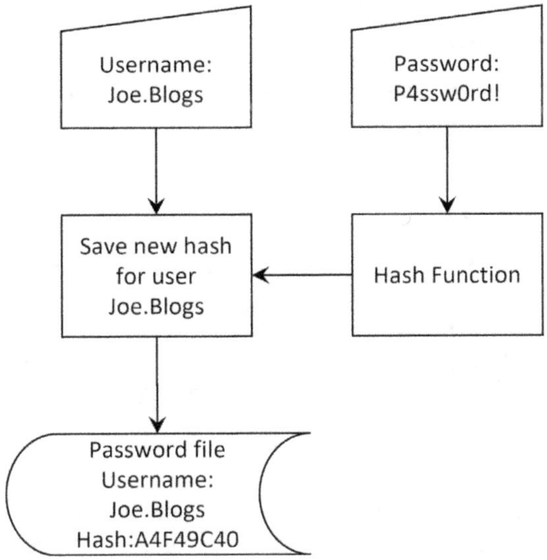

Figure 6.7 *Creating a new password*

Box 6.4 Hash functions

A hash is a fixed size representation of a password, or any other data, that is created using a hashing function. This function takes input data of any size and produces a short 'signature', 'digest' or 'checksum' which is intended to be a unique identifier of the input*. There are different ways to achieve this, including

- Dividing the data into blocks and then each block undergoing a hashing function (e.g. Secure Hash Algorithm 2 [SHA-2] in variants of SHA-256, SHA-512). The block data is then combined with the output of the hash function from the preceding block before being hashed, effectively linking or 'chaining' each block to the preceding ones. Hence, the value of the final hash depends on the value of data in all of the preceding blocks.

*As a simple analogy, the total of a supermarket receipt is a checksum of all of the items in the basket. The total can then be used to verify the set of items purchased. However, different sets of items may have the same total, so totalling the items is not a unique verification. Hence, more complicated functions are needed to create hashes with a very high probability of being unique.

(Continues)

Box 6.4 (Continued)

This means that even a small change in the input data can result in a very different output.

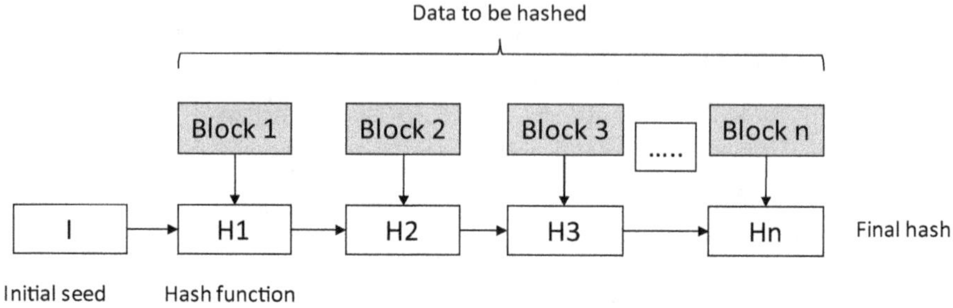

- Perhaps a more readily understandable, but uncommon approach, is to use an encryption function that produces a larger encrypted code (cypher text) than the input plain text [53] and effectively apply this backwards. In this case, the text or data being hashed is run through the decryption function. Each time the data is passed through the decrypt function, the resulting text is reduced in size. This can be repeated until a hash of the desired size, for example, 80 characters long, has been produced.

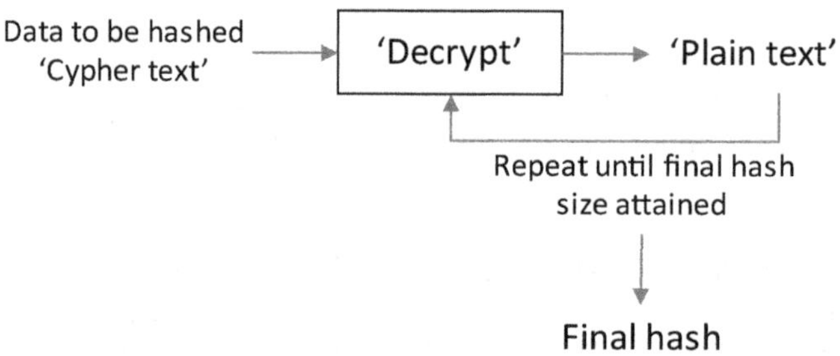

In the above descriptions we have not included a description of the actual hash functions as these are beyond the scope of this book. Essentially, they 'mix' the two inputs (block data and output of previous blocks) through a series of Boolean operations and an encryption function. A clear explanation and example Javascript code for SHA-1 and SHA-2 is provided by Movable Type [54].

Hashes have a wide variety of uses because they can show whether the original data have been changed; if it has then the result is a completely different hash, even if the change is small. Another important requirement of hash functions is that there should be a low risk that the same hash will arise from different input data. When such instances arise these are known as 'collisions'.

The location of the password file depends on the type of operating system or application. Typically, the username is stored along with the password hash. The password file may also contain information such as when the password expires, the number of 'rounds' and the 'salt', which are discussed below.

- A salt is a randomly generated string of characters that is stored with the hash of the password. When a new password is created, the randomly generated 'salt' is appended to the password before hashing. This means that, when 'salted', the same password used in other applications would have a different hash. Hence, if a user's account had been compromised in one application, an attacker would not have access to other applications unless they also had access to the password file which identifies the salt used.
- Rounds are the number of times the password is processed through the hash function. Processing the hash multiple times slows down the process of logging in, which can extend the time required for attacks to the extent that they are unworkable [55].

6.6.1 Offline password cracking

So-called 'offline' attacks are based on the attacker having access to the hash of a user's password, or more typically the hashes of millions of users, such as from the 2012 LinkedIn attack [56]. The attacker is able to crack the passwords offline and then use the combination of user ID and password in other applications.

There are two main approaches for offline password cracking:

1. Simple Brute Force Attacks apply every possible combination of a sequence of letters to identify an unknown password. For example, a 7 character password containing just lowercase letters of the alphabet would mean 26^7 different combinations of letters or ~8 billion different combinations. Using upper case letters as well turns this into 52^7 or ~1 trillion combinations. However, with computing power ever increasing, these large numbers do not take long to generate.
2. Dictionary Attacks are another form of Brute Force Attack that use 'Dictionaries', which are the lists of possible passwords (dog, d0g, cat, c4t, aaaa, password etc.) that run into millions of words. Dictionaries are used with different rule sets that create alternatives for a given password, replacing letters with numbers (pa55w0rd) and appending numbers (password100) and so on.

A recent study of password enumeration [57] showed that of the 160M LinkedIn SHA-1 hashed passwords leaked, 60% could be cracked in around 4 hours. Increased speed in cracking passwords is possible by using look-up tables of pre-computed hashes of password guesses (from dictionary or Brute force methods). After generating different passwords, a hash is created for each, which can then be used to look-up the hash of an unknown password and compare to its input value, which is the password being sought.

6.6.2 Rainbow tables

A rainbow table is a special type of look-up table where a series of hashes are calculated but only the first and last values in the series or 'hash chain' are compared to see if there is a possible match. Given a hash of an unknown password, the chain is recalculated in steps and at each step the result is compared to the hash chain [58]. If a match is found, this indicates that the password may be found from a particular pre-computed chain. Rainbow tables are a specific type of hash-chain that reduce the number of 'collisions' between hashes. According to Ars Technica [59], rainbow tables become unwieldy for passwords of 9–10 characters and have become less useful given the rise in powerful graphics processor units.

6.6.3 Key derivation functions

Hash functions are intended to be fast to compute, which makes them vulnerable to brute force attacks. A further improvement in password security is the use of 'key derivation functions', which use encryption functions rather than hash functions. With key derivation functions, encryption and decryption consume a high amount of CPU time and memory, thus being known as 'memory hard' functions. Examples of functions are bcrypt [60], PBKDF1 and PBKDF2, and scrypt.

As a consequence, they are harder to crack than hash-based passwords as each computation, such as the encryption of a password guess, may take around 0.5 ms. This computation time is unobservable to a user, but makes brute force attacks extremely slow.

6.6.4 Password policies

Different organisations invoke different password policies, and these have evolved over time and continue to evolve. The conventional advice has been to generate random passwords of a specified minimum length (10–12 characters) and then change them frequently. Such advice has generally been developed from an assessment of the length of time that passwords could be 'cracked' in.

Computer users may also establish their own patterns, using lower quality passwords for accounts that are of little value. Florêncio *et al.* [61] describe five categories of accounts from 'don't care' to low/medium/high consequence accounts and 'ultra-sensitive' accounts. They argue that it should not be a surprise if a user uses 'password' as the password on a 'don't care' account such as one used to access a media article. Organisational password policy can also conflict with user needs, particularly where the usability of password protocols conflicts with users' tasks [62] and result in short cuts such as sharing of passwords, writing passwords down, reusing passwords etc.

The quality of a password is indicated by its 'entropy', which is a measure of uncertainty. The entropy of a password can be thought of as a measure of how unpredictable it is. To increase entropy we can increase the number of characters in a password. This is why password advice is to use the whole of a keyboard's available characters. The ASCII character set supports: lowercase, uppercase, numbers and special characters such as $, & etc. However, of the 95 printable characters, only 84 are typically allowed for passwords as some are excluded such as space and punctuation [63].

6.6.5 Multifactor authentication

To increase authentication security, there has been a general move to multifactor authentication (MFA), typically two-factor. Readers will likely be familiar with this from their experience in online banking and more recently retail. MFA typically combines information from one of the following categories:

- Something you know (e.g. Username and PIN)
- Something you have (e.g. Security Dongle/RFID Chip)
- Something you are (e.g. Biometrics – fingerprints, voice)

By mixing the types of authentication together, it is much harder for an attacker to gain access to a system. Access to many websites, especially for banking, now requires a combination of username and password with a second security key generated from a dongle or sent by SMS text, known as a 'one time password'. Any attack would require the username, password and physical access to the dongle, which is itself password protected. Advice on MFA may be found from the UK NCSC [64].

While MFA has now become a key control in cyber security, there are vulnerabilities depending on how it is implemented. Computer Weekly reports the use of the 'pass-the-cookie' attack [65]. Here, the user has authenticated a session with two factor authentication, which is stored in a 'cookie'. However, an attacker may subsequently steal the temporary session cookie for their own malicious purpose.

6.7 Malware protection

Malware is a generic term for different types of malicious code, including viruses, 'worms', 'trojans' and so on. Viruses are a self-replicating type of malware. Virus and malware are terms that are often used interchangeably. However, malware has come to be used to express the more sophisticated end of the malware spectrum, and viruses as the less sophisticated end, those that can be identified with unique signatures [66].

Once malware is activated it executes a malicious payload, causing damage and typically replicates itself to inflict onward harm to other computers. Patil and Joshi [67] describe malware/viruses as typically containing three functions:

- 'Infect-executable', which finds executable files and infects them by copying its code into them.
- 'Do-damage', also known as the payload of the virus, is the code responsible for delivering the malicious part of the virus.

- 'Trigger-pulled', which checks if the desired conditions are met in order to deliver its payload.

There are various reports of the first virus, even as early as 1959, but classification of the exact type of malware is unclear and computer viruses appear to have generally taken hold in the 1980s [68]. The rise of Anti-Virus (AV) software likewise began in the 1980s [69, 70].

Protection against malware is largely through Anti Virus (AV) and anti-Malware software.

AV scans and inspects files and objects on a device (endpoint), detecting and blocking known malware or viruses. Viruses typically infect an endpoint through files, emails or software installation. Viruses can lay dormant on a host for a significant period of time and many users may not be aware that the endpoint is infected. This is because many symptoms of viruses may be subtle and may just make the computer seem slower or 'crash' occasionally.

AV software typically employs a database of virus signatures known as a 'virus dictionary', which includes the lines of codes or signatures of a large number of known viruses [71]. The signatures may be selected pieces of code (in hexadecimal or binary form) taken from the known virus or may be created by running a 'hash' algorithm on parts of the virus code, for example, using the MD5 or SHA256 hash functions. The signature is then composed of, for example, the file size, the hash and the file location [72]. The signature of the suspected virus is compared with the signature database and if there is a match the AV program will 'quarantine' it, making the virus unable to execute and cause damage to the computer system/files. The virus can then be deleted or 'cleaned' from the system by the AV software.

An issue with AV software is that viruses are constantly changed by attackers and therefore may not be recognised by the signature dictionary. The same virus software can be 'packed' with additional random data so that it creates a different signature that will not be recognised by the AV software. Viruses may also be coded to change themselves to avoid detection, which are known as polymorphic viruses [73]. The virus will modify itself or encrypt part of the virus in order to evade detection.

Nowadays, for all intents and purposes, AV and anti-malware software perform the same functions. However, a general distinction that has been made is that anti-malware does not rely on a signature database but rather looks at unusual or anomalous behaviour, using legitimate functions of a machine, such as attempting to connect to an external command and control server, or accessing a computer's camera or microphone.

Some issues with AV and anti-malware software are as follows:

- Legitimate functions, such as software and applications, can be flagged as a virus or malware depending on the type of product deployed on end points.
- While AV and anti-malware applications are similar, where they do address different threats they may both be installed on an endpoint. This risks, however, conflicts that may crash computers and be less effective [74].

- AV software will not help with 'zero-day' threats [75], which exploit zero-day vulnerabilities, i.e. vulnerabilities that may not have been discovered or publicly disclosed by the software producer. As a consequence, the producer will not have created a software patch and the AV software will not be able to detect the attack through signature based scanning.

6.8 Firewalls

A firewall protects a system or network from unauthorised access from another network, typically where a network is connected to the Internet or for segmenting networks within an organisation. It is designed to monitor and, if necessary, restrict network traffic between a trusted network (e.g. a company or home network) and an untrusted network like the Internet. Firewalls are not simply plug and play devices but are individually tailored according to the local security policy. Firewalls can be separate network devices or a software programme running on a device; readers will be familiar with software implementations such as Windows Defender. Dedicated hardware devices may also be placed at the perimeter of a network, between the network and an external unrestricted network such as the Internet. Some firewall products include additional functions, such as wireless access points, network management protocols, content filtering, anti-virus or Virtual Private Network services.

Firewalls use different methods to identify harmful Internet traffic, for example:

- Packet filtering which checks who is being communicated with.
- Application layer filtering, which attempts to understand what the communication is about. By operating at the application layer, the filtering can inspect the contents of traffic and apply rules based on it. For example, an email containing a specific type of file that is not permitted by the organisation's security policy.
- Proxys, which act as an intermediary between the two sides of a connection, preventing direct connection between two hosts (computers) but allowing traffic according to certain rules which again reflect the organisation's security policy. In addition, proxies can require authentication of the network user, such as username and password. Proxies also work at the application layer and support content based filtering, including the ability to decrypt certain types of traffic [76].

A thorough explanation of all of these methods is beyond the scope of this book, but in the following paragraphs we explore packet filtering.

6.8.1 Packet filter firewalls

A packet filter firewall (or a network layer firewall) examines packets of information sent with Internet communication protocols (TCP/IP) and compares it to predefined rules. If a packet of data does not comply with the rules, it is not permitted through the firewall. The packet comprises header information and the 'payload', which is the actual data. The rules are generally based on the source and destination IP addresses and the port numbers referenced in the individual packets.

The rules for the firewall are defined by the 'firewall administrator' and may be quite complex; as the administrator seeks to balance the interests of stopping malicious activity against normal Internet use. Because of this complexity, the configuration of firewalls is error prone [77] and therefore a source of vulnerability. On a basic level, the firewall does not examine the content of the packets and is not able to make decisions based on the information being exchanged or where it actually came from. There are three ways in which a packet filter can be configured, once the set of filtering rules has been defined:

1. The filter accepts only those packets that are safe, rejecting all others. This is the most secure mode, but it can cause user frustration if legitimate packets are inadvertently rejected.
2. The filter drops only those packets that are unsafe, accepting all others. This mode is the least secure but is less inconvenient to the user.
3. The filter encounters a packet for which its rules do not provide instructions and quarantines it or queries the user on what should be done with it. This can be inconvenient if it causes numerous dialog boxes to appear, for example, during web browsing. In proxy firewalls, the packet is pushed to the proxy to decide what to do with it.

Packet filter firewalls are classified as stateless or stateful, in the sense that the firewall depends on a previous state based on data stored in its memory:

- A 'stateless' firewall examines network traffic and restricts or blocks packets based primarily on their source and destination addresses. It has no knowledge of the content or context of the connection between two systems. These types of firewall run on basic rulesets for packet filtering, making decisions based on static information such as destination/source addresses and ports, which is provided by the packets themselves. The set of rules are known as an 'Access Control List' and the order in which they are applied has an influence on performance [78]. Stateless firewalls are considered vulnerable to spoofing attacks, where the source IP address is faked, since they consider each packet in isolation without contextualising the connection or previous packets. Stateless firewalls are, however, considered useful in terms of their speed and efficiency when working with high traffic loads.
- A 'stateful' firewall is 'session based' and allows or blocks traffic based on state, port and protocol used between applications, monitoring all activity from the opening of a connection, such as TCP, until it is closed, i.e. the 'session'. Once a trusted connection is established, the firewall does not examine any more packets as long as they are part of the same connection [78].

6.8.2 Deep packet inspection

Conventional packet filtering reads the header of the packet. Deep packet inspection (DPI) 76 is a type of network packet filtering that evaluates and examines the contents of the packet as well as the source address. Similarly to AV programmes, the packets are compared against, for example, malware signatures using fast 'string matching' algorithms to see if any packet of data contains anything that might be harmful. This is not effective on packets with encrypted data as any inspection of packets needs to occur after the data have been decrypted.

6.8.3 Application-layer firewalls

Application-layer firewalls work on the application level of the TCP/IP stack (e.g., browser or FTP traffic) and may intercept all packets travelling to or from an application. Application firewalls work much like a packet filter but apply filtering rules (allow/block) to each process rather than each connection. Application firewalls further filter connections by examining the ID of data packets against a rule set for the local process involved in the data transmission. There are thousands of varieties of applications/software available and therefore application firewalls have more complex rule sets.

As security moves away from traditional perimeter security and purely hardware firewall devices, Web Application Firewalls (WAFs) have become increasingly important in protecting distributed networks and ecosystems. These are application firewalls that reside in front of the web server [76].

6.8.4 Implementation considerations

Detailed guidance on firewalls is provided by Amoroso [79]. Two key points to note though are as follows:

- Firewalls use a pre-defined set of rules to make decisions about whether a connection should be allowed or rejected. It is essential that the rule set is maintained and kept up-to-date. If the rule set is compromised, then the firewall could be vulnerable to attack.
- All firewalls have a software component which can be compromised and should therefore be regularity updated as any other network device.

Finally, we note that attackers have found ways to circumvent firewalls through commonly used ports that are unlikely to be blocked by the firewall. Due to this, organisations are deploying intrusion detection systems to monitor traffic that has been allowed by the firewall but may in fact be malicious. This subject is covered in Chapter 7.

6.9 Email security

6.9.1 The email problem

6.9.1.1 Introduction

Email security is an important component of preventative security, given its universal use, and that it so successfully exploited by attackers [79]. Historically, email was first introduced without any security [80], meaning that emails could be intercepted, read and modified with ease. Since then organisations have been playing catch-up on email security, with a large variety of security add-ons, some of which we discuss in this section.

Until recently, spammers had an easy time with email, as they could send millions of unwanted emails with abandon, containing nuisance marketing or malicious content. This also did not require high sophistication from attackers, who could easily carry out high volume attacks with email addresses purchased from criminal 'dark web' market places. By sending in high volume, only a small percentage of attacks need to be successful to be worthwhile. The motivations for these attacks are many and include financial gain through fraudulent activity, intellectual property theft, extortion and blackmail.

The objectives for improving email security can be characterised as a need to preserve the confidentiality, integrity and availability of emails:

- Confidentiality, so that none but the sender and recipient(s) are able to view an email.
- Integrity so that the email is known to be (i) unaltered from the sender and (ii) not carrying any malicious content, or attempting to orchestrate a business email compromise event [81].
- Availability, so that streams of spam or 'unsolicited bulk email' do not cause a loss of service [82].

6.9.1.2 Email as a common point of entry

Most email providers offer some protection by filtering 'junk' email and applying automated malware detection. However, attackers are finding ways to circumvent controls and continue to leverage phishing emails to distribute malware. For example, the Verizon Data Breach investigations Report [83] shows that 23% of malware was delivered by email.

Ransomware, spyware and other malware are often delivered through emails, which trick users into clicking on an attachment or a shortcut in the email. Attachments may be program (executable or .exe) files, 'zip' files, Word/Excel files etc., and users will likely be asked to enable macros on opening.

'Spearphishing' emails, which are targeted to 'phish' a specific person, often use social engineering to manipulate the email recipient into clicking on the email attachment or link and downloading files or providing information such as login credentials. Attackers have become increasingly sophisticated in their design of the email to make it seem wholly legitimate. This could include using an organisation's logo, using similar domains to that of a legitimate sender, or spoofing the legitimate names of individuals or companies that the recipient may communicate with. Emails have been the attack vector for numerous attacks in the transport sector, and some examples are given in Box 6.5.

Box 6.5 Open Source examples of attacks on transport systems with email as the attack vector

Sophisticated advanced persistent threat (APT) groups are reported in open source to have used spear phishing emails as a core feature of their tactics, techniques and procedures in targeting Critical National Infrastructure sectors, including aviation and their associated third-party companies [84]. These emails contained malicious attachments which could use native Microsoft Office functions to make connections with an external server. The attackers used password cracking tools and masqueraded as a legitimate user on the network. Further research showed the threat actor group then used falsified emails claiming to be a contract agreement, with an attached PDF. The PDF would contain malicious URLs or embedded links to redirect recipients and thereby steal their credentials, thus enabling the attackers to gain access to the network and exfiltrate data or connect remotely to operational systems (Industrial Control Systems).

Other sophisticated APT groups have been associated with targeting the aerospace sector through the use of malicious and heavily targeted emails directed at employees with military and commercial aviation sector roles [85]. A Mandiant report by O'Leary *et al.* [85] provides their insights into an APT group reportedly targeting an aviation organisation in order to gain insight on Saudi Arabia's aviation capabilities. The emails in question used generic recruitment themes targeted at individuals within the industry, to lure them into clicking on seemingly legitimate links to employment websites. If the link was clicked, a '.hta' (executable HTML) file containing embedded code was automatically executed to download an unauthorised method of access, also known as a 'backdoor'.

With respect to threats facing the maritime sector, Proofpoint researchers have analysed spear phishing activities aimed at naval defence contractors and associated ship, vessel and submarine building or research institutions [86]. These emails use techniques such as attached Excel and Word documents to distribute malware. The Proofpoint research also found that the attackers were using compromised email accounts to send further waves of emails and, once established in the network, attempting to compromise the organisation's servers and use them as command and control servers to serve further malware.

6.9.2 Key components of email communication

Email software enables senders to compose, receive and store emails and attachments. The software may be installed on a user's computer, as an instance on a virtual machine, as a web application (e.g. Hotmail and Gmail), and on devices such as multifunction printer-scanners (so that a scan can be emailed to the user).

Figure 6.8 illustrates the passage of an email, where it can be seen that there are four computers (or more with mail relay) and numerous software components (the Agents) involved. The 'Mail User Agent' sends and receives emails to/from that user's 'Mail Server'. The Mail User Agent (MUA) communicates with the 'Mail Submission Agent' when sending email and the 'Mail Delivery Agent' when receiving email. 'The Mail Transfer Agent' sends the email from one mail server to another.

Each part of the process is governed by protocols that describe the individual steps and parameters involved. These protocols involve an exchange of information between one computer and another, instructing the email transfer and informing when it is complete.

The protocol that defines how mail is transferred (or 'pushed') from the sender to the mail server and between mail servers is the SMTP [87]. Transfer of messages occurs by establishing and then terminating a connection between SMTP client and server. Email messages may be transferred in a single connection or a series of hops through intermediary systems. SMTP transports what is referred to as a 'mail object', which comprises the email header (like an envelope) and the email content (the message and any attachments).

Other protocols define how messages are 'pulled' or downloaded from the recipient's mail server; these are Internet message access protocol (IMAP) or POP3. IMAP has more functions than POP3, such as enabling it to flag the state of an email (read or not) and allowing users to create folders and move emails between them.

A further protocol to be aware of is multipurpose Internet mail extensions (MIME), which supports a wider set of characters (beyond ASCII) and non-text components (multimedia files) that are attached to an email [88]. There also exists a secure version of this, known as S/MIME [89].

To translate or 'resolve' an email address into an IP address, the Domain Name System (DNS) is used to look up the latter part of the email address (after the '@'), for example, 'theiet.org'. Additional information is associated with this domain name through what are known as 'resource records'. These include the mail exchange (MX) records, which identify the email servers which should be used

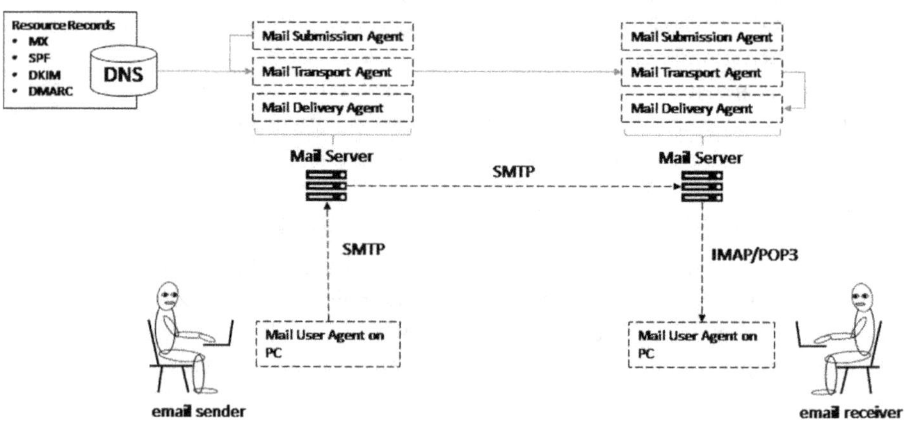

Figure 6.8 An email's passage from sender to receiver

during email communication [90]. Put simply, an MX record gives the IP address of where your email server resides and therefore where your email traffic should be directed. Organisations may use multiple MX records in case a mail server fails. When an email is sent, the sender's email server queries the DNS server to return the MX record of the recipient's domain, which gives the IP address to which the email is to be sent. This is somewhat like looking up a post box number for an organisation before posting a letter.

Readers may be familiar with these email protocols when manually setting up their own email client program, although increasingly this is automated and requires only entering an email address and password. Examining the 'properties' of an email you have received will show the sort of information exchanged. What is shown in the 'in-box' is mostly the message itself with some header information (from, to, subject). The email properties will show additional information and, significantly, the actual email sender's address; this arises because the header information in the in-box can be spoofed. According to Media Temple, 'It is important to know that when reading an email header every line can be forged, so only the "Received:" lines that are created by your service or computer should be completely trusted' [91]. Table 6.1 gives an example of what can be learned in comparing the 'From:' address in an email as presented to the user and what is seen in the 'properties' of the email.

Table 6.1 Comparison of email information seen by recipient and available in 'properties'

SMTP field	Content	Comment
Email as shown to recipient:		
From:	**'Apple'< service-appleid@intel.com>**	Seemingly authoritative sender, but this field is forged, i.e. not the actual address of the sender.
To:	< john.smith@example.me >	
Date:	Tue, 25 Oct 2016 05:03:42–0000	
Subject:	verify your account information !	
Email fields (in 'email properties'):		
Return-Path:	**< myintl@vps326718.ovh.net>**	
Delivered-To:	< john.smith@example.me >	
Received:	from haggis.mythic-beasts.com ([2 a00:1098:0:86:1000:0:2:1]) by onza.mythic-beasts.com with esmtp (Exim 4.84_2) **(envelope-from< myintl@vps326718.ovh.net>**) id 1c3 × 17–0007JC-8R for john.smith@example.me; Tue, 08 Nov 2016 03:24:09+0000	Shows the trail of the email when received by the recipient's server. The 'envelope-from' shows a different email address to that displayed to the recipient. Normally there are several 'Received:' entries recording the path of the email as it is relayed between different servers.
Message-ID:	< E1byttm-0007uR-Ht@**vps326718.ovh.net**>	A unique string for the message, which can be forged but not in this case.

Table 6.2 Example of XOR operation to create cypher text from plain text XOR with a key stream

Plain text	0110 1010 0100 1011
Key stream	1101 0110 1001 0010
Cypher text	1011 1100 1101 1001

6.9.3 Securing email

Emails are generally sent in plain text and are laden with security issues. TLS addresses this, applying encryption to emails between servers. Similar to the transmission of IP packets, email is sent in relays or 'hops' between servers. While TLS may be applied by the servers (nodes) between these hops, these data packets will need to be decrypted for the mail server to take action on them (e.g. by the message transfer agent and message delivery agent). Each node must therefore be secure for the email to be considered secure.

Because TLS provides security between servers at the transport layer, it provides partial security only. Complete confidentiality between the sender and receiver is known as 'end-to-end' security and is only achieved by encrypting the message at the sender MUA and decrypting it at the receiver MUA. There may then also be security issues with how the received emails, and any attachments, are then stored; in plain text or encrypted form.

The different parts of the email's journey which need to be secured are illustrated in Figure 6.9, with the security boundaries shaded, highlighting three sections of TLS and other security required around the client PCs and servers. The use of TLS grew rapidly in 2014–2015, reaching 80% of outbound messages and 60% of inbound messages. This was largely driven by the most popular providers such as Yahoo and Outlook, leaving a long tail of 700,000 SMTP servers where 'only 82% support TLS, of which a mere 35% are properly configured to allow server authentication' [92].

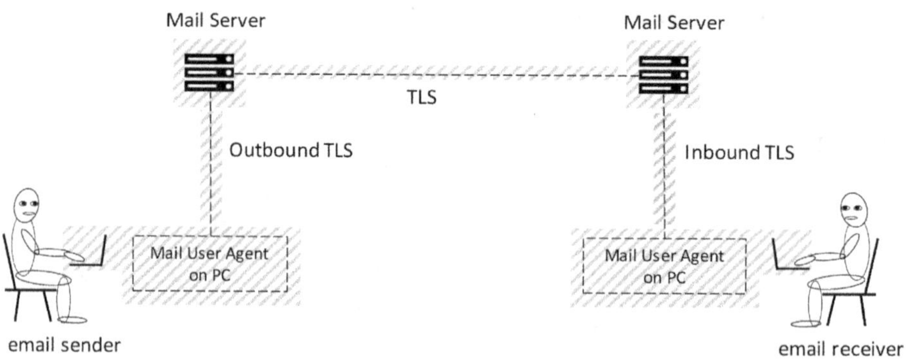

Figure 6.9 Securing the email's passage

More recent work in 2018 [93] analysed the adoption of a range of protocols (StartTLS, x509, SPF, DKIM, DMARC, DANE and DNSSEC). In mapping these to metrics of confidentiality, integrity, phishing and identity theft, it was found that only 26% of domains analysed had applied a minimum recommended set. However, 69% of email providers tested were found to protect the confidentiality of the emails that they send and receive.

Beyond encryption with TLS, a number of protocols have been developed and have been widely promoted for adoption [15, 94, 95]. These include:

- STARTTLS is an enhancement to the SMTP protocol, which has been standardised since the early 2000s [96]. On initiating sending or receiving email, the mail server advertises whether it supports TLS with the keyword 'STARTTLS'. On recognising this keyword, the mail user agent then uses the command 'STARTTLS', which begins the process of securing the exchange of messages between them.
- S/MIME (Secure/Multipurpose Internet Mail Extensions) adds authentication and encryption to messages. However, it relies on digital certificates which may not be supported by the email clients.
- DNSSEC stands for 'Domain Name System Security Extensions'. It was developed to authenticate DNS records.

There are various methods whereby attackers can intercept emails, for example:

- **DNS hijacking**. The DNS server maps IP addresses to human readable names, for example, 'theiet.org' is mapped to 45.60.154.129. This is called 'DNS resolution'. If an attacker has access to an organisation's router, they can provide a false DNS server to resolve the domain name from 'theiet.org' to an IP address of their choice. The attacker is then able to view emails before forwarding them to the correct recipient.
- **TLS downgrades**, where the STARTTLS discussed in Section 6.9.1 is deleted or replaced, disrupting the request for encrypted transmission.
- **Man in the Middle Attacks (MitM)** allows attackers using the same Wi-Fi network as the user to intercept traffic by using a proxy or traffic/packet 'sniffer' such as the legitimate 'Wireshark' network analysis tool [97]. This could allow an attacker to view credentials entered into a browser by tricking the user into running an HTTP session rather than HTTPS; this is also known as 'TLS/SSL stripping' [92].

6.9.4 DMARC, SPF and DKIM

There are three main protocols that are widely promoted to play a part in strengthening email security. Together they act to check that when an email is received it comes from the domain it claims to be from:

1. **SPF** (sender policy framework) could be summarised as 'My domain uses these IP addresses only to send mail – that is, no other IP addresses are authorised so beware if you receive an email originating from elsewhere'.

2. **DKIM** (domain keys identified mail) could be summarised as 'I certify that this exact email has come from my domain – I have used my private key to sign the email's contents so you can tell if it has been tampered with'.
3. **DMARC** (Domain-based Message Authentication Reporting and Conformance) was conceived to standardise coordination actions around SPF and DKIM between sender and receiver.

When SPF and then DKIM were first deployed, it was expected that the amount of spam and other malicious email (phishing, etc.) would be greatly reduced. However, there were some missing elements at the receiver end of emails, such as feedback to senders on the effectiveness of their actions [98]. This led to the creation of a standardised approach known as DMARC. In the following paragraphs we discuss these protocols in more detail, starting with SPF, then DKIM and DMARC. These protocols, however, do not solve the problem of email security and there is a continuous creation of attack methods, used individually and in combination [99].

6.9.4.1 Sender policy framework

Sender policy framework (SPF) is an authentication protocol which defines which mail servers are authorised to send mail for an organisation's domain. To enable SPF, the organisation publishes an 'SPF record' of the IP addresses to be used to send mail from. The receiving mail server then applies SPF checks to ensure that the outgoing or incoming email is being sent to or coming from an authorised domain. If you view the properties of one of your emails you can see various uses of 'spf' checks and 'mailfrom' in the email header, for example:

> "spf=pass
> ...
> smtp.mailfrom=preventativecontrols.com; needsomecontrols.org; dkim=pass"

SPF allows flexibility in how the rules are configured and the results of checks are published in the email header. However, there are some drawbacks to SPF [100]:

- Attackers can still use a similar domain to the one they are aiming to spoof and then configure it with SPF, which would mean malicious emails could pass SPF checks;
- SPF reads the header 'mailfrom', which can be spoofed.

6.9.4.2 Domain keys identified mail

Once SPF records have been set up, the authentication mechanisms of domain keys identified mail (DKIM) and DMARC may be configured. DKIM [101] provides protection against domain spoofing by identifying the domain that the email is coming from and confirming the validity of outbound mail. It is considered a stronger authentication mechanism than SPF [102].

DKIM works by the sending mail server (the organisation's or mail service provider's server) signing each email with a unique identifier. This identifier is based on a hash of the email being sent, which is then encrypted. The encryption is done with the private key of the organisation or mail service provider. The corresponding public key is lodged with the DNS record.

The resulting DKIM signature is then added to the email header. When the recipient mail server receives the email, it decrypts the hash using the public key. It then creates its own hash of the message and compares them. If the hashes are a match, then the DKIM check is found to be successful and the email is authenticated to the sender's organisation; otherwise, the email will be directed to the spam or junk folder.

Encryption with the private key does not serve any confidentiality purpose as it is readily decrypted with the public key. However, it does serve an authentication purpose as it matches the public key to the unknown private key, i.e. the public key can only decrypt messages encrypted by the private key.

6.9.4.3 DMARC

DMARC creates a feedback loop from the email receivers back to legitimate email senders [103]. It is a reporting, policy and authentication protocol that was conceived in 2010 and is increasingly deployed by email providers to protect against fraudulent domains.

DMARC applies to incoming email messages. A DMARC record is stored as a text file in the DNS record. It instructs the mail servers on what action to take with emails if the SPF or DKIM checks fail. Hence, the baseline requirement for DMARC is to first implement SPF and DKIM, then to supplement these by listing 'policies' in a text file that informs the receiver what actions to take if an email does not meet the policies. These policies could

- reject an email and send a report to the sender (at a specified email address)
- quarantine an email for further checks by the receiver

Fine tuning can be applied to the policies, such as applying the rules for a certain percentage of the time, although this creates work for the receivers of those emails.

Policy in this sense means the actions taken on receipt of an email, including checking for blacklisted IP addresses, SPF and DKIM checks. DMARC can therefore be seen as a checklist for each email and criteria as to what actions to take following failure of a check. Once SPF records have been set up, the authentication mechanisms of DKIM and DMARC may be configured.

When SPF and DKIM were introduced, they were intended to authenticate the sender and lead to a decrease in phishing emails. A variety of reasons are cited for this not happening [104], including:

- The complexity of the email environment of organisations, which may have different domains, subdomains and email systems, including third-party providers, and is regularly changing.

- Errors in authentication algorithms, so that some authentic emails are not authenticated and some spam emails are.
- Poor feedback to senders on these errors, so that there is little information to improve authentication.
- The bias of receivers towards reducing 'false positives' (not rejecting legitimate email) which allows through more spam email.

In addition, email administrators believe that the slow adoption of the protocols is primarily due to the lack of a critical mass [105]. Like many network protocols, the benefits of the anti-spoofing protocols come into existence only if a large number of Internet domains adopt the protocols to publish their authentication records. Currently, the incentive of adoption is not strong, especially for Internet domains that do not host email services.

6.9.5 Email security awareness

Training and awareness in identifying a malicious email is extremely important, particularly when encouraging staff to practice scepticism when opening emails and clicking on links and attachments. Although there are many tools on the market which attempt to block malicious emails, attackers are becoming increasingly sophisticated in creating emails which circumvent detection.

Fagerland [106] explores how technical controls fail and claims that promoting email security and making users aware of the threat is the only truly effective solution. Fagerland's research explored the characteristics of malicious emails, by analysing a set of them which circumvented preventative controls. This led to the proposal to strengthen email security by motivating users through awareness and human observations of risk alongside automated system processes.

The UK's NCSC provides guidance on social engineering attacks based on a digital footprint. Many organisations now use phishing scenario campaigns to test staff awareness of malicious emails following training and awareness sessions. This will allow organisations to look at click rates of staff in various echelons of an organisation, and it may help in identifying where more training resource is needed, based on likelihood of click through.

The concern is that if an employee can be tricked into running code on behalf of an attacker, sometimes as simple as clicking a web link, then many layers of security can potentially be circumvented without any 'hacking' taking place. This can be mitigated in several ways, including the principle of least privilege, i.e. give users only enough digital privileges as is required to perform their job function. However, training users how to spot potential phishing attacks or malware is an essential foundation of cyber defence.

A further concern for organisations is the use of personal webmail access from enterprise networked devices, since an employee using their webmail account and clicking on a malicious email presents the same risk as the corporate email.

6.10 Conclusion

This chapter began by considering software and why vulnerabilities arise through the process of developing code. These vulnerabilities enable attackers to enter and exploit systems and so require remedial action, such as software patching. Additional controls are needed to protect systems against unauthorised access and the transfer of data between systems, typically across the Internet. We therefore considered a range of controls, which will largely be familiar to readers, including patching, firewalls, encryption, passwords, AV and email security.

A common factor for the controls discussed is that they were first introduced to mitigate a particular vulnerability but then needed to evolve with increasing sophistication and to work with additional controls. Email is an excellent case study in how legacy systems were first developed without any security, which was then added later. Security experts now speak of 'secure products' versus 'product security', and we reflect back to Ed Amoroso, and other security thought leaders, who have long been emphasising the need for better software.

Because no control can be relied upon on its own, several layers of security are needed. Even with these layers, security practitioners cannot provide complete protection, and attacks will very likely be successful in penetrating systems. Organisations therefore apply controls to detect and manage such penetrations, which is the subject of the next chapter.

References

[1] National Cyber Security Center. *Reducing your Exposure to Cyber Attack* [online]. Available from https://www.ncsc.gov.uk/information/reducing-your-exposure-to-cyber-attack.

[2] Cloud Security Alliance *Software-defined perimeter (SDP) and zero trust*. 2020 May 27. Available from https://cloudsecurityalliance.org/artifacts/software-defined-perimeter-and-zero-trust/

[3] Puthal D., Mohanty S.P., Nanda P., Choppali U. 'Building security perimeters to protect network systems against cyber threats [future directions]'. *IEEE Consumer Electronics Magazine*. 2017;6(4):24–7.

[4] Greenwald G. 'No place to hide: Edward Snowden, the NSA, and the US surveillance state'. *Macmillan*. 2014.

[5] Reason J. 'Human error: models and management'. *BMJ*. 2000;320(7237): 768–70.

[6] Meijer C., Van Gastel B. 'Self-encrypting deception: weaknesses in the encryption of solid state drives'. *2019 IEEE Symposium on Security and Privacy (SP)*; San Francisco, CA, 2019. pp. 72–87.

[7] Cleghorn L. 'Network defense methodology: a comparison of defense in depth and defense in breadth'. *Journal of Information Security*. 2013;04(3): 144–9.

[8] Conklin W.A., Dietrich G. 'Systems theory model for information security'. *Proceedings of the 41st Annual Hawaii International Conference on System Sciences (HICSS 2008)*; Waikoloa, Big Island, Hawaii, 2008. pp. 265–265.

[9] Khouzani M.H., Malacaria P., Hankin C., Fielder A., Smeraldi F. Efficient numerical frameworks for multi-objective cyber security planning. *European Symposium on Research in Computer Security*; Springer, Cham, September 26; 2016. pp. 179–97.

[10] Amoroso E. *Cyber security*. Summit, NJ: Silicon Press; 2006.

[11] CCV details [online]. Available from https://cve.mitre.org/blog/https://www.cvedetails.com/browse-by-date.php.

[12] *The state of vulnerability reports: What the CVE surge means* [online]. Available from https://techbeacon.com/security/state-vulnerability-reports-what-cve-surge-means.

[13] Open Source Security [online]. Available from https://opensourcesecurity.io/2021/03/30/its-time-to-fix-cve/.

[14] de Vicente Mohino., Bermejo Higuera., Bermejo Higuera., Sicilia Montalvo. 'The application of a new secure software development life cycle (S-SDLC) with agile methodologies'. *Electronics*. 2019;8(11):1218.

[15] Stokes J. 'An illustrated introduction to microprocessors and computer architecture' in *Inside the machine*. San Francisco, CA: No Starch Press; 2007.

[16] Petzold C. *Code: the hidden language of computer hardware and software*. Washington: Microsoft Press; 2000.

[17] Guðmundsson A. *32-bit virus threats on 64-bit windows. symantec*. Wallingford, OX: Virus Bulletin;

[18] Ousterhout J.K. 'Scripting: higher level programming for the 21st century'. *Computer*. 1998;31(3):23–30.

[19] Perl [online]. Available from https://www.perl.org/about.html.

[20] Who is the OWASP® Foundation? [online]. Available from https://owasp.org/ [Accessed 30 Jul 2021].

[21] Edsger W. Dijkstra [online]. Available from https://en.wikipedia.org/wiki/Edsger_W._Dijkstra#cite_note-22.

[22] Ferguson N., Schneier B. *A cryptographic evaluation of ipsec*. Schneier Bruce; 1999.

[23] The Hacker News *The continuing threat of unpatched security vulnerabilities*. 2022 Mar 8. Available from https://thehackernews.com/2022/03/the-continuing-threat-of-unpatched.html

[24] National Cyber Security Centre. *Vulnerability management*. Available from https://www.ncsc.gov.uk/guidance/vulnerability-management

[25] CVE [online]. Available from https://cve.mitre.org/cgi-bin/cvename.cgi?name=CVE-2003-0264.

[26] Malladi S.S., Subramanian H.C. 'Bug bounty programs for cybersecurity: practices, issues, and recommendations'. *IEEE Software*. 2019;37(1):31–9.

[27] Chrome Vulnerability Reward Program Rules [online]. Available from https://www.google.com/about/appsecurity/chrome-rewards/.

[28] CVE. CVE-2018-10088. Available from https://cve.mitre.org/cgi-bin/cvename.cgi?name=CVE-2018-10088#:~:text=Buffer%20overflow%20in%20XiongMai%20uc,than%20CVE-2017-16725

[29] Dissanayake N., Jayatilaka A., Zahedi M., Babar M.A. 'Software security patch management – A systematic literature review of challenges, approaches, tools and practices'. *Information and Software Technology*. 2022;144:106771.

[30] Gentile U., Serio L. 'Survey on international standards and best practices for patch management of complex industrial control systems: the critical infrastructure of particle accelerators case study'. *International Journal of Critical Computer-Based Systems*. 2019;9(1-2):115–32.

[31] Singh S. *The Code Book: The Evolution of Secrecy from Mary, Queen of Scots, to Quantum Cryptography*. New York: Doubleday; 1999.

[32] Piper F., Murphy S. *Cryptography: A Very Short Introduction*. Oxford OUP; 2002.

[33] Shannon C.E. 'Communication theory of secrecy systems*'. *Bell System Technical Journal*. 1949;28(4):656–715. Available from http://doi.wiley.com/10.1002/bltj.1949.28.issue-4

[34] Diffie W., Hellman M. 'New directions in cryptography'. *IEEE Transactions on Information Theory*. 1976;22(6):644–54.

[35] Rivest R.L., Shamir A., Adleman L. 'A method for obtaining digital signatures and public-key cryptosystems'. *Communications of the ACM*. 1978;21(2):120–6.

[36] Rose S., Nightingale S., Garfinkel S., Chandramouli R. 'Trustworthy Email. National Institute of standards and technology'. *NIST Special Publication*. 2017:800–177.

[37] Apaza R.D., Popescu L. 'The way to the future has already started: ICAO aeronautical telecommunication network (ATN) using internet protocol suite (IPS) standards and protocol evolution update'. *2018 IEEE/AIAA 37th Digital Avionics Systems Conference (DASC)*; London, UK, 2018. pp. 1–6.

[38] Narten T. 'Internet routing'. *ACM SIGCOMM Computer Communication Review*. 1989;19(4):271–82.

[39] Braden R., Postel J. Requirements for Internet Gateways. ISI, USC Information Sciences Institute; 1987.

[40] Huurdeman A.A. *The Worldwide History of Telecommunications*. London: J. Wiley; 2003.

[41] Shah M.N. 'Implementation of graph theory in computer networking to find an efficient routing algorithm'. *International Journal of Innovative Research in Computer and Communication Engineering*. 2018;5:325–30.

[42] *How Browsers Work: Behind the scenes of modern web browsers* [online]. Available from https://www.html5rocks.com/en/tutorials/internals/howbrowserswork [Accessed 5 Nov 2019].

[43] Socolofsky T., Kale C. *A TCP/IP tutorial. network working group. RFC 1180*. 1991. Available from https://www.rfc-editor.org/rfc/rfc1180

[44] GGP, *EGP and 25 years of BGP, a brief history of internet routing, Router Freak.*

[45] Severance C. *Introduction to networking*. 2015. Available from https://do1.dr-chuck.net/net-intro/EN_us/net-intro.pdf

[46] Forouzan B.A. *Data communications and networking*. 5th Ed. Huga Media; 2012.

[47] Alshamsi A., Saito T. 'A technical comparison of IPSec and SSL'. *19th International Conference on Advanced Information Networking and Applications*; Taipei, Taiwan, 2005. pp. 395–98.

[48] Thiruvasagam P., George K.J., Arumugam S., Prasad A.R., Corporation, Japan N.E.C., NEC Corporation, Japan. 'IPSec: performance analysis in IPv4 and IPv6'. *Journal of ICT Standardization*. 2019;7(1):61–80.

[49] Draper-Gil G., Sanchez I. *My email communications security assessment (MECSA): 2018 results*. The Publications Office of the European Union;

[50] Thomas S.A. *SSL and TLS essentials: securing the web*. John Wiley & Sons; 2000.

[51] Bauman E., Lu Y., Lin Z. 'Half a century of practice: Who is still storing plaintext passwords?'. *International Conference on Information Security Practice and Experience*; Springer, Cham; 2015. pp. 253–67.

[52] Plain Text Offenders [online]. Available from www.plaintextoffenders.com. www.plaintextoffenders.com [Accessed 4 Oct 2019].

[53] WeSecure.net [online]. Available from http://wesecure.net.http://wesecure.net [Accessed 3 Oct 2019].

[54] SHA-256 Cryptographic Hash Algorithm [online]. Available from https://www.movable-type.co.uk/scripts/sha256.html.https://www.movable-type.co.uk/scripts/sha256.html [Accessed 3 Oct 2019].

[55] Salted Password Hashing – Doing it Right [online]. Available from https://crackstation.net/hashing-security.htm.https://crackstation.net/hashing-security.htm [Accessed 4 Oct 2019].

[56] Ars Technica reported 6.5 million LinkedIn password hashes stolen and that more than 90 percent of them were cracked just 6 days later [online]. Available from https://arstechnica.com/information-technology/2012/08/passwords-under-assault/2/.

[57] Valois M., Lacharme P., Le Bars J.M. 'Performance of password guessing enumerators under cracking conditions'. *IFIP International Conference on ICT Systems Security and Privacy Protection*; Lisbon, Portugal, 2019. pp. 67–80.

[58] Rainbow table [online]. Available from https://en.wikipedia.org/wiki/Rainbow_table [Accessed 6 Oct 2019].

[59] Goodin D. Why passwords have never been weaker—and crackers have never been stronger [online]. 2012. Available from https://arstechnica.com/information-technology/2012/08/passwords-under-assault/2/ [Accessed 6 October 2019].

[60] Provos N., Mazieres D. A Future-Adaptable Password Scheme. *USENIX Annual Technical Conference*; FREENIX Track; 1999. pp. 81–91.

[61] Florêncio D., Herley C., Van Oorschot P.C. 'An administrator's guide to internet password research'. *28th Large Installation System Administration Conference (LISA14)*; Seattle, WA, 2014. pp. 44–61.

[62] Brostoff S., Sasse M.A. 'Safe and sound: a safety-critical approach to security'. *Proceedings of the 2001 Workshop on New Security Paradigms*; Cloudcroft, New Mexico, 2001. pp. 41–50.

[63] Shelton B. Use non-ASCII characters in your password [online]. Available from www.infosecurity-magazine.com [Accessed 6 Oct 2019].

[64] NCSC *Multi-factor authentication for online services*. Available from https://www.ncsc.gov.uk/guidance/multi-factor-authentication-online-services

[65] Scroxton A. 'Should I be worried about MFA-bypassing pass-the-cookie attacks?' *Computer Weekly*. 2021;20.

[66] Malware and Antivirus: What's the Difference? [online]. Available from https://www.vipre.com/resource/malware-antivirus/.

[67] Patil M.B., Joshi M.J. 'A study of Past, Present Computer Virus & Performance of Selected Security Tools'. *Journal of Advanced Computer Science and Technology*. 2012:316–24.

[68] Hamlen K.W. 'Stealthy software: next-generation cyber-attacks and defenses'. *2013 IEEE International Conference on Intelligence and Security Informatics*; Seattle, WA, 2013. pp. 109–12.

[69] Kline J.S., McAfee Associates *Competitive strategies for the computer antivirus industry*. Phoenix, Arizona: MediaNet Solutions, Inc;

[70] Beblavý M., Kureková L.M. 'Into the first league: the competitive advantage of the antivirus industry in the Czech Republic and Slovakia'. *Competition & Change*. 2014;18(5):421–37.

[71] Haidar A.M. 'Common network security threats and counter measures'. *Proceedings of the International Conference on Security and Management (SAM) 2011 (p. 1). The Steering Committee of The World Congress in Computer Science, Computer Engineering and Applied Computing (WorldComp)*; Las Vegas, Nevada, 2011. pp. 1.

[72] Ralinov R. *Anti-virus tool: using anti-virus techniques for malware detection*. The University of Manchester; 2016.

[73] Hamlen K.W. *Stealthy software - next-generation cyber-attacks and defenses*. IEEE; 2013.

[74] Running More Than One Antivirus Program [online]. Available from https://www.kaspersky.co.uk/resource-center/preemptive-safety/running-more-than-one-antivirus-program.

[75] What is a Zero-day Attack? – Definition and Explanation [online]. Available from https://www.kaspersky.co.uk/resource-center/definitions/zero-day-exploit.

[76] NIST SP 800 41. *Guidance on firewalls and firewall policy*. Gaithersburg, MD: National Institute of Standards and Technology (NIST); 2015.

[77] Voronkov A., Lindskog S., Martucci L.A. 'Challenges in managing firewalls'. *Nordic Conference on Secure IT Systems*; Stockholm, Sweden, 2015. pp. 191–96.

[78] Sudarsan A., Vasu A., Ganesh A., Ramalingam D., Gokul V. 'Performance evaluation of data structures in implementing access control Lists'. *International Journal of Computer Networks and Security*. 2014;24(2):1303–8.

[79] Amoroso D.E.G. 'Tag cyber security annual'. *Outlook For Fifty Cyber Security Controls*. 2019;1.

[80] Postel JB. RFC 821: Simple mail transfer protocol [online]. 1982. Available from http://www.ietf.org/rfc/rfc821http://www.ietf.org/rfc/rfc821.

[81] Cidon A., Gavish L., Bleier I., Korshun N., Schweighauser M., Tsitkin A. 'High precision detection of business email compromise'. *28th {USENIX} Security Symposium ({USENIX} Security 19)*; 2019. pp. 1291–307.

[82] Chandramouli R., Garfinkel S.L., Nightingale J.S., Rose S.W. 'Trustworthy email'. *NIST Special Publication*. 2016:800–177.

[83] Verizon. *Data breach investigations report*. New York, USA: VERIZON; 2021.

[84] US-CERT. *Alert (TA18-074A): russian government cyber activity targeting energy and other critical infrastructure sectors*. Cybersecurity & Infrastructure Security Agency; 2018.

[85] O'Leary J. Mandiant: Insights into Iranian Cyber Espionage: APT33 Targets Aerospace and Energy Sectors and has Ties to Destructive Malware [online]. Available from https://www.mandiant.com/resources/apt33-insights-into-iranian-cyber-espionage [Accessed 15 Feb 2018].

[86] Axel F., Pierre T. Leviathan: Espionage actor spearphishes maritime and defense targets [online]. Available from https://www.proofpoint.com/us/threat-insight/post/leviathan-espionage-actor-spearphishes-maritime-and-defense-targets [Accessed 15 Feb 2018].

[87] Kensin J. *RFC 5321-simple mail transfer protocol*. Internet Engineering Task Force; 2008.

[88] Freed N., Borenstein N. Multipurpose internet mail extensions (MIME) part one: Format of internet message bodies [online]. 1996. Available from http://www.hjp. at/doc/rfc/rfc2045.html [Accessed 5 Dec 2019].

[89] Schaad L.J. 'Secure/multipurpose internet mail extensions (S/MIME) version 4.0' in *Message specification draft-ietf-lamps-rfc5751-bis-12*. Brute Squad Labs, Inc;

[90] Mockapetris P. 'Information sciences Institute'. *Domain Names-implementation and Specification, RFC1035*. 1987.

[91] Available from https://mediatemple.net/community/products/dv/204643950/understanding-an-email-header#5 [Accessed 23 Mar 2021].

[92] Durumeric Z., Adrian D., Mirian A, *et al*. 'Neither snow nor rain nor MITM: an empirical analysis of email delivery security'. *Proceedings of the 2015 Internet Measurement Conference*; Tokyo, Japan, 2015. pp. 27–39.

[93] Sanchez M.J., Draper Gil G. My Email Communications Security Assessment (MECSA): 2018 Results. *EUR 29674 EN,* Publications Office of the European Union; Luxembourg; 2019.

[94] Email security and anti-spoofing [online]. Available from https://www.ncsc.gov.uk/guidance/email-security-and-anti-spoofing.

[95] Malatras A., Coisel I., Sanchez I. 'Technical recommendations for improving security of email communications'. *2016 39th International Convention on Information and Communication Technology, Electronics and Microelectronics (MIPRO)*; Opatija, Croatia, 2016. pp. 1381–86.

[96] Hoffman P. *SMTP service extension for secure SMTP over TLS: RFC 2487*. Internet Mail Consortium; 1999.

[97] Wireshark [online]. Available from https://www.wireshark.org/ [Accessed 10 Feb 2021].

[98] Kucherawy M., Zwicky E. 'Domain-Based message authentication reporting and conformance (DMARC)'. 2015.

[99] Shen K., Wang C., Guo M., *et al.* Weak Links in Authentication Chains: A Large-scale Analysis of Email Sender Spoofing Attacks. *30th {USENIX} Security Symposium ({USENIX} Security 21)*; 2021.

[100] Whalen S. *Demystifying DMARC: A guide to preventing email spoofing*. 2018 Jun 15. Available from https://seanthegeek.net/459/demystifying-dmarc/

[101] DKIM Frequently asked questions [online]. Available from http://dkim.org/info/dkim-faq.html [Accessed 9 Feb 2021].

[102] Email security and anti-spoofing [online]. Available from https://www.ncsc.gov.uk/guidance/email-security-and-anti- spoofing [Accessed 9 Feb 2021].

[103] Barrett P. *Testimony on cyber R&D challenges and solutions. United States house of representatives committee on science, space and technology subcommittee on research and subcommittee on technology*. Washington, USA: Committee on Science, Space, and Technology; 2013.

[104] DMARC overview [online]. Available from https://dmarc.org/overview/ [Accessed 12 Dec 2019].

[105] Hu H., Peng P., Wang G. 'Towards understanding the adoption of anti-spoofing protocols in email systems'. *2018 IEEE Cybersecurity Development (SecDev)*; Cambridge, MA, 2018. pp. 94–101.

[106] Fagerland V. *Automatic Analysis of Scam Emails. Norwegian*. Trondheim: University of Science and Technology. NTNU; 2017.

Chapter 7
Threat identification, monitoring and detection

Lucy Caldwell, Andrew Watson, Andreas Wehowsky[1], *and Patrick Mana*[2]

7.1 Introduction

The preceding chapter focused on pre-event or 'preventative' controls, which reduce the likelihood of an attack being successful. In this chapter, we delve into a further category of control, 'detection', which assists in the discovery of both attempted and successful cyberattacks. Furthermore, we explore how threat intelligence and threat data can be used to model, understand, detect and address malicious or anomalous behaviour within networks.

This chapter aims to help readers understand how attacks are identified and the methods and models that security teams use to monitor systems, detect attacks and inhibit them. As in the previous chapter, we do not cover an exhaustive set of controls or methodologies, and those discussed are also not specific to the transport sector. By gaining an understanding of how they work in general, we hope to broaden knowledge and understanding around how detection controls can aid in securing systems.

This chapter does not endeavour to provide guidance on configuration or implementation of such detection tools, technologies, or methodologies – but rather aims to equip the reader with an overarching understanding of the concepts discussed and explore how they can be used to augment defensive needs. The chapter draws on examples to enhance the descriptions; however, these should not be considered exhaustive or definitive use cases. Implementation and utilisation of such tools require expertise and methodologies can be widely nuanced; the use of such depend heavily on a wide array of criteria, including an organisations' risk appetite as well as their unique environment or network topology.

The underlying assumption for this work is that attacks can no longer be stopped at the perimeter of a system or organisation, so to detect attacks and defend against them, we need to understand the attacker's motives, objectives and methods. This has led to informing strategies and technologies that deter and defend against

[1]Founder & CEO, Muninn, Denmark
[2]Cyber Security Program Manager and EATM-CERT Manager, EUROCONTROL, Belgium

attacks, as well as the continued development of security knowledge, expertise, technologies and services with an aim to defend networks and ecosystems.

7.1.1 Overview

The chapter addresses a range of technologies, services and methodologies that may appear to have very similar objectives and approaches. To help navigate these, Figure 7.1 shows the broad relationship between the main topics discussed.

Supporting the technologies and services are threat data and threat intelligence, which is discussed first in the chapter. Monitoring systems both consume and generate threat data, which can be analysed, correlated and turned into threat intelligence

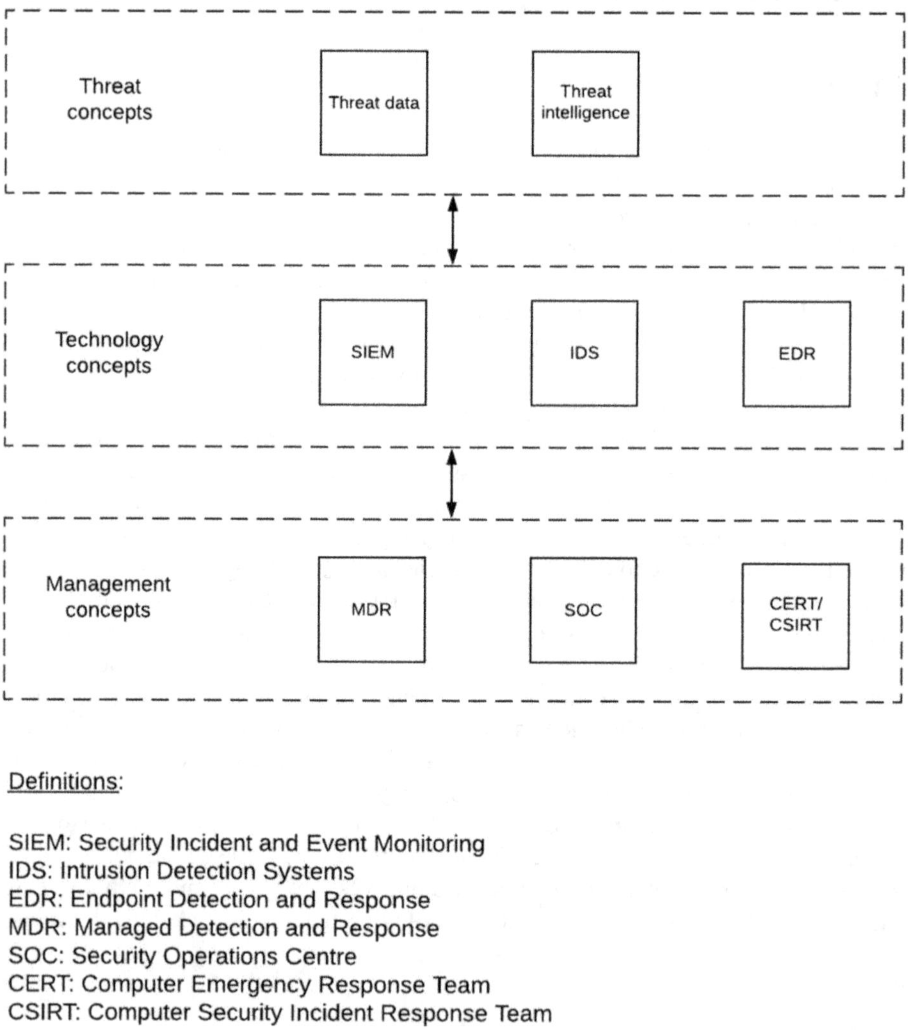

Figure 7.1 Relationships between key topics in the chapter

for security management, business stakeholders and sharing with others. Threat intelligence can inform the configuration of defensive technology based on the current threat landscape.

The chapter then explores monitoring and detection technologies, focussing on 'security information and event managers' (SIEM), 'intrusion detection systems (IDS), and 'end point detection and response' (EDR) as three non-exhaustive examples of such technology:

- Most larger organisations will collect log information pertaining to the network infrastructure and devices. This provides an abundance of information that can inform monitoring and help organisations distinguish between normal and abnormal behaviour. A commonly used solution to collate logs is a 'security information and event manager' (SIEM), which is a software system that collects and aggregates log data generated across an organisation and supports security and information technology (IT) teams in monitoring behaviour across a network.
- The final technology covered is 'end point detection and response' (EDR). This is a software solution that logs all activity across all endpoints running the software and maps this to a central location for data analysis and correlation to identify potential patterns of malicious behaviour.

The final part of the chapter covers services leveraged or outsourced by organisations to monitor networks and detect unusual behaviour, respond to cyber security incidents and respond to threat intelligence - often using data derived from the technologies explored above. The chapter covers 'security operations centres' (SOCs), 'managed detection and response' (MDR), and 'computer emergency response teams' CERTs.

7.1.2 Why securing the perimeter is not enough

7.1.2.1 The permeable organisation in the current threat landscape

In the current cyberthreat landscape, attackers (groups or individuals) have a distinct advantage. An attacker only needs to leverage a single vulnerable entry point into an organisation's network in order to compromise a network and gain unauthorised access to a network or system.

While detection technologies are increasingly seen as a pivotal part of any cybersecurity strategy, many organisations have yet to adopt them, focusing on traditional preventative technologies. This means many cyberattacks or network intrusions remain undetected. Reference [1] summarises the key issue: 'Adversaries are developing new methods for embedding malware in networks, remaining undetected for long periods, and stealing data or disrupting critical systems'. The Ponemon Institute surveys the time taken to identify and contain a breach; in 2021 this was 212 days and 75 days respectively, a total of 287 days [2]. Such a delay to identify and contain a breach could cause a series of serious consequences for organisations, depending on attacker motivation.

The use of purely preventative cyber security controls is no longer conducive to adequate protection for modern organisations' ecosystems and operating models. Most organisations will operate as part of a complex and interconnected supply chain, with dependencies on third parties both for key services, as well as for supply of equipment or software. A cyber-attack on just one organisation within these complex supply chains can have a significant cascading impact on many other organisations – whether due to the intent of the attacker, or through collateral damage. A new paradigm of global interconnected devices and networks, remote working and cloud computing drives the necessity for a defensive strategy within organisations to mitigate attacks earlier on in the attack chain to minimise impact to organisations.

While an abundance of global cyber attacks are indiscriminate or opportunistic bulk attacks, often featuring commodity malware or leveraging social engineering techniques such as large scale phishing or other fraudulent campaigns, attacks are also becoming more targeted. Threat actors increasingly leverage leaked credentials or personal information of victims to curate their lures; this coupled with increasingly bespoke tooling, advanced social engineering techniques, and sophisticated technical exploits allow attackers to specifically target individuals, or vulnerabilities in organisation and global supply chains.

The IBM Security X-Force Threat Intelligence Index 2022 report [3], which analysed trends and attack patterns from their data sources throughout 2021, observed that ransomware was the top attack type in 2021, with global supply chains being heavily impacted – particularly in the manufacturing sector. The analysis observed a 33% increase in attacks which leveraged vulnerability exploitation of unpatched software, further highlighting the importance of robust vulnerability management processes, as explored in chapter 6.

While there is a wide expanse of entry vectors available for adversaries to leverage, commonly this entry point is via malicious email communication. A ransomware attack on a US maritime facility in 2019 originated from an employee clicking on a malicious link in an email. The ransomware reportedly not only impacted the IT network but also infiltrated cargo transfer industrial control systems and shut down critical processes [4]. As emails are essential for business communication, attackers have increasingly circumvented email security controls, making it a commonly successful entry vector.

The IBM Security X-Force Threat Intelligence Index 2022 report's [3] analysis of the incidents they remediated in the transportation sector in 2021 found that half were initially caused by a phishing email, with stolen credentials and vulnerability exploitation closely following as the cause of the incident at 33% and 17% respectively. The report highlights how an array of attack types and vectors were used against the transportation sector in 2021 including malicious insiders, ransomware, remote access trojans (RATs), server access attacks, credential harvesting and data theft.

Hence, it is increasingly accepted in the security community that it is not possible to stop all adversaries at the perimeter of a system or a network, emphasising the importance of a defensive strategy and detection technologies. While traditional preventative controls such as firewalls are in common use, as discussed in Chapter 6, they have limitations, including risk of misconfiguration and common challenges with encrypted traffic inspection. Furthermore, attack tactics and techniques increasingly attempt to circumvent firewall controls. This means using a defence in depth approach, detecting and responding to malicious activity inside internal networks, and on network attached devices (known as 'end points'), is an important consideration for all organisations.

As organisations' defensive needs increase in line with the increasing sophistication and motivation of attackers, we also look at the importance of threat intelligence. The fast evolving threat landscape has triggered a more dynamic approach to security management, from compliance-based towards risk-based approaches. In short, organisations can fairly assume that their networks are permeable, and should therefore seek to understand associated threats relevant to their organisation, and be prepared to monitor, detect and respond to inevitable network intrusion attempts.

7.1.2.2 Cyber resilience

Cyber resilience, covered in Chapter 4, is increasingly a strategic focus for organisations, which refers to an organisation's capacity to continuously deliver the intended business outcomes, despite adverse cyber events [5]. NIST defines cyber resiliency as *the ability to anticipate, withstand, recover from, and adapt to adverse conditions, stresses, attacks, or compromises on systems that use or are enabled by cyber resources* [6]. The UK's National Cyber Security Centre (NCSC) promotes 10 steps to Cyber Resilience [7], which includes; risk management, engagement and training, asset management, architecture and configuration, vulnerability management, identity and access management, data security, logging and monitoring, incident management and supply chain security. While all are fundamental aspects of overall cyber security and will be covered to varying degrees in this book, this chapter will explore the importance of 'logging and monitoring' for organisations to be able to detect and investigate incidents.

Recent attacks on transport systems exhibit how attackers can render online systems and key transport services inoperable, such as the ransomware attacks on Metro Vancouver's transportation agency Translink [8] and Montreal's Société de transport de Montréal public transport system [9]. A review of attacks to the aviation industry is provided by Ukwandu *et al.* [10] and a recent ransomware attack on Bangkok Airways is covered in Reference [11]. Maintaining cyber-physical resilience is therefore paramount [12]. Recent work in this area includes the ResiCAV project [13], which identifies the need to 'detect, understand and respond to emerging cybersecurity threats in real time across the mobility ecosystem', thereby enhancing cybersecurity and resilience.

Distributed Denial of Service (DDoS) attacks are a common attack type to cause significant disruption to the resilience of an organisation or their services and have increased in both frequency and volume in recent years; research by F5 [14] found that between 2020 and 2021, DDoS attacks increased by 55%. Furthermore, the research found that the attacks were becoming more complex with multiple attack vectors being used, underscoring the importance of being able to detect DDoS attacks in a timely manner in order to mitigate the attack and minimise impact; for example, routing, blocking, or scrubbing traffic after a DDoS attack is detected, to minimise impact to key services.

Undiscovered threats are particularly a problem in the light of how rapidly cyberattacks propagate through a network and can cause operational impact. CrowdStrike [15] reports that the 'breakout' times of cyberattacks attributed to certain adversaries were an average of 4 hours 37 minutes. Breakout time is defined as the time taken for the adversary to move laterally on a network after the initial compromise. Moreover, analysis of advanced threat actors saw breakout times of 19 minutes. These findings demonstrate the importance of detection and effective, *timely* response and support. CrowdStrike has defined a '1–10–60' rule of thumb, recommending that organisations should detect intrusions in under 1 minute, perform full investigation in under 10 minutes and eradicate the adversary in under 60 minutes. Achieving such performance may be an ideal situation, depending on the environment, the sophistication of the attack and whether eradication is feasible in the context of the ongoing operation; for example, if the part of the system or end point of concern is part of central infrastructure, it may need to be managed until a triage plan has been identified.

The problem in discovering breaches is often exacerbated by a lack of visibility of the extent of networks. Hence, a process of 'network enumeration' is needed to discover, among other things, details about where the connected devices sit; what services are running and which operating systems and which network resources are in use. If the networks are not fully enumerated, and not all assets are known, organisations are unable to recognise or baseline what 'normal' behaviour or traffic looks like on the network and end points [16].

Furthermore, as attacks become increasingly targeted, frequent and sophisticated, it is important from a resiliency context to understand why the organisation's assets and data may be targeted and by whom so to layer controls accordingly to detect and respond to attacks in a timely manner. This forms part of the activity known as 'threat assessment', which guides organisations to understand how they might be targeted by which adversaries, which also informs how to allocate security budgets.

7.1.2.3 The increasing capability of threat actors

The development of detection tools has led threat actors to respond by continuously changing their tactics.

It is commonly understood amongst cyber security professionals that threat actors are increasing in sophistication, with attackers working on increasingly innovative ways to circumvent security controls. Microsoft's Digital Defense Report 2020 [17] explores this increase in sophistication by Nation State actors, noting that the most common attack techniques included 'reconnaissance, credential harvesting, malware and virtual private network (VPN) exploits'.

In the cybercriminal space, the ransomware threat environment has increased in sophistication and prevalence at an exponential rate, with the NCSC Annual Review 2021 [18] noting that

> *Ransomware became the most significant cyber threat facing the UK this year. Due to the likely impact of a successful attack on essential services or critical national infrastructure it was assessed as potentially harmful as state-sponsored espionage.*

The increase in sophistication of the ransomware groups' techniques, tactics and procedures (TTPs*) including more sophisticated exploits, the exfiltration of data prior to encryption of files and systems, as well as an array of extortion tactics, further emphasises the importance of early detection of attempted ransomware attacks to minimise impact earlier on in the attack chain.

When it comes to detection, threats actors are increasingly attempting to circumvent traditional security controls and diversify their tactics to avoid being detected. For example, many antivirus tools use signature-based detection, which match character strings in the software code of a suspected virus to a database [19]. If there is a match, the code is suspected as a virus and quarantined. In this way of working, a virus has to be first identified and then its signature captured and uploaded to the database. To respond to this, polymorphic viruses have been developed to change their code as they replicate, just sufficiently, to create a different signature, and so evade detection.

Attackers also use the native and legitimate functions of computers to evade detection, by using operating system shell scripts (e.g., Windows PowerShell), Java or macros. These attacks, known colloquially as 'living off the land' [20], may pass security controls as there is no software in a computer's file system to detect. Attackers have also increasingly been observed leveraging 'fileless malware' which resides only in the computer's memory [21].

Arguably, the changes in threat actor techniques could also be a sign of improved detection capabilities. These issues will be explored later in the chapter in regard to how detection capabilities need to be regularly configured to align with the changing threat landscape.

7.1.2.4 Chapter contents

The main elements addressed in this chapter are:

- **Threat and threat intelligence**: We look at what threat intelligence is, where it comes from and how it is used.

*TTPs are defined in NIST SP 800-150 as follows: The behaviour of an actor. A tactic is the highest-level description of this behaviour, while techniques give a more detailed description of behaviour in the context of a tactic, and procedures an even lower-level, highly detailed description in the context of a technique.

- **Monitoring and detection technologies**: We describe an array of technology that can be leveraged for defensive purposes and may already feature in your organisation's suite of cyber defences such as SIEM software.
- **Monitoring and detection services**: We look at how organisations can leverage SOCs or managed services such as MDR services and computer emergency response teams (CERTs) as part of their security strategies and operations.

7.2 What are threats and how do we detect them?

7.2.1 Threats and Threat Intelligence

There are various definitions of what a threat is and in Chapter 2, we drew on ISO 22399:2007 [22] for a definition of a threat as 'a potential cause of an unwanted incident which could result in harm to a system or function'. Threat intelligence, according to Mavroeidis and Bromander [23], 'is the provision of evidence-based knowledge about existing or potential threats'. Information on a cyber threat could broadly encompass (list not exhaustive):

- the method of attack, such as the attack vector(s) or details of specific malware or family of malware
- the intent and capability of the attacker [24]
- the potential for exploitation of specific vulnerabilities in systems (software and/or hardware), defences or people

Threat intelligence is gathered, shared or purchased by organisations because it can give insight into these aspects and help them prioritise and direct defensive actions.

Threat intelligence plays an important role in augmenting an organisation's defences. It provides organisations with information on the threats which could potentially impact them as well as on the adversaries orchestrating those threats, allowing organisations to make informed decisions on cyber defence accordingly. According to Gartner [25], threat intelligence is 'evidence-based knowledge, including context, mechanisms, indicators, implications and actionable advice, about an existing or emerging menace or hazard to assets that can be used to inform decisions regarding the subject's response to that menace or hazard'.

We can distinguish between raw threat *data* and *intelligence*. Threat *intelligence* provides information on threats, which is both contextualised and actionable, following analysis and enrichment. It is also timely and should be able to be understood and consumed by decision makers or the intended stakeholders [26]. Threat *data*, while often marketed as intelligence, are mostly indicators of compromise (IoCs, discussed in section 7.2.3) that have been associated with nefarious behaviour, but without further context or enrichment. Such threat data can be used by technical controls to augment detection.

There are various forms of threat intelligence (expanded in 7.2.2). For example, strategic intelligence products tend to provide qualitative analysis on trends within the cyberthreat landscape, including adversary behaviour or geopolitical events, which may have implications for an organisation. By comparison, technical or tactical threat intelligence tends to provide detailed technical analysis of threat actors' infrastructure, IoCs, exploited vulnerabilities, as well as the tactics, techniques and procedures (TTPs) of key adversaries. This allows security teams to augment their defensive posture or conduct threat hunting to assess potential exposure or compromise. Threat intelligence can also consist of forward-looking analysis to allow organisations to augment their preventative posture, as well as provide analysis on new threats observed 'in the wild', so that security teams can review if a compromise has occurred or whether their controls will defend against the new threat.

Gathering, creation, development and evaluation of cyberthreat intelligence may follow a traditional intelligence process, such as defined by the 'Intelligence Cycle' [27]. This conceptual model describes how intelligence can be produced from raw data or information. The cycle is somewhat perpetual and describes how intelligence is ideally driven by requirements from stakeholders, which then informs information collection plans in order for the processing and analysis phases to take place, leading to dissemination and subsequent evaluation of requirements. The above elements are a continuous process. The idea is to provide cyber threat intelligence teams with priority intelligence requirements against which to align their plan for intelligence collection and analysis, and to ensure that through thorough review at each stage of the cycle, 'the intelligence consumer's requirements are constantly at the heart of the process' [28].

In Reference 29, Kime outlines a process of continuous analysis for systematic review and management of cyberthreats and associated risk, 'Intelligence Preparation of the Cyber Operational Environment', comprising four main elements:

1. Define the operational environment
2. Describe the operational environment's effect on network defence
3. Evaluate cyberthreats (evaluating the adversary)
4. Developing cyberthreat courses of action

Organisations should adopt a requirements driven framework which promotes a coordinated and continuously evaluated approach to cyber threat intelligence processes within their organisation. This should align with their operating model, organisational priorities and resource. Later in this chapter, more methods and frameworks to provide a structured approach to threat analysis are explored.

7.2.2 Types of threat intelligence

Threat intelligence, also known as 'cyberthreat intelligence', comes in many forms depending on requirements and consumption across an enterprise. Threat intelligence can consist of both open and closed source information. Threat intelligence has been categorised by various sources, including Chismon and Ruks [30] who define four distinct categories: Strategic, Operational, Tactical and Technical. Definitions of the four areas vary depending on the source, but may broadly be described as follows:

- **Strategic threat intelligence** often covers high-level information on adversary groups and their motivations, as well as broader trends being observed in the cyberthreat landscape. This category requires analysis by human experts, such as trend analysis, geopolitical analysis or profiling of particular threat groups to derive potential insight into future attack scenarios. This type of intelligence is often discussed at board level and may inform key decisions around security strategy and resource allocation based on geopolitical conflict and how that may impact the cyber domain.
- **Operational threat intelligence** includes information on impending attacks or future campaigns. It may include information on the capabilities of specific adversaries and is often used to inform security teams how controls could be augmented to prevent and detect impending attacks. Recent studies indicate the potential to use artificial intelligence to collate operational intelligence and to protect networks by attributing tactics to adversaries and maintaining an up-to-date contextual view of the cyberthreat landscape, to help determine mitigating controls [31].
- **Tactical threat intelligence** is generally information on the tactics, techniques and procedures (TTPs) of threat actors. This type of intelligence relies on security personnel understanding attacker tools and methodologies, as well as the associated modus operandi of the threat actor in order to augment defenses.
- **Technical threat intelligence** consists mainly of IoCs (see 7.2.3), and is often in a form that can be ingested automatically by security controls or policies allowing organisations to detect and monitor for malicious activity. This information can come from organisations' internal systems, such as firewalls, IPS (Intrusion Prevention System)/IDS (Intrusion Detection System) and email filters, as well as external sources. The lifespan of technical intelligence is often short where it includes information such as URLs (specific web pages), 'bad' Internet Protocol (IP) addresses (specific servers) and malware hashes. These IoCs can become redundant in a relatively short time frame as attacks change; hence, threat intelligence and protective monitoring are time-sensitive.

Organisations may choose to use a Threat Intelligence Platform (TIP) to store and manage threat data and threat intelligence, which can also integrate with other security controls such as a SIEM, covered later in this chapter. However TIPs require careful configuration, management and evaluation, which is important to consider to ensure the TIP provides value [32].

7.2.3 Indicators of compromise

As referred to briefly above, IoCs are technical threat data which can indicate an attempted compromise, or that a compromise has occurred [33]. IoCs are identified from careful monitoring of network information through network telemetry† (gathering end point, network, log or other data) and threat hunting activities, and can also be used by network defenders or Security Operations Centre (SOC) teams to investigate unusual activity or attempted network intrusions.

IoCs are referred to often in this chapter, as they can be used by security controls such as SIEM software, which collects and aggregates data across an environment for subsequent human analysis. If the IoC data is highly reliable, and less prone to give false positives, it may be integrated into active network tools [34], such as email gateways and intrusion prevention systems (IPS), which block malicious activity at an early stage. A note of caution with IoCs is that attacker infrastructure changes regularly, and some malware types mutate so fast that the IoCs can become obsolete in minutes or even seconds.

Table 7.1 provides some examples of basic IoCs, which are a fraction of the total amount [38, 39].

7.2.4 Threat hunting

Threat hunting is a human process of searching for past or current IoCs or unusual behaviour to detect and isolate threats. Threat hunting aims to detect unusual or sophisticated threats, which may not be picked up through rule-based security components. It is usually based on a hypothesis in order to understand attackers' TTPs and improve security controls [40, 41].

Threat hunters assume an attacker mindset and have specialist skills and knowledge to both detect and respond to attacks. Through continuous identification and testing of hypotheses and use cases, threat hunting can improve an organisation's detection capability by looking at how attackers behave and how controls can be deployed to defend. Because attackers evolve their techniques as defenders' capabilities improve, some IoCs may not provide actionable insight without supporting contextual information. Threat hunters therefore use threat intelligence to explore evidence of compromise in an environment, and subsequently recommend updates to detection controls.

When automation and alerts are appropriately configured on detection controls (explored below), threat hunters can focus their expertise on hunting for unknown threats and consider how detection could be improved to capture attacks at an earlier stage. CrowdStrike have coined the term indicators of attack [42], which focuses on detecting attackers in real time by focussing on the tactics and behaviours of an attacker rather than evidence left through IoCs.

For further reading on how much 'pain' it would cause an attacker should analysts and threat hunters detect malicious activity, we recommend exploring Bianco's Pyramid of Pain; which outlines how not all indicators of activity are equal, and explains that, for example, detecting attackers' TTPs causes more issues for attackers than hash values [39, 43].

†Network telemetry refers to automated measurement of network-related information from devices across the network.

Table 7.1 Example indicators of compromise

Indicator of compromise	Details/Examples
File hashes (e.g., MD5, SHA-1, SHA-256)	Hashes provide unique signatures of a set of data such as a file of malware code. By running a (cryptographic) hash algorithm on a suspicious file, the resulting output ('hash') can be compared with a database of previously identified malware hashes. Any match indicates the potential existence of malware on the infected computer or end point.
IP addresses	An IP address provides information on the infrastructure used by the attacker. For example, IP addresses leveraged in attacks such as Distributed Denial of Service attacks, while having the potential to be spoofed, can provide insight into attacker infrastructure. Connections, or attempted connections, to suspicious IP addresses may indicate malicious activity. These addresses sometimes have a short lifespan as attack groups may move to new servers or use legitimate servers belonging to others to carry out attacks.
URLs	Uniform Resource Locators (URLs) are the addresses of web pages (e.g., https://www.theiet.org). A website's address can be checked against databases of known malicious websites.
Host and network artefacts such as registry changes	Registry keys could be added or modified by an attacker as a means to keep a persistent presence on a computer and, for example, trigger malware to run at a particular time if not already active. This is a technique often used by attackers when writing trojans [35], malicious code that masquerades as, or attaches to, legitimate software.
Unusual DNS requests	The Domain Name System (DNS) translates domain names (e.g., 'theiet.org') to IP addresses. If a computer or other device is under the control of an attacker, there may be an unusual amount of traffic between the device and the attacker's IP address. The Internet traffic can be analysed to establish evidence of Command and Control traffic beaconing from malware on a host out to the master server for attack instructions [36] using a domain name. However, attackers may also generate random domain names using Domain Generation Algorithms to hide the actual Command and Control addresses in use and evade detection [37]. Examples of other indicators are unusual domain name requests, unusual DNS logs showing requests to unusual geographical locations, abnormal volumes of queries and out-of-hours requests.
File name, location or size changes	Malware may be embedded in malicious file extensions - for example, an '.exe' attachment on an unexpected email. However, '.exe' is used to indicate legitimate program files that are 'executable'. Changes to a file, its size, or its location could indicate malware.
Changes in privileged account activity	Indicators may include accessing a large number of files, exfiltration of data, writing to a removable device, password changes or failed logons.

7.2.5 Threat actors

'Attackers' or 'threat actors' may be broadly categorised into the following groups:

- **Nation State**: Sophisticated threat actors can be funded by their State to carry out cyber operations in alignment with geopolitical agendas (see Box 7.1 on geopolitical threats). Often referred to as advanced persistent threat (APT) actors, various Nation State groups have been classified with, for example, a group number or name - often based on security vendor nomenclature. Their objectives will differ depending on motive, but their activities range from espionage, destruction and disruption, to the exfiltration of data. Nation State adversaries are commonly made up of highly skilled attackers who deploy a range of methods to reach their goal. There are clear trends of cyberattacks increasingly being fuelled mainly, but not only, by geopolitical tensions [47].
- **Criminal Groups**: Cybercriminals are motivated by money and their primary aim is to gain financially from cybercrime through ransom payments, fraud and theft.
- **Insider**: An insider can be malicious or non-malicious (under duress, in error) but is essentially somebody who has access to files and systems within an organisation or who can exploit their access to sensitive data, escalate their privileges to conduct nefarious activity or compromise files, systems or data, leading to exfiltration or disruption. Motivations of malicious insiders vary.
- **Hacktivists**: Politically or ideologically driven, hacktivists aim to disrupt an organisation or sector to promote their cause, usually through causing disruption to services.
- **Cyber terrorists**: Cyber terrorists use technology or computers as a means to inflict terror, fear and cause widespread disruption and harm – usually politically motivated or aligned with extremist ideals.
- **Opportunists**: Opportunists, often referred to as 'script kiddies', are essentially any person or group who exploit or take advantage of a vulnerability they may have observed, to see how far they can go to reach an objective and test their expertise.

Attributing attacks to attack groups is possible by analysing malware, attack indicators and the tools, tactics and procedures (TTPs) of the attackers. For example, WannaCry, the ransomware which impacted the UK's NHS [48], was attributed to the Lazarus group, associated with North Korea, due to commonalities in the tools, techniques and code previously used by the group. Frameworks and taxonomies, such as the MITRE ATT&CK framework outlined later in this chapter, can be used to map attack attributes of attackers and explore mitigating controls. Attack attribution allows organisations to understand their adversaries from the archetypical behaviour of a potential threat actor and deploy controls to mitigate attacks. For example, by configuring patch updates for commonly exploited vulnerabilities, disabling admin shares and changing account credentials.

Box 7.1 Geopolitical threats

Geopolitics increasingly drives offensive cyber operations; the ability for opposing countries and adversaries to leverage computers to advance their interests, and destabilise opponents continues to grow. Cyber operations are now inherently intertwined in modern international relations; offensive cyber capabilities are part of a state's strategic toolkit. Buchanan [44] states that 'cyber attacks have become a low-grade yet persistent part of geopolitical competition. They happen every day. Government hackers play an unending game of espionage and deception, attack and counterattack, destabilization and retaliation'.

Political rhetoric sheds light on the concern of asymmetric conflict, where there is a clear disparity between the ability of attackers and defenders. In the translation of 'Unrestricted Warfare' [45] in 1999, Liang and Xiangsui, two members of the People's Liberation Army, outline how China and developing countries could engage in technological warfare to compensate for traditional military downfalls. They explore an amalgamation of warfare techniques, specifically through the use of technology. This text was significant in shifting global recognition of the threats faced from other nations and how threats were now manifesting.

Over recent decades, it has become apparent that threat actors have new tools and methods to carry out attacks and that Nation States had alternative ways to exert power or advance their national interests through cyber operations. Nation State orchestrated cyber-attacks are increasingly reported on in open source media, with objectives and tactics broadly varying depending on motivations. Microsoft's Digital Defense Report 2021 [46] outlines the increase in sophistication and scope of disinformation campaigns. Here false information is deliberately leveraged to influence public opinion, and is an emerging threat leveraging the cyber domain as part of information warfare tactics. Disinformation campaigns can be orchestrated through a broad range of techniques and for varying motives (including cybercriminal campaigns), however Nation State actors have increasingly used disinformation campaigns aligned with their geopolitical agendas to manipulate narratives, influence masses, and to cause disruption to democratic populations.

While actors' objectives to destabilise opponents, cause disruption and conduct offensive operations has not changed since purely kinetic warfare, in an increasingly interconnected world, the method(s) by which this can be carried out has expanded, and offensive cyber toolkits remain very much a core component of a nation's arsenal.

An early adoption of threat actor identification was carried out in 2007 by Intel, which defined a Threat Agent Library [49]. They defined 22 standard archetypes of threat groups with 8 attributes in order to help organisations identify relevant threat actors with standard terminology to aid communication. In 2015, the original author of the Intel White Paper refined the Threat Agent Library by adding the motivations behind threat actor activity [50].

It is important to note that derivatives of code and variants of tools used throughout an attack can be used by actors of many capabilities, so it is often difficult to infer attribution with absolute certainty. Through the collective experience of multiple cyberattacks, it has become clear that sophisticated malware and tools created by actors such as Nation States, are subsequently leveraged by less sophisticated groups [51]. This creates a perpetual cycle of ever-increasing threat actor capabilities. Furthermore, attribution deflection and different attack groups sharing tools, code, techniques and infrastructure continues to complicate the attribution picture.

7.2.6 Threat analysis methods and frameworks

Using threat data requires a structured approach, so that the information gathered is translated into intelligence products that inform decisions, that is, 'actionable intelligence'. The time interval between identifying, using and sharing threat intelligence is also critical.

In this section, we cover examples of methods or frameworks developed to provide a structured approach to threat analysis,

- Threat modelling
    - STRIDE
- Adversary models
- The Lockheed Martin Kill Chain®
- MITRE ATT&CK Framework

7.2.6.1 Threat modelling

Threat modelling was developed primarily as a software development tool but, as defined by the Open Web Application Security Project®, known as OWASP, has wider application to systems and business processes. OWASP defines threat modelling in the context of the software development life cycle as working to 'identify, communicate and understand threats and mitigations within the context of protecting something of value' [52].

For software development, threat modelling is achieved through a variety of formal methods, such as attack/threat trees, data flow diagrams and misuse cases, which may be modelled in ‡Unified Modelling Language [53]. Threat modelling may also drive software testing to ensure a systematic approach to identifying threats [54], which are then tested against, for example, tests for cross-site scripting or Structured Query Language (SQL) injection attacks.

‡Unified Modelling Language.

Considering how threat modelling might be applied at a high level, such as the system or business process level, Ma and Schmittner [55] describe threat modelling as addressing:

- the vulnerabilities of the system and its assumptions of trust
- the nature of adversaries (motivations, capabilities and TTPs)

They further describe, in the context of automotive systems, a progression from concept (high-level security requirements) to product development (technical requirements and specifications), to production (penetration testing[§] [56]). Digital Shadows [57] widen this view of threat modelling into a process of defining critical assets, assessing attacker motivation, understanding threat actor capability and intent, and developing scenarios to evaluate the effectiveness of controls.

There is therefore some overlap between threat modelling and risk assessment, particularly in defining the system and its boundaries, the critical assets and the relevant threats. Hence, the approaches to threat modelling vary more with the objectives and technical detail of the exercise being carried out rather than the underlying principles. The second dimension, however, is also to consider the threat landscape and ask questions about the nature of adversaries. This can inform the types of threat that organisations may face, including whether they may be directly targeted. This links to an adjacent area known as adversary models, briefly covered below.

7.2.6.2 STRIDE

Threat modelling is a core component of Microsoft's Security Development Lifecycle [58]. STRIDE is a mnemonic for 'Spoofing, Tampering, Repudiation, Information disclosure, Denial of service, and Elevation of privilege'. Shostak [59] outlines the STRIDE methodology, which is a well-known tailorable threat model for basic threat assessment developed by Loren Kohnfelder and Praerit Garg in 1999. The STRIDE model can be used more broadly than for software development, such as a threat elicitation tool to aid a holistic assessment of a network, or more granularly for each system. Microsoft provides a comprehensive description of STRIDE threat modelling and access to its tool on line in Reference 60.

The STRIDE threat modelling methodology is a formalised process of decomposing the software design into functional elements, and the data flows between them with a 'Data Flow Diagram'. Each element may be considered vulnerable to one or more of the STRIDE threats, which allows the designer to consider appropriate controls (mitigations) to add to the design. The link between threats and security properties are shown in Table 7.2 (various sources including references [55, 59 and 60]), which also shows the security properties affected and examples of types of controls which can be used (non exhaustive) to mitigate the threat.

[§]Penetration testing is described by the UK's NCSC as 'A method for gaining assurance in the security of an IT system by attempting to breach some or all of that system's security, using the same tools and techniques as an adversary might'.

Table 7.2 STRIDE threats mapped to security properties

Threat	Property affected*	Definition	Example	Typical control(s)
Spoofing	I, C, A	Pretending to be someone or something you're not.	Brute force attack on user credentials to spoof identity of user.	Strong authentication controls (e.g., multifactor authentication).
Tampering	I	Modifying artefacts, data, memory, network, code.	Malicious modification of a database.	Digital signatures, access control lists.
Repudiation	I, NR	Claiming action was or was not attributed to yourself or your action.	Attacker (insider or external adversary) uses someone else's network account to perform a malicious action, which is subsequently hard to attribute to the attacker.	Robust audit logging and aggregation to identify unusual behaviour.
Information disclosure	C	Enabling unauthorised access to data or leakage of data.	Unauthorised access to files or databases, which could result in data exfiltration.	Proper encryption standards, avoiding self-signed certificates. Robust data loss prevention security policies and controls to limit data exfiltration (e.g., email gateway rules, firewall rules, data upload rules, file repository site rules). Robust access control practices.
Denial of service	A	Causing essential resources needed for a service to be rendered unavailable, e.g., system flooded with continuous requests impacting operations.	Distributed Denial of Service attack, crypto mining attack.	DDoS mitigation services. Network layer and application layer controls. For example, detecting traffic patterns, traffic scrubbing, diversion, filtering, load balancing.

(Continues)

Table 7.2 *Continued*

Threat	Property affected*	Definition	Example	Typical control(s)
Escalation of privilege	I, C, AA	Changing authorisation level of access rights in order to carry out actions that would otherwise be denied, e.g., remote users having the ability to execute code.	Adversary may use tools and techniques to escalate their privileges to an admin user after compromising a standard user device by exploiting operating system vulnerabilities or through obtaining credentials.	Role-based access control, privileged access management, vulnerability management.

*Key: (I) Integrity, (C) Confidentiality, (A) Availability (AA) Authentication/authorisation (NR) Non repudiation

7.2.6.3 Adversary models

Adversary models allow organisations to understand threat actors and associated attack paths, to anticipate attacks from specific types of attacker and to put controls in place. Do *et al.* [61] describe adversary models as a formalisation of an attacker, either as statements on the adversary's capabilities and goals or, with a more extensive formulation, as an algorithm. They regard adversary models as more generalised models for understanding attacks on a system as they represent the 'complete' attacker, compared with threat models, which tend to model distinct attacks in detail. The 'complete' attacker here refers to information on the attacker's goals (intent), capability and assumptions (whether the attacker is internal or external, has access to networks, access privileges, etc.).

The concept of an Attacker Behaviour Model, proposed by Moskal [62], considers the interaction between a network and an attacker by utilising cyberattack scenarios and network defence simulators. The idea was to create more realistic attack paths using a 'multistage attack scenario simulator', for which they simulated various cyberattack scenarios. Once an exploit or technique has been 'fingerprinted' or understood by defence teams, it can quickly be shared among collaboration groups, rendering the exploit far less likely to succeed or remain anonymous on the next occasions it is used.

There have been various attempts to create models that are flexible enough to be utilised by multiple organisations, pre-empting attacks and assigning controls to prevent and detect attacks of a similar nature. In reference [63], Sonalke and Griffor explore how, due to the interconnectivity of critical systems, adversary modelling is an essential exercise for organisations to understand their security posture, associated vulnerabilities and their inherent and residual risk (following the assignment of appropriate controls).

Sonalke and Griffor [63] point out, however, that traditional adversary models do not take into account the levels of connectivity and automation between systems and, with modern intertwined systems, air gaps [64] may ensure separation in one dimension (mechanical, digital or pneumatic) but may be penetrable in another. The authors propose an alternative 'New Adversary Model' to consider how adversaries can pivot between interconnected systems. Their model involves the following:

- Identifying potential adversaries and their capabilities
- Prioritising assets
- Weighting attack surfaces through considering multiple attack entry points
- Strategising the attacker's operating methods and return on investment

7.2.6.4 The Cyber Kill Chain® as a framework for understanding cyber attacks

The Lockheed Martin Cyber Kill Chain® [65] has become the de facto framework for analysing and understanding the various stages of cyberattacks. The concept of a kill chain derives from conventional warfare stages and is a useful framework to understand key stages of cyberattacks. The stages described by the Lockheed Martin Cyber Kill Chain® are: Reconnaissance, Weaponization, Delivery, Exploitation, Installation, Command & Control (C2), and Action on Objectives.

The United Kingdom's National Cyber Security Centre (NCSC) has produced a simplified version of the Lockheed Martin Kill Chain® to help organisations understand a cyberattack, compressing the kill chain into four stages: survey, delivery, breach and affect [66].

There have been some criticisms of the kill chain, particularly that it is based on the assumption of a traditional perimeter security infrastructure, and also that it does not acknowledge the insider threat [67]. However, its widespread use indicates that it is still regarded as a useful tool to understand attacks.

7.2.6.5 MITRE ATT&CK Framework

The MITRE ATT&CK Framework [68] is a well-regarded free resource providing information on adversary groups and TTPs that have been observed targeting multiple sectors. Defined as 'ATT&CK' (Adversary Tactics, Techniques and Common Knowledge), the model uses the 'Cyber Kill Chain' (©Lockheed Martin) to portray adversary TTPs used in cyber operations. ATT&CK is characterised as a behavioural-based threat model [69], 'to identify relevant defensive sensors and build, test, and refine behavioural-based analytic detection capabilities using adversary emulation'.

The model comprises a matrix of various 'tactic categories' to describe the stages of a cyberattack and the methodologies used by attackers, from Reconnaissance to Command and Control. Each category contains between 10 and 40 different 'techniques', such as 'Phishing for Information' in the Reconnaissance category. MITRE also provides information on detection, potential mitigations and attributed threat actor groups. Within the tactic categories, MITRE outlines techniques leveraged by adversaries at each phase of the Cyber Kill Chain®.

ATT&CK takes a behaviour-based approach, and MITRE encourages sharing of threat intelligence into a knowledge base to continuously leverage benefits for all users. The framework is regularly used by analysts in threat intelligence reports to map patterns across a common taxonomy in order to assess adversary behaviours categorised as malicious, and to understand TTPs observed at each stage of a cyberattack. This helps foster information sharing initiatives and the operationalisation of threat intelligence so that organisations' detection and response capabilities can correlate with information extracted from the MITRE repository.

MITRE also provides guidance on how to tailor detection methodologies to identify specific tactics and techniques. The ATT&CK matrix is flexible, allowing organisations to prioritise detection and mitigation techniques based on their network architecture and specific threats. There are different matrices for different types of organisation or industry: Enterprise, Mobile and ICS [70]. Furthermore, organisations can use the matrix to structure red and blue team exercises and create heat maps to assess their security posture with respect to detecting attacks [68]. EUROCONTROL has developed a specific aviation heatmap using MITRE ATT&CK, which highlights techniques used to conduct attacks against aviation.

7.3 Monitoring and detection technologies

7.3.1 Introduction

As discussed previously, the acknowledgement that purely perimeter based, preventative security is no longer adequate has driven the adoption of monitoring and detection technologies. Monitoring and detection technologies identify malicious activity and look for anomalous behaviour across an environment. As well as identifying possible threats and attacks, the monitoring information they yield provides data to assess the risks facing an organisation.

A prerequisite is to establish a clear picture of the network, to limit the number of false positives or false negatives and ensure that alerts are relevant to the environment. Capturing the network picture includes, but is not limited to:

- Network discovery and enumeration, which is a process of identifying all of the resources on a particular network. This leads to a clear picture of the network, its connections, attached devices and other data that may indicate potential vulnerabilities [71].
- Obtaining an inventory of the different technologies, software and applications deployed throughout an environment, so as to tune data monitoring to them.
- Ensuring sufficient network telemetry is generated and stored from network infrastructure to allow for log analysis, correlation and alerting.
- Understanding supply chain/third party connections.

Many organisations will have cloud computing infrastructure to contend with and consider when adopting detection technologies and establishing network visibility. This creates additional complexity due to varying cloud computing service models, such as 'Infrastructure as a Service', 'Software as a Service' and 'Platform as a Service'. These require detailed consideration of the scope of monitoring and the interaction with the hosting/third party's security components. Furthermore, the dynamics of these service environments may pose greater challenges in configuration for effective monitoring and third-party assurance. Establishing a baseline of alerting for monitoring purposes is an important exercise, although configuration may be complex [72]. Reference 72 also describes how in public cloud infrastructure, resources may be shared by organisations introducing potential risk, for example, Distributed Denial of Service attacks on one resident within a shared infrastructure environment could also impact those who share resources. Thus, it is increasingly important to understand the cloud environment and configure effective monitoring.

7.3.1.1 Associated standards and regulation

There are many security standards, guidance and control sets that refer to detection and monitoring practices: ISO27000, CREST (the Council for Registered Ethical Security Testers¶), National Institute of Standards and Technology (NIST),

¶CREST is an 'international not-for-profit accreditation and certification body that represents and supports the technical information security market'.

UK NCSC guidelines, IEC 62443 (Industrial Control Systems) and the UK Government Minimum Cyber Security Standard [73]. These standards and guidelines emphasise the importance of an organisation's monitoring capacity and the ability to identify known threats within a network. Of particular relevance to the transport sector is the European Union Directive on the security of Network and Information Systems (NIS Directive), which came into force in May 2018. The legislation is relevant to operators of essential services and to digital service providers [74], where Member States are responsible for identifying the operators to which the Directive will apply, noting that Annex II of the regulation refers to air, rail, water and road transport.

The NCSC has specifically developed principles and guidance under the objective of 'Detecting Cyber Security Events' under the Cyber Assessment Framework (CAF) [75]:

- **Security monitoring**. The UK's NCSC advises that an organisation should monitor ['the security status of the networks and systems supporting the operation of essential functions...']. Amongst a wide range of guidance, the following are noted as important: threat intelligence, generating appropriate alerting, the monitoring of user activity, the ability to monitor both current and historical known and unknown threats, as well as the necessity to gather and aggregate logs from all key systems, and the ability to review IoCs. Furthermore, the NCSC emphasise the importance of defining governance processes and roles in order to continuously evaluate the effectiveness of security controls.
- **Proactive security event discovery**. The UK's NCSC also advises organisations to maintain the ability to review anomalous events, where deviations from normal network behaviour and evidence of potential compromise should be captured, assessed and analysed [76]. This type of monitoring should build on security monitoring and requires expert analysis, network knowledge and business risk understanding. The aim is to move beyond assessing events and behaviour against attack signatures or known attack paths towards identifying unknown threats and attack paths. The guidance refers to the growing use of machine learning to identify clear deviations in network behaviour patterns. However, the guidance also warns of the difficulties of using such 'intelligent security tools', which are very much in their infancy and can result in a large number of false positives. Machine learning also relies on enough data for the algorithms to recognise what 'bad' looks like, a constant challenge given the evolving techniques of attackers.

In the remainder of this section, we cover the following technology concepts:

- Log management
- Security information and event manager (SIEM)
- Network monitoring and intrusion detection systems
- Anomaly detection
- End point detection and response (EDR)

7.3.2 *Log management*

Most sizeable organisations will collect log information pertaining to the network infrastructure and devices. This provides an abundance of information that can inform monitoring and help organisations distinguish between normal and abnormal behaviour. NCSC guidance on log collection for Security Operations Centres (SOCs), suggests log sources should include: application, host, network and cloud logs, as well as collate user authentication logs, security control logs, and DNS logs [77]:

Log source	Rationale
IP connections to the Internet from the network.	To give analysts and IT teams visibility of connections, alert them to potential Command and Control sessions and potentially highlight any unusual web traffic.
Web traffic to the Internet, ideally full HTTP(S) header information and, as a minimum, URLs and domain names.	This is important as malicious and persistent connections are often through HTTP(S) connections from an end point.
Email traffic and, in particular, send and receive metadata. Email contents and headers provide further clarity and visibility.	Emails are a common attack vector, and this information is therefore important for analysis in alignment with web traffic.
IP connections between IT and OT networks.	This data will contain evidence of actions taken to penetrate and disrupt cyber-physical systems.
Host and end point activity.	To detect unauthorised activity on computers in regard to anomalous software behaviour and use. This is important because if the sources only focus on network-level monitoring, some advanced attacks may not be detected.

Log data can be analysed and compared against IoCs, from threat intelligence or data feeds. It is important that logs can be easily queried, correlated and mapped back to particular assets in a timely manner, with bespoke alerts configured for further protective monitoring; a SIEM is often used for this purpose.

Log data needs to have sufficient security and storage itself and it should be stored in a database isolated from other internal domains, as sophisticated attackers have been known to modify or delete logs to obfuscate their presence [78]. This also supports privacy protection in the case that some personal data may reside in logs. The United Kingdom's NCSC has developed a free and open-source software, 'Logging Made Easy' [79] to assist organisations in gathering logs and network telemetry. This also helps organisations gather logs and query evidence of attacks through reviewing TTPs. This software is not seen by NCSC as a replacement SIEM, but rather as a tool for organisations with few resources to have greater oversight over their environment with respect to patching levels, device usage, administrative commands and IoCs.

7.3.3 Security information and event manager (SIEM)

A SIEM is a software solution that collects and aggregates log data generated across an organisation's network and allows security and IT teams to monitor behaviour across a network. The logs are generated from network components (with the SIEM software agent installed on it) and sources such as network switches, routers, firewalls, servers, applications and user devices/end points. All of these devices continuously create 'events', such as a log-in failure, and the SIEM is used to aggregate and consolidate the information. SIEMs also gather audit information from operating systems, networks and applications. Logs are transferred from network components using protocols such as the syslog protocol [80], and the SIEM's centralised logging receiver will receive the logs and consolidate the data. These logs are pushed to the SIEM tool in a continuous stream, producing a large amount of data, and therefore requiring a large amount of storage. Most SIEMs allow for real-time information to be pulled into a single dashboard. Reports can be pulled from the SIEM to look at specific variables and correlate data across devices.

One of the key issues associated with a SIEM is how it is used and configured. Many organisations deploy a SIEM tool as a key security component in order to achieve accreditation and meet recognised standards, but unless it is carefully configured with a series of rules there is a risk of the SIEM underperforming. Example configurations are correlating events, including all the relevant sources, assessing the criticality of those sources and so on [81]. While all networks differ, and therefore configurations will be different, key aspects to consider include:

- If the SIEM is monitoring various geographical locations, then a log in at 2 a.m. from a remote location in one country may either signify unusual behaviour or normal usage, depending on the location of the user. Hence, the time zone context should be acknowledged when setting up alerts. Network Time Protocol [82] can be used to ensure all devices synchronise their clocks to ensure the logs correlate appropriately in real time.
- Feeding threat data into the SIEM, including malicious indicators, artefacts and signatures for cross-correlation and actionable insight.
- Alerts should be customised to an organisation's environment. For example, an alert from a user who does not have privileged access rights may trigger higher priority alerts if their user account is seen escalating privileges or accessing unusual places associated with their profile.
- Events should also be correlated against what could be legitimate behaviour; for example, if a privileged user changes a domain setting, this should be checked against a change request or configuration management process. Geographic correlation is also important, for example, a user logging in from two different geographic locations at the same time.

Finally, it should be mentioned that there are some negatives with SIEMs: a risk of too many alerts ('alert storms'), false positives, and challenges in integrating all devices (especially in cloud environments).

7.3.4 Network monitoring and intrusion detection systems (IDS)

The previous section described SIEM systems that receive log feeds from various sources, but only cover managed devices, that is, devices where a software agent can be installed to record data and forward logs. This section addresses intrusion detection systems that monitor the network for malicious behaviour. When monitoring networks, IDS are also referred to as Network Intrusion Detection Systems (NIDS).

There are an increasing number of network-connected devices in corporate networks; smartphones, tablets, and 'Internet of Things' devices such as video cameras. Hence the ability to have full insight of all network traffic is critical. IDS have been developed to do more in-depth analysis by virtue of feeds from raw network packets. The raw data is often Internet Protocol (IP)-based traffic, which could include traffic from supervisory control and data acquisition (SCADA) systems, and physical security data such as electronic locks and alarm systems. Transport systems often use SCADA technology for key processes such as the automation of traffic lights, the monitoring and control of underground and overground rail networks, vehicle tracking, and controlling cyber-physical systems such a rail crossings. Hence, IDS have a key role to play in transport infrastructure.

Examining raw packet data may also be necessary for forensics[**] or hunting for threats on a network at packet level. Free tools such as Wireshark [83] or NetworkMiner [84] are quite efficient for this purpose. However, looking for threats in raw IP packets would be the same as looking for a needle in a haystack. In a 10-Gbps infrastructure, up to about 14 million packets/second can be expected [85]. Powerful software is needed to automate the task for detecting and even preventing threats in real time.

7.3.4.1 IDS Implementation considerations

IDS are typically computers running software that captures and analyses network traffic. As a simple example, we can consider a single network switch, a server running IDS software and multiple client nodes or PCs. The network switch is configured to mirror all traffic on all network ports to a single port connected to the IDS via a Network Interface Card in monitor mode. This allows all network traffic being sent and received across all network nodes to be replicated to the IDS for inspection – whether traffic remains internal to the network perimeter or is routed out via the Internet. This way the IDS can inspect all traffic for IoCs by performing some, or all of the following:

- Pattern matching: matching known IoCs, and other potentially damaging patterns, against specific rule sets, for example, port scans, SQL injections and exploit code

[**] 'Forensics' is used in the general sense as used by security teams. Digital forensics as applied by law enforcement agencies is a subject in its own right and concerns the preservation and analysis of data as evidence for criminal law cases. See, for example, Casey E. Digital Evidence and Computer Crime. 3rd Ed. Elsevier. 2011.

- Logging and categorising: the output of which can be displayed and manipulated via graphical user interfaces and dashboards allowing for quick dissemination of the data and, depending on the product, allow for other useful features like alerting, management reports and integration with other security systems
- Monitoring potentially unwanted actions: for example, accessing file repository sites or communications platforms, which could be a policy violation
- Checking for security violations: for example, detection of credentials such as unencrypted usernames and passwords, attempted access of blacklisted URLs or their IP addresses
- Monitoring for vulnerabilities: for example, out-of-date software versions, commonly exploited vulnerabilities and broken cryptography

Similar to other monitoring tools, the IDS needs to be security hardened, with strict access control, and backed up to prevent manipulation of this important data. To be effective, IDS also need the supporting processes, infrastructure and human expertise. Effective operation means human operators should be able to read the logs, configure the IDS and its reaction to certain events, respond to events and maintain the system (keeping it patched, running efficiently and ensuring it does not slow down any part of the work system).

A particular note of caution concerns the existing network components. IDS generally make comparisons against the operating norm of the network when the IDS is first installed. Hence, if there was some malware or embedded component present prior to installing the IDS, it is very likely that the embedded malware will remain active and deemed to be the norm.

An IDS provides security teams insight into network communication, and communication (internal and external) is often an integral part of an attack, whether a hacker is executing commands and exfiltrating data, or whether malware is propagating or communicating.

An IDS should logically be placed on the network where it can gain the most information to analyse, and in the best case, this means sending the IDS everything. How this is achieved depends on the particular network environment. This should be assessed and defined by network experts who will provide insight on how the data ingestion will minimise impact to key systems and cost to the organisation. Of course, there are always concessions to consider and more data that is sent to the IDS means:

- more storage space is required to store captured data
- more random access memory and processors required to process this data
- increased load on network switches and IDS Network Interface Card(s)
- a greater amount of data for security teams to analyse
- potential for more false positives
- more fine tuning required to negate false positives
- increased management overhead

There are many variations of IDS, which vary in complexity and features. Free open-source software is available, such as Security Onion [86], which incorporates a set of tools into a security hardened Linux operating system for intrusion detection, enterprise security monitoring and log management. Often, tools require prior expertise, knowledge and know-how. New commercial systems aim to reduce the management and expertise required, such as MUNINN™ [87], which uses artificial intelligence (AI)-based anomaly detection to detect and block unknown threats.

Essential requirements include the following:

- overview of the traffic and alerts, showed graphically
- relevant alerts with a minimum of false positives
- good reporting function
- forensics and auditing (who did what when)
- updates via cloud services
- integration with other systems, for example, SIEM

The traces that malware and malicious users leave in the network can be found at different abstraction levels. The most basic low-level stateless procedure is to look for known threats by inspecting each packet. However, more advanced analysis, such as described in Box 7.2, is required in order to analyse the states, content and purpose of network connections.[††]

Box 7.2 Protocol analysis

Protocol analysis is very useful in giving a broader and more detailed picture of the current events in the network traffic. Protocol analysis refers to the protocol stack [88] in network communication and involves building stateful session information that represents the communication between two devices. For example, when a user copies a file from a server in a Microsoft network share, thousands of TCP (Transmission Control Protocol) packets flow back and forth. The output of a protocol analysis could be 20 key value pairs describing the event, including the originator's and the responder's IP addresses, that the protocol used was the file sharing protocol 'Server Message Block' [89], the filename was 'my passwords.txt' and the remote folder name was '/shared/passwords/'. Other details could be the file hash sum, file size and timestamp. All this information, which is also referred to as network metadata, is useful when looking for indicators of malicious activity.

[††]Network states meaning the life cycle of a connection, from opening to ongoing to closed or closing.

Malicious network activity can be categorised into known and unknown threats. Section 7.3.6 discusses anomaly detection for identifying unknown threats. Known threats can be categorised in blacklists, whitelists, signatures and static traffic usage patterns. Some IDS can receive feeds of IoCs from cloud services and thus be able to detect the latest blacklisted IPs, domains, certificates, email addresses, websites, file hashes and more that have been attributed to malicious behaviour. However, some malware types mutate so fast that these IoCs can become obsolete in minutes or even seconds. Using IoCs is, however, an efficient way of detecting if a network has been infected with known malware or if devices have been historically or are currently infected with malware that leaves traces of known network traffic.

Detecting known malicious traffic usage patterns that happen over time can be achieved by scripts or software programs, which may also use probabilistic models and statistics. A probabilistic model uses stochastic, or random, variables to describe normal and expected behaviour and can thus pinpoint anomalies. These programs do more than just examining a packet at a time, forgetting about the previous packet. For example, ransomware that encrypts files on a network share can be detected by looking at its pattern, that is, renaming and/or creating new files at a very high frequency. By setting the threshold to be very sensitive, this kind of ransomware can be detected in milliseconds.

7.3.4.2 IDS limitations

As with any security technology, one size does not fit all, and it is important to understand the limitations of an IDS. For example, if an IDS analyses the virtual local area network (VLAN) used for 'data', but the voice over IP (VOIP) VLAN is omitted from scrutiny, then a compromise via the VOIP VLAN could allow for undetected movement via this network.

Another problem with IDS, particularly when first deployed, is false positives. These can be reduced by skilled security practitioners with a good understanding of the particular network, so as to discount irrelevant detections. Over time, experience will be gained to decrease false positives, although there is a balance to ensure that true positives are not missed. For example, IDS may categorise the use of a file repository as high risk due to the potential of data exfiltration by a malicious actor (whether an internal employee or an external attacker). This depends on the policy towards file repository sites, and if used company-wide, this does not mean that one of the thousands of uploads was not an attacker stealing data; such a scenario should be addressed based on available resource and risk appetite.

IDS cannot stand alone in protecting the IT systems of an organisation. It is often necessary to examine the cause of certain network behaviour by investigating processes on the host that caused traffic. Therefore, having an agent installed on the host that tracks system processes and other operating system events is helpful or even necessary to find the root cause, as the case study in Box 7.3 illustrates.

Box 7.3 IDS forensic analysis case study

An IDS also provides forensic ability in the case of a compromise. Sometimes a 'perfect storm' of circumstances will allow an attacker access to a network completely undetected. A remnant of this compromise may be detected in future, long after the hacker has completed their objective, but it may not be known what has been accessed, changed or stolen. This is where the IDS is invaluable in a forensic capacity, as its extensive data capture allows security teams to trace the attacker's movement starting from the original IoC and working backwards or forwards (depending on what was detected).

Andrew Watson describes his use of a new implementation of Security Onion on a network that was considered highly secure, including, among other things, the following controls:

- Redundant firewalls with Stateful Packet Inspection and in line AntiVirus (AV)
- Third-party email AV and malware scanning
- Host- and server-based AV with hardened detection configuration
- Host- and server-based firewalls
- Regular penetration testing and vulnerability scanning
- Default hardened O/S configurations
- Up-to-date perimeter IPS

Despite the layers of security that were implemented, the Security Onion IDS logs showed a connection to a URL that was marked as malware related. Open-source intelligence revealed only one Google search result, which indicated that the URL was related to a Windows Management Instrumentation (WMI)-based malware. In searching the WMI for anything of note, Andrew discovered a JavaScript payload hidden in the database and undetected by AV. He reverse-engineered the code and discovered that the payload was attempting to allow Remote Desktop Protocol access to the infected machine, set guest and admin passwords and receive new commands from the Command and Control server. While these actions were defeated by multiple layers of the 'defence in depth' strategy, without the IDS, it would not have been possible to discover and remove the infection.

7.3.4.3 Summary

In summary, the most important advantages with network monitoring systems are their ability to:

- Detect malware automatically and in real time as it propagates within the network
- Detect malicious activity such as data breaches
- Gain insight of the network, including assets or devices, traffic flows, services and protocols and use it to harden the network

Box 7.4 IDS/IPS case study

Andrew Watson describes his experience with a computer that was infected with ransomware: The ransomware had established a connection to a Command and Control server to exchange encryption keys with the aim of encrypting all of the files on a network. In this instance, the IDS detected the IoC and alerted the security practitioner, but it took no action by itself, whereas the IPS detected and blocked the exchange based on its rule sets. In this scenario, the IPS was preferable as it prevented the spread of infection.

Conversely, if the same computer had been infected with ransomware while off site, and where keys were already exchanged, on reconnection to the internal network, it would begin to infect every system on the LAN. While an IDS would detect this and alert the security team, an IPS would not detect this as there is no LAN to Internet communication. In this scenario, an IDS is preferable as it detects the infection before all nodes are compromised.

7.3.5 Intrusion prevention

IPS are perimeter-based systems but are worth mentioning here as they are based on similar technology to IDS. IPS typically examine one network packet at a time and do not see the bigger picture of multiple interrelated connections stemming from devices within the perimeter. As many types of attacks start with a device on the LAN being compromised, the traditional IPS no longer plays such a strong defensive role. Box 7.4 provides a case study of IPS and IDS operating together. The example demonstrates the importance of a defence in depth strategy using multiple layers of controls.

7.3.6 Anomaly detection

7.3.6.1 Baselining the network

A 1987 paper by Dorothy Denning was pioneering for its time and was a pivotal step towards acknowledging the importance of analysing abnormal patterns within a network, by positioning a real-time IDS to detect an amalgam of security violations. Denning described a model intrusion detection expert system as a 'rule-based pattern matching system', based on looking for deviations away from baseline behaviour. Baseline behaviour is established by analysing anomaly records through audit data and statistical modelling to alert the security team to potentially nefarious activity. In conclusion, Denning foresaw adversaries escaping detection in future by gradual changes in their behaviours [90].

There is no exact definition or requirement specification for a network baseline. It is intended to be a description of the normal behaviour and traffic observed in a network. In its simplest form, it could be a list of accepted network ports, services and devices. But the more accurately the baseline is described, the better the prediction of anomalies leading to fewer false positives. A baseline could include probabilistic models of behaviour over time, describing as many features in the traffic as possible – for example, observed domains, software, users, MAC addresses, user agents and geographical location.

As an example of a baseline property, the user 'Smith' interacts with the ticketing system database at specific times during the week and normally performs specific types of queries, file transfers and operations, whereas user 'Roberts' accesses the jump box that connects to the OT interface for rail crossing control panel.

Anomalous behaviour can be defined as behaviour that deviates from that described in the baseline. For example, user Smith starts logging on at different times than usual and performs different queries. An anomaly is not necessarily malicious – a user may have taken up more responsibility or changed their shift patterns, for example. In order to determine if it was malicious, a user (or the software program) needs to investigate what exactly triggered the anomaly and why. If it was caused by a user on a given machine, one needs to examine what other alerts or events stem from this user and machine to understand the broader picture.

7.3.6.2 Anomaly detection algorithms and machine learning

Security teams are often arbitrarily deriving intelligence, exploring security events and correlating data across network devices manually. This is not necessarily efficient or effective, so it is preferable for baselines to be defined or learned by an algorithm, particularly as all networks are different and change over time. Today, a user of an IDS, typically a system or network administrator, will expect that the baselining and anomaly detection needs to constantly learn and fit its model to changing network features.

The advent of high-performance computing, particularly based on specialised graphic processing units, has enabled a new breed of machine learning algorithms. Network baselines can be represented in a machine learning model; for example, a deep learning neural network can describe normal behaviour in the network as observed over a recent period. It is very much up to the software designer or mathematician behind the algorithm to make the machine learning model and corresponding algorithm useful in representing the baseline and predicting anomalies. Sarker *et al.* [91] discusses popular machine learning models in the context of developing a generalised intrusion detection model. These models are probabilistic and need tuning to reduce the number of false-positive alerts to a manageable level and help users assess the severity of an anomaly for prioritising forensics and investigation tasks.

New machine learning approaches have the potential to aggregate and share information to promote adaptive learning [92]. Fraley and Cannady [93] researched a deep neural network to orchestrate analysis of a high volume of security events. Their research saw an organisation's security team consolidating security event categories from 20 categories into 5: Malware, Reconnaissance, Denial of Service, Policy and Exploit, with associated subcategories. This was done with key stakeholders and subject matter experts. The defined categories allowed the machine learning system to process a large number of daily alerts and common threat vectors and allow the human analysts to focus on more complex cyber events associated with zero-day vulnerabilities. The deep neural network technology was also able to produce reports to share IoCs. The generated neural model was used on a test dataset of over 9 million alerts associated with previous human decisions; the model classified the alerts with 99% accuracy and the study claimed that nearly 80% of personnel in the Security Operation Centre could save time using a machine learning model.

7.3.6.3 Monitoring decentralised networks

Most organisations have computer networks and services spread over multiple locations with servers and services running on premises as well as in the cloud, whether a virtual private cloud or third-party cloud services like Amazon Web Services. At the same time, people and vehicles use a multitude of network-enabled devices, connecting to various parts of the network via different communication channels, such as Wi-Fi, 4G or 5G cellular networks, or satellite communication. Several shipping companies, airlines or freight companies using ground transport have connected their ships, aircraft and trucks to their headquarters via cellular and satcom networks. The interconnectivity comes with an abundance of advantages but also introduces security risks. Malicious actors could theoretically hack a ship's network from remote locations. Furthermore, vehicles have an increasing number of devices with connectivity installed, for example, to enable remote maintenance and reduce support costs.

This new paradigm of global networking creates a whole new challenge of maintaining control of the networks and being able to monitor them. Usually, the essential traffic flows through main hubs and gateways, that is, switches, which is where IDS should be deployed. Often, several IDS are required to cover the entire network. Therefore, to comprehend the bigger picture, it is important that a cluster of IDS can exchange threat information as well as being managed from a central point. An example of an IDS deployed in the transport sector is provided in Box 7.5.

7.3.7 End point detection and response

While AntiVirus software is still an important part of defence, attackers have evolved their techniques to bypass such controls. In response, a new breed of end-point security software, called 'endpoint detection and response' was developed to overcome the limitations of signature-based detection. End points are the devices that connect to the network, such as computers, peripherals, mobiles and IoT devices.

This chapter has already introduced several related but different concepts, so before discussing EDR technology, we provide a comparison with SIEM. Although there are overlaps in their functions, an EDR is generally considered to complement a SIEM by providing another log source with end-point-centric data.

Box 7.5 Example IDS system deployed in the transport sector

MUNINN is a commercial NIDS and IPS from Wehowsky.com Cyber Security Systems [87]. MUNINN performs full packet capture and protocol analysis to feed metadata into AI engines. MUNINN learns the normal internal network traffic usage patterns by baselining the network and uses both machine learning and static methods to detect anomalies and other IoCs.

The following example occurred at a company specialising in ground transportation. The company has several geographical hubs and a relatively large and complex computer network. This transport company, like many others, is highly dependent on high availability services and system integrations with partners. A cyberattack can cause havoc and huge losses, as a result of disruption of IT services, which can stall the flow of transport of goods, and even espionage from competitors.

MUNINN was deployed at the company to monitor traffic and detect and prevent cyberthreats. The system detected several compromised work stations with installed malware that was interacting with Command and Control servers. The malware was performing lateral movement, trying to identify weaknesses in other devices in the network by the means of address and port scans, and password guessing on file servers and SSH (Secure Shell) servers. At the same time, MUNINN's AI engine detected an anomalous data transfer to a blacklisted site. The identification of these malicious activities was backed up by other static models in MUNINN that triggered alerts concerning invalid self-signed certificates. The transport company was able to identify the compromised work stations, clean them up and use the reports generated by MUNINN to harden their network to stand against future cyberthreats.

EDR [94] is a software solution that logs all activity across all end points that are running the software, and sends this to a central location for data analysis and correlation. This type of data can then be further analysed by security operations centres (SOCs) and threat hunting teams.

The EDR software continuously monitors end-point system-level events and correlates this information in a central location to detect any significant deviations from the norm or potential patterns of attacker behaviour. The information collated by the EDR software may include activity such as IP addresses that the end point is interacting with, identification of beaconing (when malware communicates with the Command and Control server), filesystem alterations, system registry alterations and anomalous activity in the end-point memory, which can provide analysts with insight of malicious behaviour.

EDR represents a new approach that no longer only relies on signatures, but instead examines the core behaviour and functions within an end point to highlight anomalies that could indicate exploitation. Such behaviours could be genuine, creating false-positive alerts, and therefore thresholds and the resulting confidence in alerts are a factor. End-point telemetry is not the only data source required for comprehensive detection and investigation. As discussed previously, network data, application logs and other data sources contribute to investigations; hence, EDR systems should allow for integration and correlation with other data sources.

A key consideration is whether the EDR solution examines for artefacts already present on an end point from a current attack, or a prior attack. Some EDR toolsets only start recording data from end points from the moment they are installed. If an attack is taking place, or has already taken place, this model will not allow the defensive teams to identify historic compromise. For this reason, a good EDR toolset will also allow users to search through end point data for 'artefacts' left from prior attacks, which then allow for appropriate response.

EDR is used by human specialists, sometimes as part of a Managed Detection & Reponse (MDR) service, to query end points and cross-correlate the results with other data to detect attacks. This approach, if implemented alongside the right defensive teams, offers the best chance of detecting and investigating attacks. Using EDR, analysts are able to carry out damage limitation and containment techniques at different stages of an attack. These include:

- isolating affected devices or servers
- shutting down processes to mitigate persistence[‡‡] [95] and stop lateral movement in the network
- view network activity to identify any outbound connections from an end point, such as Command and Control, and blocking these connections

The above techniques are particularly relevant to organisations whose offices span large geographical areas and may be subject to sophisticated threats from sophisticated threat actors, where signatures and prior IoCs are irrelevant. As discussed earlier in the chapter, with attacks occurring rapidly, attackers can reach their objective in minutes or hours. Hence, EDR solutions should allow incident response functions to collect and analyse data rapidly to keep pace with and respond to the attack. A good EDR software agent will allow defensive teams the ability to remotely copy data from the end point's memory for analysis and isolate machines or restrict network connections.

7.3.7.1 Practical considerations with EDR

Depending on the organisation, there are different considerations for EDR technology implementation. Some key factors are:

[‡‡]'Techniques used for persistence include any access, action, or configuration changes that let them maintain their foothold on systems, such as replacing or hijacking legitimate code or adding startup code'. Source: https://attack.mitre.org/tactics/TA0003/

- the ability to control how much bandwidth the EDR solution consumes
- whether the EDR solution interferes with the stability of certain end points
- the amount of memory needed

An EDR solution is typically designed for standard operating systems only, so where end points or OT use non-standard operating systems, organisations may have to rely on more readily available data such as system logs [96].

7.3.7.2 Adding threat intelligence feeds into EDR

Threat intelligence or threat data feeds are often an input into EDR solutions allowing security teams to assess characteristics of attacker behaviours, specific IoCs, or link multiple isolated events that may suggest a previous or current attack. Organisations usually connect to a threat intelligence or data feed via an Application Programming Interface and/or use their own intelligence. This aids with continuous update and iteration of threat intelligence and threat databases and the EDR's capability to detect and monitor potential unknown malware. However, against advanced attacks with new bespoke malware, some threat data may have little value.

Some EDR solutions use machine learning to detect malware strains, or those deliberately modified to avoid signature-based detection, that is, 'unknown malware' [97]. The aim is for machine learning algorithms and mathematical models to attempt to identify malware based on large data sets of known malware to identify particular characteristics and therefore detect and classify new strains.

EDR and fileless malware

As discussed earlier, increasing sophistication by attackers has led to fileless malware, which does not require any data writing to the hard drive [98] on first infection and subsequent propagation to other systems. A disadvantage for the attackers, however, is ensuring that the malware persists once the end point is rebooted or shut down, as anything in main memory (random access memory or 'RAM') is lost once the machine loses power. The attacker can either re-exploit the end point, which is time-consuming, or ensure that before shutdown occurs, the malware ensures it persists by writing itself to disk and becoming resident in memory again once the end point has rebooted. Once written to the hard disk drive, the defensive teams have an opportunity to detect the malware's presence if memory analysis is not present. This, however, provides just a fleeting opportunity for detection and highlights why memory resident malware is so effective.

7.4 Services

Larger organisations deploy or outsource a variety of services, notably MDR services, SOCs and CERTs, which we discuss in this section.

7.4.1 Managed detection and response

MDR stands for managed detection and response – an MDR service provider is an outsourced detection and response service.

There are multiple service providers offering MDR services, with varying models [99]. The approaches vary, from provision of detection and response using only EDR data, to combining network, log and EDR data. The former service presents a risk to organisations that rely solely on their outsource detection partner during an incident, as coordinating between network and log output, and the outsourced EDR provider in a real-time attack can present issues. These can be amplified further if the incident response team is outsourced, resulting in much slower response between all parties.

Depending on the EDR solution used, the MDR service can be more focused towards automated alerting or can consist of teams of experienced and specialist threat hunters in combination with more sophisticated EDR solutions on a 24/7 basis. The response capability when outsourced will be based on service-level agreements and scenario planning, which should revolve around the organisational risk appetite.

A comprehensive MDR service would incorporate data from network, end point and logs and use threat hunting techniques to carry out real-time analysis and bespoke alerting to organisations, based on their particular threat landscape and risk appetite.

7.4.2 Security operations centre

A SOC is an internal or outsourced function that carries out various operational security activities such as security monitoring, detection, analysis, reporting, vulnerability management and/or incident response. A SOC could encompass all of the functions discussed in this chapter or focus on a small set, such as basic logging, supported by outsourced services. SOCs can also provide key insights on the security posture of the organisation's infrastructure. A SOC can encompass most of these activities or have a narrowly defined function, such as threat hunting for advanced attacks only. As a centralised unit, SOCs use information from the organisation's infrastructure and from much of the technology explored in this chapter alongside external sources of information to investigate alerts, protect networks and derive intelligence.

7.4.2.1 Implementing a SOC

When designing a SOC, the input of telemetry data (e.g. logs, network data) are critical factors. For example, the telemetry that is pulled into the SOC or security components could prioritise the more critical assets, which may have a significant impact if compromised, including public safety, personal data breaches, system malfunctions, financial impacts and so on. However, one workstation could become a single point of failure if compromised, and as such, workstation telemetry can provide a wealth of information to detect and stop attacks at

Table 7.3 *Factors to consider in implementing a SOC*

	Key factors to consider
People	• Ensuring that people with relevant skill sets are deployed, for example, those with offensive knowledge as well as incident management and forensic experience
Process	• Ensure threat cases are considered that are specific to the organisation the SOC is operating within (or for). Use common frameworks (e.g., ATT&CK) and sharing standards • Ensure the SOC upholds relevant SOC standards and regulations such as SCADA (Supervisory Control And Data Acquisition) and STAR (Security Trust And Risk), and adhering to regulations such as PCI DSS (Payment Card Industry Data Security Standard), GDPR (General Data Protection Regulation), ISO standards etc. • Ensure processes and procedures are in place – e.g. Incident response plans, processes for forensic analysis • Have defined service-level agreements and key performance indicators with other members of security and the broader organisation • Ensure requirements are clearly defined
Technology	• Have in depth knowledge of the environment. Have complete visibility of assets and the environment and knowledge of the network and its systems to identify anomalous behaviour – Understand the environment, technology leveraged and other connectivity; for example; third-party connectivity, Internet connectivity, data centres, virtual private networks (VPNs) – Consider interoperability of systems • Ensure proactive monitoring and data capture of all events within the environment Ensure preventative tools are not operating in silos in order to improve data collection and visibility of unusual activity across a network [100]

earlier stages. Therefore it is important that telemetry is carefully considered and that the design of the SOC depends on the priorities and risk appetite of the organisation.

The following factors (Table 7.3) could be considered when implementing a SOC or procuring SOC services [101–103].

7.4.3 Computer emergency response teams

A computer emergency response (or readiness) team (CERT), or a CSIRT (computer security incident response team) is a specialist group that handles cybersecurity incidents, analyses threats and often act as a key stakeholder in sharing cybersecurity information. CERTs are typically country or sector based, but there are smaller-scale CERTs emerging on an organisational level. Box 7.6 describes the recent implementation of the EUROCONTROL EATM-CERT, a real-life example of work within a transport sector CERT.

Box 7.6 EUROCONTROL EATM-CERT

Author: **Patrick MANA**, EUROCONTROL, Cyber-Security Programme and EATM-CERT Manager

Among the many factors that go towards building cyber resilience into Europe's air traffic management (ATM) network, there are three in particular that need urgent consideration: the need to *change management culture* around managing cyberattacks, a willingness to *share information* on challenges and solutions and an understanding that *problems must be tackled well in advance of deployment.*

Europe's ATM system is under a constant threat of cyberattack from an operational point of view and from threats to the administrative network – bank transfers, fraud, confidential information that can be disclosed and so on. However, there is still an *illusion* among too many in the industry that the cyberthreat is not serious.

The EATM-CERT publishes quarterly an aviation cyberthreat landscape report for senior management, which lists cyberattacks on aviation systems/services and also on other domains that are using same software or products.

The focus of ATM has been traditionally on safety – building safety resilience on a solid experience of incident trends and analysis. But the cyberthreat is entirely different – what has happened in the past is entirely unrelated to what will happen in the future. When political tensions have flared between two countries, government-backed hackers have been able to attack the aviation infrastructure of their rivals in unprecedented ways – including disrupting the flight message boards in airports. The rate of cybercrime-related attacks is not linear but is increasing and with new types of attacks. Hacktivists will target more and more aviation especially because of aviation-negative impact on the environment/climate change.

The message we try to convey is 'open your eyes'. It will happen and you will never be 100% protected (we have no control on the type of attacks as mentioned above), but if you are cyber resilient, you will be able to respond quickly and minimise the damage. This can also be achieved by carrying out penetration tests where vulnerabilities of the systems are identified.

EUROCONTROL

In order to help the European aviation community in better managing the cybersecurity challenge, EUROCONTROL hosts the European Air Traffic Management Computer Emergency Response Team (EATM-CERT) providing services of common interest to aviation stakeholders of EUROCONTROL Member States.

EATM-CERT has a duty to promote sharing of cyber (security) information and does this through alerts sent by email or by pushing them into its MISP (Malware Information Sharing Platform), which is connected to CERT-EU, a ~40 strong group of national CERTs and aviation stakeholders. EATM-CERT shares cyber information not only with the Aviation ISAC but also with the European

(Continues)

Railway, European Energy ISACs and the Operational Technology ISAC. An important prerequisite to share cyber information is to establish trustworthy relationships. This can take time, and there are further complications faced, such as establishing a legal framework – a key challenge.

De-identifying or anonymising cyber information can help show that most of cyber information is mainly technical and can be shared without disclosing who was impacted. Most of the time, there is no added value in disclosing it. As an example, EATM-CERT was informed by a Civil Aviation Authority, an Air Navigation Service Provider and an Airport Operator that they were hit by an attack using an EMOTET Trojan. After analysis, EATM-CERT concluded that these attacks had the same IoCs and consequently sent an alert to all stakeholders of EUROCONTROL Member States, identifying the IoCs associated with the attack but not the organisation that was initially hit. Cyber information can be shared and 'sharing means caring'!

Finally, senders of information use traffic light marking to help limit the further distribution of information through appropriate classification.

EUROCONTROL also works with national cybersecurity agencies to make sure the European network meets the various national requirements. EUROCONTROL supports its Member States by providing cybersecurity-related policy, guidance material, workshop, training courses and a combined ATM/cyber expertise to national CERTs. This contributes to build relations based on trust and opens the door to share cyber threat intelligence (CTI) with national CERTs and aviation stakeholders, for example, advanced persistent threat targeting aviation, including their TTPs as domain of interest or critical infrastructures of a specific State including aviation. IoCs (e.g., malicious IP addresses or URLs, vulnerabilities, malware hash, etc.) are also shared.

Examples of EATM-CERT work – credential leaks

EUROCONTROL has started to test a cybersecurity service, which detects when credentials (email address and password) are known by hackers due to data leaks. Credentials and other information can be stolen by hackers and sold to other hackers or criminal groups. The financial gain may be significant to the original attacker (recently a hacker was selling 620 million credentials for 20,000 dollars) and of greater value to criminal groups.

How is this used by hackers?

It is well known that people reuse passwords for different logons, and emails have become the default username for many websites. This lowers the security of their logon credentials (email and password) to the least secure website that they use. When hackers obtain these credentials from a low-security website, they have everything they need to log onto other websites or even company networks. Hence, it is very important to educate staff and stakeholders of this and other basic security hygiene. Another example of how hackers use this information is through 'sextortion' – a campaign that hit some EUROCONTROL staff. Hackers send an email

(Continues)

Box 7.6 (Continued)

and put the recipient under threat by disclosing their password and making them believe they have access to their sensitive information. EATM-CERT is offering a service to detect the leaks and have signed up over 70 stakeholders (CAAs, airlines, airports and air navigation service providers). As our role is to protect the European ATM Network, we are intending to provide it to stakeholders for free, as part of EUROCONTROL's service. The rationale being that for aviation stakeholders, *we are as strong as the weakest link* as we are all interconnected. Hackers target the most vulnerable, which means targeting neighbours who have not joined up, so by offering this as a free service, we aim to make it as widely used as possible.

Phishing Protection:

The EATM-CERT team protects aviation stakeholders, for example, from route charge–related phishing attacks, or fraudulent attempts to obtain sensitive information such as usernames, passwords and financial details. We have already taken down domains responsible for hundreds of these fraudulent bank transfer attacks (see Table 7.4) and are extending this activity by working with Europol and conduct phishing awareness campaigns – especially to airline staff involved in route charge payment.

Data Leak Protection:

EATM-CERT also helps the aviation community to discover what sensitive documents have leaked and are known to the hackers, for example, electrical installation drawings of an airport, surveillance video camera system drawings of an air traffic control centre site, including the login and password to control the system and site map of an air traffic centre (Figure 7.2). The first lesson learned was that in 90% of the cases, the leak came from a supplier.

Another EATM-CERT activity will be to increase the level of information sharing (as automated as possible) with national CERTs and with ATM stakeholders, for example, through automated means to share cyber information such as using MISP (Malware Information Sharing Platform).

Table 7.4 Fraudulent domain names being suspended

Domain name	Domain closure: status	Attempts count
eurcontrolint.net	Suspended	50
eurocontroladmin.net	Suspended	29
euro-control-int.org	Suspended	13
euro-control.net	Suspended	8
eurocontrolint.net	Suspended	5
euro-control.org	Suspended	3
euro-controlinc.com	Suspended	2
eurocontrotint.net	Suspended	2
eurocontroint.net	Suspended	1
eurocontrolints.net	Suspended	1

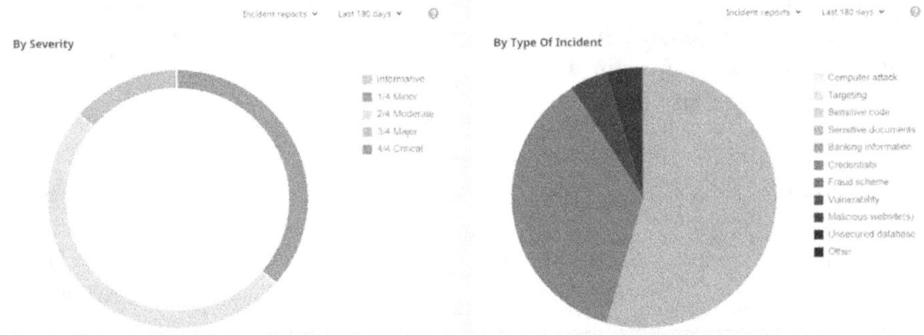

Figure 7.2 Detected leaks of sensitive documents

7.5 Conclusions

We have shown that organisations' networks and systems are permeable, and that while we can hope to reduce successful attacks, some will always get through. We therefore need tools to help us detect attacks in order to respond accordingly. We also know that if undetected, malware can reside on networks for months, so we need to be constantly monitoring networks for the presence of malware or malicious and anomalous behaviour. There is no single solution to this problem, but rather a set of strategies to be tailored to each organisation's risk and risk appetite.

The chapter covered how threat intelligence can help inform organisations of their risk profile, and how information on adversaries and the cyber threat landscape more broadly, can in turn inform how defensive technology is configured or leveraged in order to protect increasingly interconnected ecosystems from increasingly sophisticated attackers. The chapter covered how, through the use of modelling and common taxonomies, organisations can greater understand the threats facing them, and how there is great potential for organisations to increase the impact of their efforts by sharing threat data and intelligence.

We covered a subset of defensive technologies used for monitoring and detection purposes. While not an exhaustive list of technologies, the use of them is increasingly recognised as a core component of a defensive strategy. This is particularly the case when considering a defence in depth approach alongside traditional perimeter preventative controls.

We also addressed the 'arms race' issue; as we increase our defensive technologies, attackers also evolve their capabilities. For example, signature-based defences are useful but ineffective against more advanced attackers. While detection controls increasingly use machine learning and artificial intelligence, the human stays steadfastly in the loop. While these technologies can increase productivity and augment organisations' security postures, most of these systems still require active management by security teams and expert human analysis is still key.[§§]

[§§]CREST is an 'international not-for-profit accreditation and certification body that represents and supports the technical information security market'.

Of course, the monitoring and detection tools discussed in this chapter are only as good as they can enable security teams to respond to attacks in time. At some point, an attack will result in damage to systems and operations. Responding to attacks is the subject of the next chapter.¶¶

References

[1] National Academies of Sciences, Engineering, and Medicine. *Update of Security 101: A Physical Security and Cybersecurity Primer for Transportation Agencies*. Washington, DC: The National Academies Press; 2020.

[2] Ponemon Institute and IBM *The cost of a databreach report 2021*. IBM Security. Available from https://www.ibm.com/uk-en/security/data-breach

[3] *X-force threat intelligence index 2022 – IBM security*. Available from https://www.ibm.com/security/data-breach/threat-intelligence/ [Accessed 26 Apr 2022].

[4] U.S. Coast Guard Says Ryuk Ransomware Took Down Maritime Facility. Available from https://www.bleepingcomputer.com/news/security/us-coast-guard-says-ryuk-ransomware-took-down-maritime-facility/.

[5] Björck F., Henkel M., Stirna J., Zdravkovic J. 'Cyber Resilience - Fundamentals for a Definition'. *New Contributions in Information Systems and Technologies*; 2015. pp. 311–6.

[6] NIST Computer Security Resource Centre *Cyber resiliency*. Available from https://csrc.nist.gov/glossary/term/cyber_resiliency

[7] NCSC Annual Review 2021. 10 Steps to Cyber Resilience. Available at: https://www.ncsc.gov.uk/collection/ncsc-annual-review-2021/resilience/10-steps-to-cyber-resilience. Accessed on: 27/04/2022

[8] Metro Vancouver's transit system hit by Egregor ransomware. Available from https://www.bleepingcomputer.com/news/security/metro-vancouvers-transit-system-hit-by-egregor-ransomware/.

[9] Montreal's STM public transport system hit by ransomware attack. Available from https://www.bleepingcomputer.com/news/security/montreals-stm-public-transport-system-hit-by-ransomware-attack/.

[10] Ukwandu E., Farah M.A., Hindy H., *et al.* 'Cyber-Security challenges in aviation industry: a review of current and future trends'. *arXiv preprint*. 2021.

[11] Williams S. IOTW: Ransomware thieves publish major airlines' passenger information. 2021. Available from https://www.cshub.com/attacks/articles/iotw-ransomware-thieves-publish-major-airlines-passenger-information.

[12] Being Cyber Resilient Is Critical for the Maritime Industry. Available from https://www.tripwire.com/state-of-security/featured/cyber-resilient-critical-maritime-industry/.

[13] *Ground-breaking ResiCAV Project Highlights 'Urgent Need' for UK Road Transport Cybersecurity Programme*. Available from https://www.horiba-mira.com/media-centre/news/2020/05/12/ground-breaking-resicav-project/

¶¶https://www.crest-approved.org/index.html

[14] *DDoS attack trends for 2020. F5.* Available from https://www.f5.com/labs/articles/threat-intelligence/ddos-attack-trends-for-2020 [Accessed 26 Apr 2022].

[15] *Global threat report.* Available from https://www.crowdstrike.com/global-threat-report/

[16] Nallaperumal K. CyberSecurity Analytics to Combat Cyber Crimes. In 2018 IEEE International Conference on Computational Intelligence and Computing Research (ICCIC); 2018. pp. 1–4.

[17] *Microsoft digital defense report.* 2020 Sep. Available from https://blogs.microsoft.com/on-the-issues/2020/09/29/microsoft-digital-defense-report-cyber-threats/

[18] *NCSC Annual Review*. 2021. Available from https://www.ncsc.gov.uk/collection/ncsc-annual-review-2021/the-threat

[19] Al-Asli M., Ghaleb T.A. Review of signature-based techniques in antivirus products. In 2019 International Conference on Computer and Information Sciences (ICCIS); 2019. pp. 1–6.

[20] Going Beyond Malware: The Rise of "Living off the Land" Attacks. Available from https://www.crowdstrike.com/blog/going-beyond-malware-the-rise-of-living-off-the-land-attacks/.

[21] Kumar S. 'An emerging threat Fileless malware: a survey and research challenges'. *Cybersecurity*. 2020;3(1):1–2.

[22] 'ISO/PAS 22399:2007, Societal security - Guideline for incident preparedness and operational continuity management'.

[23] Mavroeidis V., Bromander S. Cyber threat intelligence model: an evaluation of taxonomies, sharing standards, and ontologies within cyber threat intelligence. In 2017 European Intelligence and Security Informatics Conference (EISIC); 2017. pp. 91–8.

[24] Von Solms R., Van Niekerk J. 'From information security to cyber security. computers & security'. 2013.

[25] Definition: Threat Intelligence. Available from https://www.gartner.com/en/documents/2487216/definition-threat-intelligence.

[26] What Is Threat Intelligence? Available from https://www.recordedfuture.com/threat-intelligence/.

[27] Intelligence cycle. Available from https://en.wikipedia.org/wiki/Intelligence_cycle.

[28] *What is cyber threat intelligence and how is it used?* Available from https://www.crest-approved.org/wp-content/uploads/2022/04/CREST-Cyber-Threat-Intelligence.pdf

[29] Kime B.P. 'Cyber threat intelligence support to incident handling'. 2017.

[30] Chismon D., Ruks M. *Threat intelligence: collecting, analysing, evaluating. MWR info security*. MWR Info Security. CERT-UK. CPNI. 2015. Available from https://www.foo.be/docs/informations-sharing/Threat-Intelligence-Whitepaper.pdf

[31] Trifonov R., Nakov O., Mladenov V. Artificial Intelligence in Cyber Threats Intelligence. In 2018 International Conference on Intelligent and Innovative Computing Applications (ICONIC); 2018. pp. 1–4.

[32] Building a Security Operations Centre. Available from https://www.ncsc.gov.uk/collection/building-a-security-operations-centre/threat-intelligence

[33] Trend Micro. Trend Micro - Indicators of Compromise. 2022. Available from https://www.trendmicro.com/vinfo/us/security/definition/indicators-of-compromise.

[34] Cyber Threat Intelligence Support to Incident Handling. Available from https://www.sans.org/reading-room/whitepapers/threatintelligence/cyber-threat-intelligence-support-incident-handling-38150.

[35] Tounsi W., Rais H. 'A survey on technical threat intelligence in the age of sophisticated cyber attacks'. *Computers & Security*. 2018;72(3): 212–33.

[36] Threat hunting. Available from https://resources.infosecinstitute.com/category/enterprise/threat-hunting/iocs-and-artifacts/threat-hunting-for-unusual-dns-requests/.

[37] Akarsh S., Sriram S., Poornachandran P., Menon V.K., Soman K.P. Deep learning framework for domain generation algorithms prediction using long short-term memory. In 2019 5th International Conference on Advanced Computing & Communication Systems (ICACCS); 2019. pp. 666–71.

[38] Threat hunting. Available from https://resources.infosecinstitute.com/category/enterprise/threat-hunting/.

[39] *The pyramid of pain*. Available from https://detect-respond.blogspot.com/2013/03/the-pyramid-of-pain.html [Accessed 25 Oct 2022].

[40] Cyber threat hunting. Available from https://en.wikipedia.org/wiki/Cyber_threat_hunting.

[41] What Is Cyber Threat Hunting? Available from https://www.mcafee.com/enterprise/en-gb/security-awareness/operations/what-is-cyber-threat-hunting.html.

[42] IOA VS IOC. Available from https://www.crowdstrike.com/blog/indicators-attack-vs-indicators-compromise/.

[43] Available from https://www.ncsc.gov.uk/collection/building-a-security-operations-centre/threat-intelligence [Accessed 25 Oct 2022].

[44] *The hacker and the state. cyber attacks and the new normal of geopolitics*. Harvard University Press; 2020. pp. 1–9.

[45] Liang Q., Xiangsui W. 'Unrestricted warfare'. 1999.

[46] *Microsoft digital defense report*. 2021. Available from https://www.microsoft.com/en-us/security/business/microsoft-digital-defense-report

[47] Proctor P. *How geopolitics impacts the cyber-threat landscape. Gartner*. Available from https://www.gartner.com/en/newsroom/press-releases/2022-06-10-how-geopolitics-impacts-the-cyber-threat-landscape

[48] Ghafur S., Kristensen S., Honeyford K., Martin G., Darzi A., Aylin P. 'A retrospective impact analysis of the Wannacry cyberattack on the NHS'. *NPJ Digital Medicine*. 2019;2(1):98.

[49] Casey T. Threat agent library helps identify information security risks. Intel White Paper; 2007.

[50] Casey T. Understanding cyber threat motivations to improve defense. Intel White Paper; 2015.

[51] Perlroth N., Sanger D., Shane S. How Chinese Spies got the NSA's Hacking tools, and used them for attacks. New York Times; 2019.

[52] Threat Modeling. Available from https://www.owasp.org/index.php/Application_Threat_Modeling.

[53] Unified Modelling Language. Available from https://www.uml.org.

[54] Marback A. 'A threat model-based approach to security testing. Software - Practice and Experience'. *Software - Practice and Experience*. 2013;43(2):129–258.

[55] Ma Z., Schmittner C. 'Threat modeling for automotive security analysis. 2016'. *Advanced Science and Technology Letters*. 2016;139:333–9.

[56] NCSC. *Penetration Testing - advice on how to get the most from penetration testing*. Available from https://www.ncsc.gov.uk/guidance/penetration-testing

[57] Understanding Threat Modelling. Available from https://www.digitalshadows.com/blog-and-research/understanding-threat-modelling/.

[58] Hernan S., Lambert S., Ostwald T., Shostack A. 'Uncover security design flaws using the STRIDE approach'. *Threat Modeling*. 2006.

[59] Threat Modeling with Stride. 2014. Available from https://users.encs.concordia.ca/~clark/courses/1601-6150/scribe/L04c.pdf.

[60] Microsoft. *Threat modeling*. Microsoft. Available from https://www.microsoft.com/en-us/securityengineering/sdl/threatmodeling

[61] Do Q., Martini B., Choo K.-K.R. 'The role of the adversary model in applied security research'. *Computers & Security*. 2019;81(8):156–81.

[62] Moskal S., Yang S.J., Kuhl M.E., Shanchieh. 'Cyber threat assessment via attack scenario simulation using an integrated adversary and network modeling approach'. *The Journal of Defense Modeling and Simulation: Applications, Methodology, Technology*. 2018;15(1):13–29. Sage Publications.

[63] Sonalker A., Griffor E. 'In Handbook of system safety and security'. *Evolving Security*. 2017.

[64] *Air gap (networking). Wikipedia*. Available from https://en.wikipedia.org/wiki/Air_gap_(networking)

[65] *The cyber kill chain*. Lockheed Martin. Available from https://www.lockheedmartin.com/en-us/capabilities/cyber/cyber-kill-chain.html

[66] How cyber attacks work. Available from https://www.ncsc.gov.uk/information/how-cyber-attacks-work.

[67] Why the 'cyber kill chain' needs an upgrade. Available from https://www.networkworld.com/article/3104542/why-the-cyber-kill-chain-needs-an-upgradesecurity-pros-need-to-focus-more-on-catching-attackers-aft.html.

[68] MITRE ATT&CK. Available from https://attack.mitre.org.

[69] Strom B.E., Battaglia J.A., Kemmerer M.S. Finding cyber threats with ATT&CK-based analytics; 2017.

[70] Enterprise Matrix. Available from https://attack.mitre.org/matrices/enterprise/.

[71] Murali G., Pranavi M., Navateja Y., *et al.* 'Network security scanner'. *International Journal of Computer Applications in Technology*. 2011;2(6): 1800–5.

[72] Kamhoua C., Martin A., Tosh D.K., Kwiat K.A., Heitzenrater C., Sengupta S. Cyber-threats information sharing in cloud computing: A game theoretic approach. In 2015 IEEE 2nd International Conference on Cyber Security and Cloud Computing; 2015. pp. 382–9.

[73] *Minimum cyber security standard.* UK Government. 2018. Available from https://www.gov.uk/government/publications/the-minimum-cyber-security-standard

[74] Directive (EU) 2016/1148 of the European Parliament and of the Council of 6 July 2016. 'Directive (EU) 2016/1148 of the European Parliament and of the Council of 6 July 2016 concerning measures for a high common level of security of network and information systems across the Union'. *Official Journal of the European Union.*

[75] The National Cyber Security Centre. Available from https://www.ncsc.gov.uk/collection/caf/cyber-assessment-framework/caf-objective-c-detecting-cyber-security-events.

[76] *Proactive security event discovery.* UK Government NCSC. Available from https://www.ncsc.gov.uk/collection/caf/caf-principles-and-guidance/c-2-proactive-security-event-discovery

[77] *The national cyber security centre.* Available from https://www.ncsc.gov.uk/collection/building-a-security-operations-centre/onboarding-systems-and-log-sources/log-sources

[78] Yaacoub J.P., Noura H.N., Salman O., Chehab A. 'Digital forensics vs. Anti-Digital forensics: techniques, limitations and recommendations'. *arXiv preprint*. 2021.

[79] National cyber security centre. Available from https://github.com/ukncsc/lme.

[80] Nawyn K.E. 'A security analysis of system event logging with syslog. SANS Institute, no. as part of the information security reading room'. 2003.

[81] 7 tips for configuring a robust SIEM. Available from https://www.avertium.com/blog/siem-configuration.

[82] Chiradeep BasuMallick *What is network time protocol (NTP)? meaning, working, benefits, and challenges*. Spiceworks Inc. Available from https://www.spiceworks.com/tech/networking/articles/what-is-network-time-protocol/

[83] Wireshark. Available from https://www.wireshark.org/.

[84] NetworkMiner. Available from https://www.netresec.com/?page=networkminer.

[85] The calculations: 10Gbit/s wirespeed. Available from http://netoptimizer.blogspot.com/2014/05/the-calculations-10gbits-wirespeed.html.

[86] Security Onion 2. Available from https://securityonionsolutions.com/.

[87] 'WEHOWSKY.COM. MUNINN & FREKI Technical White Paper'.

[88] Protocol stack. Available from https://en.wikipedia.org/wiki/Protocol_stack.

[89] Server Message Block. Available from https://en.wikipedia.org/wiki/Server_Message_Block.

[90] Denning D.E. 'An Intrusion-Detection model'. *IEEE Transactions on Software Engineering*. 1987;SE-13(2):222–32.

[91] Sarker I.H., Abushark Y.B., Alsolami F., Khan A.I. 'IntruDTree: a machine learning based cyber security intrusion detection model'. *Symmetry*. 2020;12(5):754.

[92] Bayer U., Comparetti P.M., Hlauschek C. 'Scalable, behavior-based malware clustering'. *NDSS*. 9; 2009. pp. 8–11.

[93] Fraley J.B., Cannady J. 'The promise of machine learning in cybersecurity'. *In SoutheastCon Mar*. 2017.

[94] Gartner. 'The evolution of endpoint protection. Symantec'. *Symantec*. 2018.

[95] Persistence. Available from https://attack.mitre.org/tactics/TA0003/.

[96] Peter Lund *How to create an EDR/MDR alternative for OT systems*. Industrial Defender. 2021. Available from https://www.industrialdefender.com/blog/how-to-create-edr-mdr-alternative-for-ot-systems

[97] Wolsey A. 'The state-of-the-art in AI-based malware detection techniques: A review'. *ArXiv:2210.11239*. 2022.

[98] Saad S., Mahmood F., Briguglio W., Elmiligi H. A tale of a fileless javascript memory-resident malware. In International Conference on Information Security Practice and Experience; 2019. pp. 113–31.

[99] CrowdStrike. *What is managed detection and response (MDR)?* CrowdStrike. 2022. Available from https://www.crowdstrike.com/cybersecurity-101/managed-detection-and-response-mdr/

[100] Rethinking response: What does it take to stop attacks? Available from https://www.f-secure.com/content/dam/f-secure/en/business/campaign-mdr/collaterals/digital/f-secure-whitepaper-rethinking-response.pdf.

[101] Kokulu F.B., Soneji A., Bao T, *et al.* 'Matched and mismatched socs - A qualitative study on security operations center issues'. *In Proceedings of the 2019 ACM SIGSAC Conference on Computer and Communications Security*; 2019. pp. 1955–70.

[102] Ahmad A., Maynard S.B., Desouza K.C., Kotsias J., Whitty M.T., Baskerville R.L. 'How can organizations develop situation awareness for incident response: A case study of management practice'. *Computers & Security*. 2021;101:102122.

[103] European Union Agency for Cybersecurity *How to set up CSIRT and SOC*. 2020. Available from https://www.enisa.europa.eu/publications/how-to-set-up-csirt-and-soc

Chapter 8
Technical response and correction

David Damato[1], Jadee Hanson[2], and Andy Jones

8.1 Introduction

8.1.1 Context

The preceding chapters focused on information security controls to prevent and detect attacks. For many years, most security teams focused exclusively on prevention and detection to mitigate the risk of a successful attack, believing that major incidents were avoidable. As the number of organisations experiencing major incidents increased, particularly in the last decade, the security industry realised (a) that breaches were no longer avoidable but inevitable and (b) that although incidents will occur, the risk of major disruptions to the business can be managed by executing an effective incident response. In the transport sector, there is extensive experience of failure and even the most comprehensively risk-reduced operations can lead to severe impacts if the fallout is not managed. For example, a short-term outage in air traffic control (ATC) systems can cause days of disruption should there be no incident management. Hence, the aim of this chapter is to provide an overview of the fundamental components required to execute a successful incident response.

8.1.2 What is an incident?

There is no single definition for a computer security incident. It is, however, important to define incident response for the organisation to ensure staff can execute an effective response. There are several generally accepted definitions for an incident. NIST 800–61 [1] defines an incident as a 'violation or imminent threat of violation of computer security policies, acceptable use policies or standard security practices'. A more comprehensive definition [2] is 'any unlawful, unauthorised or unacceptable action that involves a computer system, cell phone, tablet

[1]Gemini, New York, NY, USA
[2]Code 42, Minneapolis, MN, USA

and any other electronic device with an operating system or that operates on a computer network'.

For the purpose of our discussion in this chapter, we define an incident as the unlawful, unauthorised or unacceptable event that involves an electronic device or network. This definition is a simple one and includes network communications and systems, as commonly found in the transport sector. Examples of incidents include, for example, malware that steals data, disruption of wireless network communications or theft of a laptop.

8.1.3 Types of incidents

Although there are many types of incidents, there are some common characteristics that can be used to inform better incident response planning and to enable an effective threat intelligence programme. The Vocabulary for Event Recording and Incident Sharing (VERIS) is one example of a framework that provides a common language for describing security incidents by

- actor
- actions
- attributes

8.1.3.1 Actor

An 'actor' is an entity that is responsible for the creation of an incident. The actions of actors can be intentional or unintentional, malicious or benign. Although not exhaustive, there are a few different types of actors that should concern the transport industry. VERIS categorises actors as external, internal or partner.

- External actors include those threats that initiate actions from outside the organisation. Examples include a natural event, nation states or organised crime.
- Internal actors are those originating from inside an organisation. Employees and contractors are included in this group.
- Partner actors include trusted third parties with which an organisation has a business relationship. Examples include vendors, suppliers and cloud providers.

The transport industry is impacted by external, internal and partner actors. When building an incident response programme, it is important to understand industry-specific threats. The following paragraphs give some examples of relevant threats to the transport industry.

State-sponsored actors

There are many reasons that nation states such as the United States, Russia or China would target other states, and this includes direct or indirect impact on the transport industry. The most obvious is in order to influence nations or execute a pre-emptive strike before a conventional war. In the spring of 2017, Russian state-sponsored hackers* took control of update servers for a software solution from the Ukrainian-based Linkos Group, called ‘M.E.Doc’ [3]. The malicious update pushed Russian malware to M.E.Doc installations at global companies. The end result was the destruction of millions of computer systems, including those not targeted, such as transportation giants Maersk and FedEx’s European subsidiary TNT Express. Although uncommon, these examples provide a possible blueprint as to how future wars may be fought.

Other nation states have also targeted organisations for intelligence purposes. Organisations such as Mandiant track Russian, Chinese and other nation state adversaries specifically target transportation, logistics and shipping [4]. Several large airlines and booking systems [5] were targeted by Chinese state-sponsored hackers in order to obtain access to passenger travel details. Although not destructive, these attacks impact national security and result in sizable and expensive responses.

Financially motivated criminals

Financial criminals are often opportunistic; motivated by money they search for vulnerable organisations. In September 2018, a group of criminals exploited vulnerability in Bristol airport’s computer systems in order to disable information screens [6]. The criminals used ransomware to disable the systems and demanded an undisclosed sum of money to restore operations. The incident resulted in confusion and the airport chose to rebuild the systems rather than pay the ransom – taking about two days to restore operations.

Terrorists/Hacktivists

Terrorists and hacktivists are motivated by a cause. That cause may be political, social or economic and include everything from damaging the reputation of an executive to the destruction of control systems operating a dam. A popular example that straddles the line between terrorism and activism is the Sony attack. In November 2014, a group of North Korean hackers destroyed the computer systems at Sony Pictures, in an attempt to coerce the studio into scuttling the release of a film that mocked the North Korean regime.

8.1.3.2 Actions

The actions available to attackers that cause or contribute to the incident could range from installing malware to socially engineering a user to expose their password. VERIS categorises the types of actions an actor may take as malware, hacking, social, misuse, physical, error and environmental

*‘Hackers’ are used in the sense of unauthorised use of computers but note that legitimate security researchers are also referred to as ‘white hat’ or ‘ethical’ hackers.

- ‘Malware’ is malicious code that is executed on a system. Examples include viruses, ransomware, web-shells and backdoors. Most attacks include at least some use of malware to facilitate an attack.
- ‘Hacking’ is a fairly broad category that includes any illegitimate attempts to access or impair a system. This could include SQL injection, distributed denial of service (DDoS) attacks, using malware or password guessing.
- ‘Misuse’ is similar to hacking but assumes the user already has legitimate access to a system.
- ‘Social’ includes those efforts designed to deceive or manipulate humans for malicious purposes. This includes phishing, blackmail and scams.
- ‘Physical’ includes physical events such as theft, sabotage or assault.
- ‘Error’ includes misconfigurations, malfunctions or mistakes. An example would be an administrator who configures a website insecurely, without any malcontent.
- ‘Environmental’ includes natural events such as earthquakes, floods or hurricanes.

8.1.3.3 Attributes

Actors always have an end goal or mission. These goals are referred to by VERIS as attributes. There are three attributes that all attacks can share, including the compromise of confidentiality, integrity and availability.

- Compromising confidentiality refers to an actor’s ability to observe an asset or data. For example, when a criminal steals credit card records online, it impacts the confidentiality of those records.
- Compromising integrity refers to an actor’s ability to make unauthorised changes to an asset or data. Examples include modifying a database record.
- Compromising availability refers to an actor’s ability to impact the desired accessibility of an asset and data. For example, when ransomware encrypts files, preventing legitimate access.

8.1.4 Incident response

The primary mission of a security leader is to reduce security risks to the business. NIST 800–30 [7] defines risk as ‘a measure of the extent to which an entity is treated by a potential circumstance or event and is typically a function of: (i) the adverse impacts that would arise if the circumstance or event occurs and (ii) the likelihood of occurrence’.

Every risk has an assigned ‘probability’. Sometimes this is called ‘likelihood’ or ‘frequency’. Regardless of the term, it’s clear that risks are generally not 100% avoidable. As a result, organisations must take steps to reduce either the frequency or impact of a risk.

Incident response is a documented and practiced process for managing risk, by rapidly transitioning from detection of an incident to remediation in order to minimise the impact of an incident to the business. This is especially important in the transport industry, where disruptions could easily endanger human safety or life.

8.1.4.1 Phases of an effective incident response

In the remainder of this chapter, we describe three primary phases required to perform an effective incident response:

1. preparation to ready for an incident
2. analysis and investigation into the cause and characteristics of the incident
3. remediation of the incident in order return operations to a normal state, while mitigating the risk or a re-compromise.

8.2 Preparation

An effective response to an incident must be correct and rapid. Failures or hesitation can increase the scope, extend the timeline or more deeply impact operations. As a result, preparing for an incident is arguably the most important part of incident response.

8.2.1 Incident handling policy

When forming an incident response programme, the first step is to define a policy that outlines the appropriate practices to identify, track and remediate security events and incidents that may impact systems and operations. This should include the definition of an incident, responsibility for reporting an incident, incident categorisation, and broad roles and responsibilities.

8.2.2 Definition of an incident

As stated in the introduction, defining the term 'incident' is important. Without it, the organisation will struggle to understand and execute an effective response.

8.2.3 Incident categorisation

Incidents can vary widely in their scope and severity. As will be discussed throughout this chapter, an organisation's response to incident is proportional to the risk it represents. And as a result, it is important to understand the characteristics that drive risk and define a serious incident versus a trivial incident.

The most significant driver for risk is the functional impact of the incident. For example, port scans or the occasional phishing attack may have little to no impact on the business. However, an attack that results in the disclosure of confidential data or results in the downtime of revenue generating systems will have a high-functional impact.

Table 8.1 Characteristics of minor and major incidents

Minor incident	Major incident
• Common or occasional event • Spam, isolated virus infection where individual clients are infected • Software or system misconfiguration • Security product creating instability in user services and devices	• Infrequent to rare event • Widespread malware, where more than some percentage of employees are impacted • A threat that is actively moving between devices/systems (i.e. lateral movement) • Effect on company's external reputation • Access to or theft/destruction of personal or confidential information • Reported by law enforcement or a credible third-party against a critical system

The second driver is the observed activity associated with an incident. This includes the type of attack and the systems impacted. Observed activity such as unsuccessful attempts to guess a password represents a very low risk. However, a more serious incident might be characterised by an attacker that has stolen an administrator's password and is moving from system to system.

Table 8.1 describes a non-exhaustive list of those characteristics that distinguish a minor and major incident.

Although organisations will most likely develop their own incident severity scale, many frameworks exist that may be helpful. This includes NIST 800–61 [8], which prioritises incidents based on functional impact, information impact and recoverability. The US National Cybersecurity and Communications Integration Centre (NCCIC) goes a step further, having created a Cyber Incident Scoring System [9], which maps to NIST 800–61. These and other frameworks represent may be used to distinguish what separates a serious incident from a minor incident, which will become a more important concept as this chapter delves into the incident response process.

8.2.4 Responsibility for reporting an incident

Many incidents are reported by employees. However, without a clear process or expectation, employees may not know how to effectively communicate or raise awareness about a potentially serious incident. As a result, it is very important to ensure an incident handling policy that clearly defines the mechanism and circumstances for communicating an incident, via phone, email or a ticketing system.

8.2.5 Roles and responsibilities

If and when an incident is reported, an organisation must understand the roles and responsibilities of different departments and other organisations. An incident handling policy should define high level roles and responsibilities of those directly responsible for an incident. Generally, this includes security, legal and communications. Organisations with periphery roles are often documented in the incident response plan.

8.2.6 Incident response plan

An incident handling policy establishes appropriate governance. An incident response plan describes how the incident response team will be organised and the procedures they will take in the event of an incident. The most effective plans will be cross-functional and will include a lead from each of the involved departments across the company. It is not uncommon for legal, customer support and public relations to be overlooked when building an incident response team. These three groups represent an important part of the team and will have a large role in the event of an incident.

8.2.7 Incident response team

Major incidents typically require a diverse team to investigate, remediate and recover from an event. An incident coordinator or commander typically leads this effort. Arguably the most important resource, the incident coordinator is responsible for the following:

- understanding an overview of the entire incident
- assembling the incident response team and delegating responsibilities
- drive and track timely completion of incident response tasks
- remove blockers that prevent execution of defined tasks
- create and issue periodic updates to the response team and senior stakeholders.

Incident handlers are responsible for gathering, preserving and analysing evidence at the direction of the incident coordinator. This will likely include employees with special skills including security and forensic analysts. Security analysts are skilled in understanding the scope of an incident, whereas forensic analysts are focused on the ability to recover key artefacts and maintain integrity of evidence in order to ensure a forensically sound investigation. In many organisations, this may be the same employee or team. Incident handlers may work with IT, system owners or other staff to remediate affected systems.

IT or engineering staff are responsible for assisting incident handlers in the gathering or preservation of evidence and to aid in the remediation of affected systems or applications. Many organisations consider IT or engineering to be part of an extended team – this is a mistake. Integrating the response and remediation teams will result in better communications and a faster overall recovery.

8.2.8 Extended incident response team

The extended incident response team is typically engaged in the most severe incidents, where external communications with customers or regulators may be required. These individuals may include general council, public relations, human resources, customer support and so on. Based on the respective industry, these extended incident response team members may vary. What is important is to identify these individuals prior to an incident taking place and ensure that they are briefed as to their responsibilities in the event they are pulled into a response situation for a significant incident.

Human resources will assist where an employee is involved in an incident. Similarly, incidents involving an insider or employees' personal devices may require specialised HR expertise.

A customer support organisation may be responsible for interfacing with customers and/or providing support to the incident response team as 'Subject Matter Experts'. Often a phone call to a customer, by a known support representative, can be extremely helpful when attempting to communicate an incident in a more personal manner.

The legal team is responsible for all legal concerns, including those related to the handling and communication of an incident. Key responsibilities include the following:

- providing advice on multi-jurisdictional notification issues, as they relate to an incident requiring disclosure
- reviewing all external communications
- interfacing with legal requests from government entities, as they relate to an incident.

The Public Relations and Communications team, in cooperation with legal counsel, is responsible for facilitating communications with customers and the media. In particular, they are responsible for the following actions:

- coordinating public relations, where required, to publicly communicate an incident to an organisation or individuals, other than customers
- reviewing and communicating an incident to customers

Computer security incidents occasionally occur through breaches of physical security or involve coordinated cyber and physical attacks. The incident response team also may need access to facilities during the incident handling process, for example, to acquire a compromised workstation from a locked office.

End users are responsible for cooperating with the response team and aiding the investigation. Additionally, end users are responsible for reporting any potential security incidents and security weaknesses in accordance with the defined incident handling policy.

8.2.9 Playbooks

Responding to incidents can take many paths. More frequent events and incidents should have standard 'playbooks' defined. It is also important that more than one analyst be trained on these playbooks. Incidents involving malware being delivered to an end user via a phishing link is something that is common and should have a set playbook. Other incidents that are more targeted and sophisticated may require a number of playbooks and may also require action be taken based on the direction of the incident response lead. Playbooks should be documented with simple steps and should be easy to follow. These playbooks should be understood by the incident response team and should be kept confidential; if an attacker could access them, they would have a distinct advantage over the response team.

8.2.10 Supporting documentation

When an incident occurs, it is extremely helpful to have direct access to all necessary supporting documentation so as to facilitate an effective response. To this end, there are some common documents/data that organisations should create and maintain, including:

- Contact information and on-call: a document containing contacts for key staff and an escalation procedure to ensure an incident can be raised to the right person, within an acceptable time period.
- Relevant third-party contracts: a list of all customer contracts and any unique terms that may impact the response (e.g. reporting procedures).
- Reporting requirements: a collection of applicable regulatory reporting requirements based on the data the organisation maintains and countries in which it operates (e.g. relevant to the European Unions' General Data Protection Regulation (GDPR)).
- Network diagrams and asset inventory: during an incident it will be important to understand network connectivity, connected devices and metadata associated with such devices. Maintaining an updated diagram and inventory will help speed an investigation and remediation.

8.2.11 Technology

In addition to people and process, successful incident response requires access to a range of technology. This includes workflow systems to assist with tracking and measuring incidents, investigative tools that speed analysis and remediation solutions to quickly recover from an incident. These technologies should exist and be ready before an incident.

8.2.12 Workflow technology

Many incidents are missed or mishandled because of a poor workflow or case management. Incident handling solutions, whether a customised ticketing system or purpose built case management system, are effective tools to drive consistent, efficient and measurable response to incidents. An effective ticketing or case management system should enable the following:

- Track all events to ensure that none are missed/not acted upon.
- Creation of metrics, such as average response time and incident type/volume, in order to enable continuous improvement.

- Customisable workflow to guide analysts through a repeatable and efficient process.
- Maintain outcomes of incidents in a manner that is easily referenced by other analysts, which can dramatically help reduce duplication of effort/ investigation research for similar incidents.
- Enrichment of events with forensic data, limiting the time an analyst must spend pulling data from other systems, in order to complete their response.

8.2.13 Investigative technology

During an investigation, analysts will need access to data to discover evidence. This includes having access to tools that can collect and store evidence. Data collection is generally performed by investigative tools that can extract forensic artefacts from systems. This includes live response scripts, forensic imaging and Endpoint Detection and Response (EDR) products, all of which are discussed later in the chapter. When selecting such solutions, it is recommended to prioritise speed, accuracy and ease of use; as these requirements will matter when attempting to quickly investigate an incident at scale.

Historical access to artefacts is also important. Many incidents start well before they are detected or are only identified in light of new threat intelligence. As a result, the ability to historically search forensic information is critical. Logs represent the most basic data, which must be available for responders. Meaningful data within logs should be collected centrally, prior to an incident. At a minimum, this should include logs containing details about system and application authentication, Domain Name Service (DNS)[†] resolution and Dynamic Host Protocol Service (DHCP)[‡] leases, network traffic flows and Network Address Translation (NAT)[§] or load balancer address translations.

8.2.14 Remediation technology

Most organisations take months to recover from a major incident. Surprisingly, the bulk of that recovery effort is not associated with the investigation but rather IT changes required for successful remediation and recovery. Technology can help speed that process, by scaling valuable IT and security resources that must plan and execute significant changes across an enterprise. At a minimum organisations should have technology that can provide the following:

- An asset inventory that contains accurate and timely information about all devices and applications, along with system owners, installed software/version information and baseline configurations.

[†]Domain Name Service (DNS) is an Internet standard that allows us to resolve names (e.g. Google.com) to an Internet Protocol (IP) address.
[‡]Dynamic Host Protocol Service (DHCP) is a service that provides IP addresses to computers when they access a network.
[§]Network Address Translation (NAT) allows many computers to access the Internet through just one publicly routable IP address.

- A system or solution for identifying and categorising data across all systems, for structured and unstructured data.
- Device management solutions that allow for fast and scalable deployment of new security configurations or software across all devices.
- Security solution that can provide near real-time prevention and/or alerting of custom indicators that can characterise a range of attacks.
- The ability to control network access, via host and network firewalls.

8.2.15 Training

A healthy approach for organisations is to conduct full-scale tabletop or war game exercises on a continuous basis. A tabletop exercise (TTX) is a disaster preparedness activity that takes participants through the process of dealing with a simulated disaster scenario. A wargame is a cybersecurity challenge and 'mind sport' in which the competitors must exploit or defend a vulnerability in a system or application, or gain or prevent access to a computer system. Both techniques allow an organisation to practice their preparedness and identify gaps with the expected outcome. TTX and wargames are also an excellent opportunity to improve engagement and build relationships with senior leadership. Such experiences can make leadership more acutely aware of information security and the challenges associated with detection and response of threats.

TTX or wargames can be led by internal teams or with the help of external firms. For those in the UK interested in leading their own TTX, the National Cyber Security Centre offers an online tool called Exercise in Box 8.1 [11], which provides exercises companies can use to assess their response to common cyberattacks.

8.2.16 Reputation

An organisation's reputation will play a significant factor in how customers, regulators and the general public respond to an incident. Despite the best executed incident response, companies with a poor reputation will receive more scrutiny and may incur greater fines. A good example is Facebook. After a series of privacy missteps, the company's reputation among the general public and regulators suffered. Any incident, regardless of size or how it was handled, was closely scrutinised by regulators and the public alike. Airlines, credit agencies and other industries with a poor reputation are also at risk. When preparing for an incident, organisations should consider how they can improve their reputation overall and as it relates to security.

8.3 Incident analysis and investigation

Every police investigation starts with an initial lead. Sometimes that lead provides little help, like an article of clothing left at a crime scene. Other times it can be more helpful, such as when fingerprints are left behind. The response to an initial lead will likely set the tone for the entire investigation. As a result, it is important to start this process right.

Box 8.1 Example of incident communications

On 25 September 2018, Facebook engineers confirmed the theft of user data. A portion of code used by Facebook contained a vulnerability that attackers used to take over user accounts, in the process gaining access to sensitive information. Because the data were covered by GDPR and there was the potential for immediate harm to users, Facebook had to report the breach quickly. This meant that investigative activities would not be completed, making it difficult to estimate the number of users who were impacted. Despite the accelerated time frame and an already battered reputation, Facebook's team produced a well-coordinated announcement, which limited the impact to the company. On 28 September, Facebook announced [10] the breach via their website. The announcement is a good example of how to communicate with customers about a breach, including the following key points:

- Although the investigation had not been completed, Facebook understood from initial analysis that no more than 50 million users had been impacted. As a result, they reported that maximum number and later revised the number to 30 million. By ensuring the number would not increase, the team avoided a potentially damaging scenario while adhering to regulatory requirements, such as GDPR.
- The team inspired confidence. The announcement included a detailed overview of the cause of the incident and described how Facebook had already taken steps to address the vulnerability and prevent further data loss.
- Facebook took ownership of the incident and was honest about the stage of their investigation, including that they were still working to determine who was behind the theft and what, if any, information was misused.
- The announcement did not include the hackneyed generalisation – 'we take security seriously' – that can be seen as disingenuous, especially during a breach. Refrain from using this or similar terms.

Investigations of reasonable size will typically progress through five stages, starting with an event triage, which performs analysis of an event to determine if an incident should be declared. Assuming an incident is declared, a larger investigation occurs – continuously iterating through adjustment of the incident's scope, analysis of evidence, forming conclusions based on analysis and developing actions based on such conclusions.

8.3.1 Event triage

An overview of the triage process is shown in Figure 8.1. Using a hospital emergency room (ER) as an example, the first person responding is likely a triage nurse. He or she would likely have asked you questions about your symptoms, the severity of pain and your medical history in order to rapidly assess your condition and prioritise your treatment. They might even order tests to validate the information you provided. A security analyst responsible for investigating a security event has similarities to a triage nurse. Their goal is to obtain enough information about an event to take informed and deliberate action. Failure to appropriately triage an incident may result in a false positive or a false negative – the incorrect classification of an incident as a false alarm. As a result, triage is an incredibly important process and the most frequent source of errors that contribute to a much more serious breach or cause a team some embarrassment.

8.3.1.1 Effective triage

Similar to the medical field, the best way to ensure fast and accurate prioritisation of events is to ensure qualified analysts follow a standardised and repeatable process to collect and analyse data required to triage an event.

Data collection requires a consistent set of information that is included when responders are alerted to a potential incident. Such data are typically captured in a ticketing system, which responders use to manage incident data and follow a consistent workflow. The speed at which responders can triage an event is directly proportional to the level of effort an analyst must expend to obtain data required to complete their analysis. The following checklist includes data that should be collected as part of logging a manual or automated event:

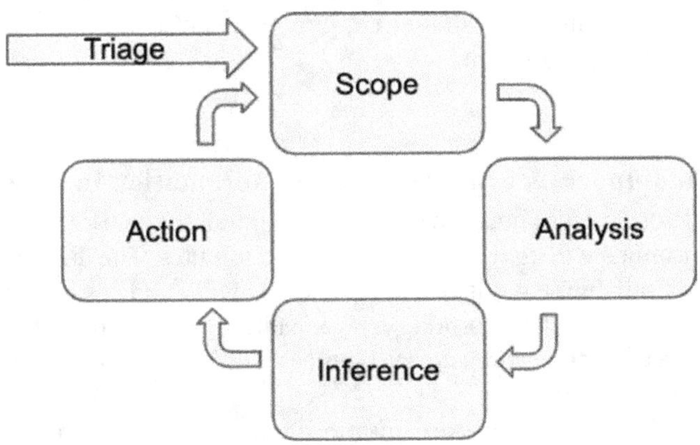

Figure 8.1 Event triage process

- Reporting a user or system and associated contact information. This may include items such as IP address, hostname, function, interconnections and other information that should already be part of the organisation's asset management database.
- Date and time of all relevant details, including incident start, activity associated with the event and actions that may have already been taken.
- Details about the incident and supporting information from relevant systems. This typically includes event logs, network traffic logs, system telemetry and information from security tools that can help recreate activity that occurred around the time of the event.

Once the analyst has access to sufficient data, they can begin analysis of the event. The analysis should attempt to answer some basic questions, which will dictate an eventual response. The following is a basic list of questions should be answered by analysing the initial data collected.

- Does the event meet the organisation's definition for an incident? For example, non-malicious spam is unlikely to be an incident, while an email attempting to steal user credentials would be declared an incident.
- How did the incident start? For example, what was the date and time a user accessed a malicious file in an email. If the start is unknown, the incident may need to be escalated for a more detailed investigation.
- Was there any lateral movement – either to or from this system – associated with this incident? If malware was pushed to or from this system, it would indicate a larger incident that requires a larger investigation.
- Is there any sensitive data on impacted systems that may dictate a specific response? The presence of specific types of data may indicate the need to take certain actions or retain evidence differently.
- Is the incident possibly ongoing? An investigation may not even be required, if antivirus blocked malware, the start of the event is well understood and no other malicious activity is present.

8.3.1.2 The importance of visibility and automation in triage

Most organisations require hours to complete the initial triage. High functioning and mature organisations will complete this process in minutes. The difference is in how much data are available (e.g. visibility) and how quickly and easily that data can be aggregated for an analyst performing triage. Mature organisations have rich logs, network captures and other forensic data sources that can be provided to analysts the moment a ticket is issued.

Another difference in mature organisations is the use of automation. Automation can help complete previously manual actions so that analysts have access to enriched data without having to take time-consuming actions. Automation also ensures that data collection is performed consistently, reducing potential errors.

8.3.2 Scope of an investigation

Assuming an analyst determines that an event meets the organisation's definition for an incident, the investigation will begin by documenting the scope. The scope of an incident is a collection of relevant leads derived from investigative analysis, which is necessary to understand the totality of the incident. The scope will be constantly updated with data from investigative analysis, until the incident is well enough understood or all investigative leads have been exhausted.

At a minimum the scope maintains the following:

- list of evidence
- list of affected systems
- timeline of attacker activity
- list of malware or other files of interest
- list of all data stolen, modified or deleted
- list of all host or network indicators
- overview of current and planned investigative tasks

The scope of an investigation is similar to the cork boards seen in law enforcement dramas, where all the evidence is visualised so that the detectives can retain an inventory of all investigative details and help prioritise follow-up on open leads. However, within security, it can be a simple spreadsheet used to prioritise investigative analysis.

8.3.2.1 Analysis

Analysis is the process of interrogating sources of evidence to discover and analyse attacker activity in order to better define the scope of the incident and drive remediation actions. Common sources of evidence that will be searched include:

- Artefacts generated by the Operating System, IaaS or PaaS, such as information recorded in Windows event logs, AWS EC2 configurations or changes to AWS Lambda function.
- Information recorded by applications or SaaS. Examples include recent history in Microsoft Word or a configuration change to a mailbox in Microsoft Office 365.
- Logs generated by network devices, which capture information about network sessions.
- Artefacts created by an attacker or malware, such as changes to local settings.

The faster the evidence is collected, the less risk of losing important information. Events generated by the operating system and other applications can overwrite forensic data and application logs may be deleted after certain size/time limits, so it is important to respond quickly or capture important forensic data for long-term retention.

8.3.2.2 What to collect

When performing analyses, there are many options as to what information can be acquired. As a minimum, the following evidence should be collected from a system of interest:

- the system name and physical characteristics
- the operating system and installed applications
- running processes
- users and group membership
- network information
- file information, including name, location, size and hash¶
- logs detailing system, user and application events
- system memory (if the system is running)

8.3.2.3 Methods of collection

There are several methods to collect evidence. This includes a 'Live Response' script, forensic imaging, network collection, enterprise services and EDR solutions.

Live Response scripts quickly and simply acquire data from a system that is currently running. There are many publicly and privately maintained live response scripts currently available. Although these can produce evidence quickly, there are a few drawbacks. Live Response scripts [12] do not permit the restoration of deleted files and as a result are not well suited for investigating systems where data may have been deleted. Data are collected on a live system and there exists the potential to erase or modify evidence on the system. This can be avoided by running a well-tested Live Response script and writing the output of the script to a remote location.

The most comprehensive method for acquiring evidence is a forensic image. A forensic image makes an exact copy of a system's disk or partition. In some cases, forensic images may just include a logical copy of files contained in a directory. Although comprehensive, forensic images require hours or days to collect. As a result, they are typically used in legal cases or in situations where an analyst needs to acquire deleted files. Two popular commercial forensic imaging tools are 'Encase' and 'FTK Imager'.

Network collection involves using enterprise network devices or specialised network security devices to capture network traffic that can help speed an investigation. There are a number of benefits to collection of network evidence. 'Network captures' may help determine systems' communications, capture files that may have been downloaded but are no longer present on a system or intercept

¶Hashes are the output of a cryptographic algorithm. Those algorithms used in forensics (e.g. SHA256) are designed to produce a fixed byte string of characters that can uniquely identify a file.

application requests. There are a number of downsides though. Network traffic is increasingly encrypted, meaning that such data may not be useful even if captured. The speed of networks has increased significantly, increasing the cost and complexity of capturing all network traffic. Networks are becoming more dispersed, meaning that it is more difficult to identify central points at which to capture traffic.

Enterprise services can be traditional services or Software as a Service (SaaS) solutions. Examples include everything from Microsoft Office 365 to Microsoft Active Directory [13]. Increasingly, these systems are either the focus of an incident or used by an adversary as part of their attack. As a result, security teams must configure these solutions to capture information that may be useful in responding to an incident.

EDR solutions, discussed in Chapter 7, are an attempt to continuously collect forensic information, in order to further speed investigations. In general, these solutions record forensic details, which can be quickly referenced and analysed by investigators. In most cases, they must be installed prior to an incident in order to work effectively.

8.3.2.4 Inference

Inference is the process of reviewing the output of multiple analyses and drawing conclusions about an attacker's activity. This stage is important, especially in large investigations, as it allows investigators to examine the bigger picture, so as to make informed investigative and remediation decisions. For smaller incidents, analysts are likely to be able to draw such conclusions, but for larger incidents inference is typically managed by an incident lead or more senior analyst.

In 2014, one of the authors investigated an attack on a Fortune 500 company. The attack had been ongoing for some time, but investigators had been unable to determine the source. Analysis suggested that the attacker was accessing target systems from non-persistent virtual desktops, protected by multi-factor authentication (MFA). Unfortunately, because these desktops were not persistent, the source of the connection and any evidence of their activity were gone. The lead investigator then correctly inferred that the attacker would have likely had to compromise the MFA system in order to gain access. This conclusion helped to prioritise the correct actions, which are described in the next section.

8.3.2.5 Action

Based on the conclusions drawn from review of the analyses, the investigators will need to develop actions. Those actions will include remediation activities and/or additional leads that need to be investigated.

Continuing from our previous example, understanding that the MFA server was likely compromised resulted in two actions. The first was prioritising the investigation of the MFA server, which confirmed a compromise of both the server and MFA seed – a secret key used to generate the one-time password (OTP). The second was that the remediation plan was updated to rebuild the MFA server and redistribute MFA tokens.

8.4 Incident remediation

Incident remediation is the most time consuming, difficult and costly component of the response process. A large remediation can involve hundreds of technical and non-technical personnel and last 2–3 months or longer. Remediation efforts can range anywhere from several thousands to tens of millions of pounds – significantly more than the investigation.

A successful remediation will ensure that the incident does not significantly expand in scope, the source of the incident is completely eradicated and normal operations are restored while reducing the likelihood of future recurrence to an acceptable level. The remediation effort will require an organisation to form a cross-functional remediation team to develop and execute a plan to contain, eradicate and recover from the incident.

8.4.1 Creating a remediation team

There is no common model for staffing a remediation team. Every organisation is unique and will need to develop a team that makes sense for their structure, function, size and geographical presence. The scope of an incident will also shape the size and structure of a remediation team – a small incident that impacts only a system may require no more than two people, whereas an enterprise-wide incident may include two hundred. Although a team's makeup and size may vary, there are several key factors that characterise successful remediation teams.

8.4.1.1 Finding the right remediation owner

The remediation owner is the single most important factor in whether a remediation is successful. It is rare to find a remediation that goes well without a strong and capable lead. The three characteristics of a strong remediation lead include the following:

1. Tenured, well respected and connected technical leader, who can successfully coordinate executives, IT, security, legal and application owners.
2. Has the support of executive leadership to quickly make the decisions necessary to resolve the incident.
3. Understands the business, internal politics and IT systems within the organisation.

8.4.1.2 Empowering security teams

The second characteristic of a successful remediation team is one that is empowered to take action, without a lengthy approval process or red tape. High-performing incident response teams typically have the power to quarantine systems, implement network restrictions or direct system owners to take immediate action. This power allows teams to shorten the response time and, in effect, limit the impact and length of a breach.

8.4.1.3 Securing incident communications from actors

Smart threat actors will target enterprise communications to better understand their environment and to determine if their presence has been detected. Incident response teams should consider operating on out-of-band solutions, including more secure or restrictive email, chat or ticketing solutions. This will ensure that should an attacker gain access to enterprise communications, information about incidents will be limited.

During large remediation efforts, where IT, legal and others are involved in a significant breach, organisations should also consider providing these users with out-of-band communications. In rare cases, it has even been beneficial to provide out-of-band devices.

8.4.2 Creating a remediation plan

Taking action against an actor, without fully understanding the scope of an incident, can expand the impact and duration of an incident. In situations where the actor is a natural event, an automated worm or a misconfiguration, this may not have much impact. But, in situations where the actor is an external or internal human threat actor, taking action before fully understanding the situation may allow an attacker to remove evidence, deploy additional malware or even execute damaging actions.

The remediation plan should capture all steps required to prepare for and execute a remediation, which includes the following phases:

- enabling the investigation and future remediation actions
- containment of the incident to prevent further damage
- eradication and recovery of the incident to return to normal operations and prevent a re-compromise

8.4.2.1 Enabling the investigation and future remediation actions

During larger incidents it may be necessary to take remediation actions that enable the broader investigation and prepare for the final eradication of the actor and recovery of operations. This includes increasing visibility to better monitor attacker activity and improving the security of systems to prevent a recurrence of compromise. Unlike other phases of the remediation, these actions appear similar to normal IT or security operations and should not alert attackers.

This phase can be largely avoided if appropriate incident planning has taken place. Incident response firms can provide organisations with a set of best practices or perform a gap analysis when delivering an incident response retainer service.

The following sections provide common recommendations, typically executed by the remediation team when attempting to enable the investigation and future remediation actions.

8.4.2.2 Logging and monitoring

System logs provide information required by investigators to quickly and thoroughly complete an investigation. Failure to configure system logs correctly, or securely store logs, may delay or disrupt the investigation. This requires organisations to:

- enable adequate system, application and network logs,
- centralise logs using a security information and event management solution,
- identify and fix systems where logging is broken or disabled,
- ensure logs include successful as well as unsuccessful actions, which will allow investigators to track all actions taken by the actor,
- limit use of known compromised credentials and systems, to make it easier to differentiate between malicious and legitimate activity.

8.4.2.3 Configurations

Deploy secure and standard configurations across workstations and servers to allow investigators to quickly identify anomalies and apply controls to prevent future compromise:

- Configure secure settings for password strength and storage, especially on legacy Windows systems.
- Ensure file system permissions are set appropriately.
- Capturing user access permissions across the organisation to clearly understand what accounts have administrative access to systems and applications – especially those with broad access.
- Passwords should be adequately protected to limit continuous compromise and make it harder for an attacker to obtain or crack passwords in the future.

8.4.2.4 Software vulnerabilities

Software vulnerabilities provide an easy path for an attacker to move to new and potentially valuable systems. To mitigate this risk, it is advisable to patch vulnerable software in timely manner to ensure that all operating system and major third-party applications are current.

- Establish a patching target, of 2 weeks or less, for high or critical vulnerabilities that could allow an attacker to gain access to systems through network connectivity or phishing attacks.

- Ensure third-party application patching is performed, and, at a minimum, includes those applications commonly exploited by attackers, including Microsoft Office, web browsers, oracle Java and Adobe Acrobat and Flash.

8.4.2.5 Limiting disruption to compromised assets

During a long remediation, there are seemingly innocent actions that may inadvertently alert attackers that they have been detected. Avoid these common mistakes in order to prevent prematurely alerting an attacker, before remediation takes place:

- Submitting a malware sample to public sites like Virus Total or to an antivirus vendor will alert a smart attacker that their presence has been detected. As a result, it is important to use private sandboxes and delay notifying antivirus vendors until remediation occurs.
- Other organisations may end up submitting a similar malware sample to antivirus vendors, which may result in the removal of malware, before remediation. For only those systems where malware has been identified, consider whitelisting the malware to avoid a situation where malware is removed before remediation, forcing the attacker to redeploy new malware and compromise additional systems.

8.4.2.6 Internal and external communications

Perhaps the most important part of a successful remediation is how organisations engage and communicate with customers and regulators. Although there is no set strategy for public communication of a breach, there are a number of principles to follow which limit negative reactions from customers, regulators and the general public:

- Engage with the public as you would want a company to engage with you.
- Do not redirect blame. Clearly state that you accept responsibility.
- Inspire confidence. Be clear and consistent about what is known and how it will be made right for impacted customers.
- Never communicate incomplete data. Only confirm facts. Revision of any details, especially an increase in the impact of the incident, will result in more attention.
- Do not continue to engage with the media if not required. After the details and planned actions have been communicated, continuing to communicate with the media will only feed and extend the news cycle.

Although the incident commander can manage communications within the incident response team, it is important that all external communications (e.g. with the company and customers) are managed by an experienced communications leader.

For larger breaches, most organisations should consider engaging a third-party experienced with managing communications for incidents.

In an ideal world, no communication to customers would be made until after an investigation has been completed, at which point an accurate announcement could be made about the impact of the breach. However, regulatory rules increasingly no longer permit companies to communicate breaches on their own terms. For example, the GDPR requires mandatory reporting of data breaches within 72 hours. As a result, many organisations are now forced to report the total number of potentially impacted records and later revise the figures down.

8.4.3 Containment

Forest fires impact millions of acres of land every year. Once ignited, fires can spread quickly, particularly with strong winds. One way to contain a forest fire is to introduce firebreaks – gaps in vegetation that help slow the spread of a fire. Bulldozers or other machinery may be deployed ahead of a fire to create firebreaks, in an attempt to contain a blaze. Containment, in the context of a remediation, is similar to a firebreak in its purpose to prevent expansion of an attack.

If applied too aggressively, containment can alert an attacker or exacerbate operational challenges. Conversely, if containment is not strong enough, an incident may expand rapidly. As a result, the goal of containment is to balance the risk of alerting the attacker and/or operational impact, with the risk of allowing the incident to continue without limits.

8.4.3.1 When to initiate containment

In 2017, Maersk shipping suffered one of the most debilitating incidents in modern transportation history. The trouble began on 27 June 2017 when employees saw their computers restart and display a blank screen (generally on laptops) or a ransomware message (generally on desktops). Employees were unsure of what to do, and some started disconnecting computers from the network, in a bid to prevent as many as possible from becoming infected; but this was not part of the containment plan. Eventually it was decided to take all systems offline, to prevent the spread of the malware, known as Not Petya. The Maersk incident is covered in detail as a case study at the end of this chapter.

Containment may or may not be necessary for an incident. The decision depends on the following factors.

8.4.3.2 Automated or human

Is the attack executed by humans or automated malware? Human attackers have the ability to adapt to changes in the network and systems. Automated attacks, once understood, are predictable and static. Containment can be applied immediately in most cases when dealing with automated attacks. However, a more careful approach is required when humans are involved.

8.4.3.3 Sophistication

If a human is behind a particular attack, it is useful to understand their sophistication. Understanding the 'who' about the attacker is most helpful when determining their capabilities. For example, an attacker that struggles with command line activity, and uses open source tools, may be inexperienced and slow to adapt to any changes. A more sophisticated attacker, like a nation state or organised criminal group, may be nearly impossible to remove with a quick and easy containment plan.

8.4.3.4 Scope

The scope of the attack is also important. If the attacker has not gained access to the entire network or critical systems, immediate steps can be taken to harden those targets, without raising the awareness of an attacker.

8.4.3.5 Timeframe

If an attack is relatively recent, there is a high likelihood of a successful and complete investigation. In the event of an older compromise (more than a few months), it is nearly impossible to ensure that all attacker access has been removed. Thus, increasing the risk of any containment action failing.

8.4.3.6 Impact to critical business functions

Many times there are environments where attackers must be quickly removed. For example, in many retail environments, it is critical that attackers do not gain access to payment systems, also referred to as Payment Card Industry (PCI) systems, that process credit cards. In the worst-case scenario, where an attacker has gained access to such systems, they may be shut down or isolated and rebuilt, so as to complete an immediate containment of the most important systems, despite making an investigation more difficult or costly.

8.4.3.7 Examples of containment

The following provides some examples of scenarios based on actual events and the containment strategy selected.

8.4.3.8 Company A

In 2013, a European company, 'Company A', was breached. They had been notified of a breach of their network by law enforcement. The initial notification included the names of files that had been stolen, which contained trade secrets. The initial investigation determined that the attack had been ongoing for more than 8 months and that hundreds of systems had been accessed. It was also determined that data theft was ongoing, occurring every evening. Key aspects were as follows:

- The attack was perpetrated by a human.
- The adversary was advanced and capable.

- The adversary had broad access to systems and had been in the environment long enough to have placed many backdoors and stolen important data.
- Some of the most important systems containing 'crown jewels' had already been accessed and data already stolen.

Given the above, the remediation team decided to perform a limited containment. Systems hosting important data, which had not already been accessed by the attacker, were secured and/or important data were selectively removed. This meant allowing data theft to continue for those systems on which the attackers were already active. It was determined that given the attackers foothold and understanding of the environment, alerting them to their detection would have resulted in a much larger breach and extended recovery.

After 2 months, Company A executed a remediation event, which involved disconnecting all network access to the Internet for a weekend in addition to removing all compromised systems. The event was successful in removing all attacker access and improvements implemented as part of the remediation not only detected an attempt to regain access but also prevented any further compromise.

8.4.3.9 Company B

Sometime in the past, a hospitality company was breached. The attack was performed by a financial criminal with moderate sophistication. The attacker had already targeted the PCI environment that processed credit cards and stolen tens of thousands of credit card numbers. The attacker was in the environment for several years and evidence of the initial compromise had been lost when logs had been overwritten. However, the attacker continued to access the network via a remote access system and valid user credentials.

The company determined that continued access to the PCI network was unacceptable and that access needed to be removed immediately without impacting operations. A plan was devised to segment the PCI environment from the rest of the network to allow for continued operations while the investigation and remediation continued. Forensic images of the PCI systems were created and a series of firewall rules were deployed to prevent the attacker from accessing the PCI environment. VPN access was also disabled.

Unknown to the investigation team, a backdoor existed in the network. When the changes were made, the attacker accessed their backdoor and regained access to the network. A misconfiguration in the firewall allowed continued access to the PCI environment and the attacker expedited the theft of more credit card numbers and distributed additional backdoors.

The containment failed due to poor planning and the presence of a backdoor that had not yet been found because the investigation was still in its infancy. Although a 'sound strategy', these are the potential risks of an aggressive containment strategy, implemented before understanding the complete scope of an incident.

8.4.4 Eradication and recovery

Eradication and recovery is often referred to as a 'remediation event' and is designed to remove the entirety of an attacker's access and restore normal business operations. For major incidents, the event may occur only after months of planning and be executed over a weekend to limit operational impact to employees.

8.4.4.1 Eliminate attacker entry vector/s and persistence

Once network connectivity has been severed, the remediation team can start working on removing the attacker's access. This ensures that when network connectivity is restored, the attacker is unable to reuse known vulnerabilities or tools to regain remote access. Key actions are:

- remove and rebuild systems that contained attacker backdoors,
- block known bad IPs and DNS,
- remove any vulnerability that resulted in the initial compromise,
- reset compromised local, domain and application accounts.

Although these tasks may seem simple, orchestrating these changes in an enterprise environment typically requires 24–48 hours. This may be shortened considerably in more modern cloud architectures, where a new environment can be prepared in parallel and then deployed with minimal disruption.

8.4.4.2 Execute recovery and prevent recurrence

Recovering from an incident requires that normal operations are restored without a short-term recurrence. Recovery from the attack requires that all business systems and processes are returned to their pre-incident status. More impactful security improvements, which may have been deferred to prevent alerting the attacker, should also be implemented. Common improvements that are implemented during the remediation event include the following:

- stronger network egress restrictions
- application whitelisting to prevent unknown binaries (computer programmes) from executing
- blocking host-to-host communications for management protocols (e.g. file sharing and remote desktop sessions) that could facilitate lateral movement if credentials are stolen
- MFA for remote access and for critical systems
- DNS or web filtering to limit malicious network traffic

8.4.4.3 Eliminate attacker connectivity

Imagine trying to clean up a messy floor, while someone continues to throw trash on the ground. This is what it would be like trying to eradicate an incident while an attacker still had access to a network. As a result, a successful eradication should require that all network access is temporarily disabled. For small incidents, this could mean quarantining an infected system or group of systems. For larger incidents, where broad access has been obtained, it may require severing all Internet access – even third-party connections – within a period of no more than 10 minutes, for the entirety of the eradication and recovery effort.

8.4.5 Post-mortem and continuous improvement

At the close of a significant incident, an organisation should collect lessons learned in order to address gaps and/or continuously improve. Such information is usually captured in a post-mortem, which captures the following details:

- What did the team do well?
- Were there any new processes or technologies developed/used that should be expanded and formalised?
- Are there things that went poorly or could have been improved upon?
- Were communications and engagements with other stakeholders clear and effective?
- Were there any skill sets or technologies that were helpful or missing?
- Was anything learned that might suggest a potential problem with a variation of the incident that was discovered? What could have made this worse?

8.4.5.1 Common lessons learned

Talk to any security professional that has gone through a breach or significant incident, the stories of lessons learned will be very similar. Preparing for these types of events is incredibly important, but with every incident the response will vary and gaps will emerge that were not initially known.

- The executive team needs to step back and let the incident team do their jobs. In many cases, the executive team will step in during a significant incident and start to give direction. This is incredibly disruptive during the critical minutes and hours after an incident has been declared. The executive team should be included as part of preparatory 'tabletop' exercises, so they have an understanding of how the process works and what their role is. Even in cases where they are part of the tabletop exercise they will still feel the need to step in and direct actions. This is not their role. If they disagree with an action taken by the response team, they have the ability to course correct. This is a support role and they need to step back and let the incident team lead the response plan as it was practiced.
- 'If you don't control the communication, it will control you'. Recent surveys have pointed to the importance of transparency and sharing information openly with those that are impacted. It is generally thought that companies that are transparent and share details of a breach of significant incident the moment they

learn about it will get a second chance to rebuild their brand and image while others may not.

- If there are different operation analysts responding to different events they need to talk. In many cases, different analysts will be assigned a specific security operations solution and they are trained as to how to respond to the events out of that application. It's only when they correlate it with other technology are they able to identify a true incident. Security incident event management solutions certainly help with this correlation but could never replace the insight and intuitiveness of security analysts. As analyst see events that are out of the ordinary or appear malicious, they need to be talking to one another.

8.5 Closing remarks

Incidents are inevitable. However, that doesn't mean that an incident needs to have a material impact on your business. Incident response is a risk mitigation strategy that when executed effectively can dramatically reduce the impact of an incident. This chapter provides the basic concepts that should enable you to plan and execute a successful incident response.

Although there is an unlimited number of incidents that could impact any organisation, leaders should aim to identify those incidents that are most likely to occur. In the transport sector, which would include automated malware, such as ransomware, which tends to thrive in large environments with outdated, legacy systems. Other likely attacks include DDoS attacks, and nation state attacks aimed at impacting system availability.

The most predictable indicator of success in responding to an incident is preparation. For a discipline focused around response, much of the result is predetermined by the amount invested into preparatory activities, such as tabletop or red-team exercises, call-trees, run books, adequate logs and other efforts designed to ensure a predictable and timely response. Organisations that continually test and improve their response capability will quantifiably fare much better in a major incident.

Finally, incident response relies on strong relationships with a range of internal and external organisations. An effective response requires strong cooperation from legal, human resources, communications, marketing and technology teams. Your organisation's reputation will play a significant factor in how customers, regulators and the general public respond to an incident. Therefore, it is important to develop strong relationships with government, regulators and your customers.

By identifying likely threats, preparing and practicing a response, and developing close relationships with both internal and external stakeholders, any organisation can be prepared to effectively respond to an incident and reduce the impact, scope and length of an incident.

To illustrate these issues further, as a case study, Andy Jones (former Chief Information Security Officer of Maersk) looks back at the 2017 NotPetya attack and introduces the concept of the extinction event.

8.6 Case Study – Surviving the extinction event – The 2017 NotPetya attack

8.6.1 What is an extinction event?

An extinction event is something that threatens the very existence of a business. It is an unexpected and unpredicted cyberattack, fast acting and devastating. While the term 'black swan event' has been used previously to describe unexpected events, the digitisation and connectivity of transport businesses, coupled with the emergence of weaponised cyberattacks – designed to destroy – takes the humble 'black swan' to another level.

Simply put, an extinction event takes a business down fast and hard – despite your best efforts.

8.6.1.1 What is the relevance to the transport sector?

Although by no means unique to the transport sector, the highly operational nature of transportation, its public visibility and a strong focus on safety makes the impact of an extinction event very apparent. Additionally, transportation's function as enabling (or critical) infrastructure will have causal impact on a multitude of other businesses and public utilities.

8.6.2 What are the causes of an extinction event?

There is no shortage of relevant examples where a cyberattack severely crippled or halted businesses or the public sector. In 2017, the Wannacry virus impacted the UK National Health Service, ransomware viruses have brought city and regional administrations, such as Atlanta (2018) and Matanuska-Susitna, Alaska (2018) to a halt, and industry impact has ranged from Saudi Aramco (2012) to Norsk-Hydro (2019). Most notably in the transportation sector, the logistical, shipping and port operations of the Maersk Group were severely crippled by the NotPetya attack in 2017.

The malicious actors responsible for the attacks can vary from criminal through to accidental, but a more recent trend is for nation states to sponsor or fund the attacks.[**] This has resulted in two significant consequences.

8.6.2.1 Technical sophistication

The involvement of nation state actors in developing new, or re-purposing existing, attacks brings significant resource to the task and can result in a very sophisticated attack or delivery method, which is capable of by-passing strong technical defences. Often, several different techniques are used in combination. For example,

[**]https://www.wired.com/story/white-house-russia-notpetya-attribution/.

the NotPetya attack used two attack vectors – Eternal Blue (also associated with Wannacry) and Mimikatz (a credential stealer) – to enable the attack to propagate even if the target has been adequately patched.

Eternal Blue is an exploit attack on Microsoft operating systems, generally believed to have been originally developed by the National Security Agency of the United States.[††]

8.6.2.2 Collateral damage

Many attacks are relatively indiscriminate in their target, such that organisations simply become collateral damage of battle. This is a consequence of both nation state and criminal engagement but has been amplified by a greater attack surface associated with the digitisation of industry. This digitisation is often associated with the network connectivity of industrial components (the term Industry 4.0 is also used). This increasingly random nature of cyberattacks renders more traditional sources of threat intelligence less effective, as existing intelligence is often sector-focused. Recent examples such as Wannacry and NotPetya are widely accepted to have caused significant damage outside of their intended targets.

These two aspects of sophisticated attacks also serve to make the transport sector more vulnerable to impacts as their extended geographic natures exposes them to multiple 'theatres of cyber war' and their cyber-physical characteristics can result in significant damage if compromised. Additionally, the criticality of transport and logistics operations to nation states makes them as attractive a target as they would be in a kinetic war.

Many nation states defence forces have a critical dependency on commercial transport and logistics providers to sustain their operations, making those providers a potentially soft target.

While there is an inevitable impulse to examine technically sophisticated attacks and to develop a technology-based defensive response to defeat them, this strategy is ultimately doomed: the resources that are available to both criminal and nation state attackers to develop new and even more capable attacks will always outweigh those resources that an organisation can afford.

Although adopting cybersecurity good practice remains important, a scenario should be envisaged where even the most state-of the-art cyber defences will fail and an organisation will be completely devastated by a cyberattack.

8.6.3 Anticipating an extinction event

Understanding how to react and recover from an extinction event can lead to a reduction in the overall impact of the event from:

- understanding the characteristics of such an event
- enhancing business continuity arrangements

[††]https://en.wikipedia.org/wiki/EternalBlue.

- identifying how to restart the business from a catastrophic event
- creating a workforce that is familiar with the scenario and confident of its ability to recover

8.6.3.1 The extinction event scenario

The impact of a potential extinction event on an organisation should be assumed to be extreme in its technical repercussions. Techniques seen in recent cyberattacks include irreversible encryption, data wiping and other destructive behaviour, such that the working assumption should be that all technology-based capability is destroyed by the attack, from the hardware layer up. Assume that there is no compute capability, no network connectivity, no communication channels and that the only information that is available is from manual written sources.

Examples of malicious attacks designed to cause irreversible damage include NotPetya, Triton, Industroyer and Shamoon.

The direct target of a cyberattack is often limited to the Information Technology (IT). However, in a modern organisation, a complex dependency will have formed, making the operational business capability completely or largely dependent on functioning IT. Hence business operations will also be critically impacted by a cyberattack even if they have not been specifically targeted.

8.6.4 Being ready

While there is an inevitability associated with the extinction event, there are a number of preparatory tasks that can either significantly reduce the impact or speed the recovery.

8.6.4.1 Do the basics really, really well

Although the sophistication of a cyberattack may overwhelm technical defences, doing the basics well can still reduce the likelihood of the event in the first place and further reduce its impact. Achieving a good level of cyber-hygiene should then remain a key activity for organisations, with an increased focus on elements that are relevant to an extinction event.

These elements include the following:

- Robust patching regimes – being able to roll out patches to the entire IT and Operational Technology estates at short notice will defend against many vulnerabilities.
- Accurate asset (physical and information) inventories – simply knowing what you have in terms of systems, computers and Industrial Control Systems is a pre-requisite.
- Active anti-virus defences – best practice would dictate that multiple products are used in combination.

- World-class backup and recovery capabilities – without comprehensive and timely back-ups, and the ability to restore from them, all will be lost.
- Resilient network design – while not a magic bullet, a layered network can slow an attack.
- Business continuity – although sometimes at the periphery of responsibility for an IT or cyber function, having confidence in business capabilities is key.

These elements are often mundane, constitute a thankless task and are also surprisingly difficult to implement in a large complex and dynamic organisation. Nevertheless, they are simply the must-dos, which must be done well.

8.6.4.2 Have an answer to the 'what ifs'

During an event, there are a number of questions that might be asked that simply may never have been asked before. Know the answers to these questions beforehand.

What if:

- We have to operate the business without any IT? Maintaining even a very basic business operational capability during an event can be a factor in boosting morale and business reputation.
- We cannot find a clean backup or restore point? Many recovery procedures assume that it is possible to recover data to a known secure point; consider how to recover if this was not possible.
- We have to restart the entire business? Know how to do this.
- The event takes 6 months to recover from? Understand how the business could operate at a reduced level if recovery is prolonged.
- It wasn't just us? If the event is widespread, causing collateral damage to a significant number of organisations in the same area, then the ability to call on other partners, such as outsourcers and recovery specialists, will be reduced or non-existent. Understand where you lie on someone else's priority list.

8.6.4.3 Practice makes perfect

Take the unthinkable and rehearse with your staff as a continuity exercise to test your plans in a business context. Go extreme, then further. A real incident is far, far scarier than you can simulate and there is a need to understand how people react in extremis to discover what is missing from your recovery tools in a safe environment and to introduce core non-IT staff (such as legal and communications) to the scenario. This will all be beneficial to reacting to the real event.

Above all, choosing an extreme event will give staff the confidence that restoring a company from an extinction event is possible – this is invaluable.

8.6.5 Managing the event

While this section does not seek to replicate existing tried and tested procedures for system recovery or business continuity, the following items are especially important in the management of an extinction event.

8.6.5.1 Reset your risk appetite

When an organisation is on its knees, the risk appetite of that organisation, and particularly the people working within it, will change. High-risk decisions will become the norm and staff at all levels will consider themselves to be empowered in ways that would never be countenanced on a normal day. For global organisations, local autonomy will reign – sometimes for purely practical reasons, such as the loss of most communication channels with the head office.

Although this must be managed, as anarchy will breed anarchy, it is important to acknowledge that high-risk decisions may be appropriate in these particular circumstances. The stakes are high.

8.6.5.2 Value your people

In an extinction event you may have little technology, but you will have plenty of people who are willing to help. Use them. Let them use their creativity. Let them use their business knowledge to keep the business at least ticking over.

Recognise that people can burn out and manage that proactively by enforcing rest breaks. Understand that some people may not rise to the challenge and will fall away – let them, but do not judge them poorly. Provide for the personal needs of people, including the families they may support, by providing the essentials, such as food, laundry and hotel space. Do this automatically such that they are not distracted.

Let people do things that you would never ever countenance on a normal day – provided it helps. After all, they can't make things worse – only better.

8.6.5.3 Communicate, communicate, communicate

An extinction event is likely to attract a high level of publicity. External cybersecurity experts will comment on and criticise the organisational response, and this can have a significant reputational impact. Knowing who these commentators are likely to be and having a pre-established relationship with them can be invaluable to managing their response. Invest in the time to network both inside and outside your organisation.

Communication with both internal staff and external parties can be a make or break factor in any recovery. There will be huge pressure for information and partnering with your communications function is fundamental.

Within the business, there will be a number of people worried about their future. Keep them informed on a regular basis using whatever channels are available. Social media can play a part here but treat those channels as public channels.

As the recovery proceeds, there will be huge interest from business functions as to when they can expect to be working again. This is perfectly understandable, as there will be preparatory tasks that the business will need to complete prior to any recovery. However, this demand for information must be constrained and managed so as not to distract from the core task of recovery.

External demands for information may come from the media and a carefully planned, but open, engagement here is key. Other demands may come from regulators, law enforcement and intelligence agencies, all vying for attention. For an organisation that operates globally, this can be overwhelming and a triaged, risk-based strategy should be adopted.

8.6.5.4 Assume that help is not coming

You may be counting on some help in your hour of need from your outsourcer, your partners, your supply-chain. You would hope that they would be able to offer assistance and boost your staffing, as your key people succumb to fatigue. However, don't count on it. You may simply be caught in the crossfire of a bigger event affecting multiple organisations who will hoover up any cavalry. Help may simply not be coming.

8.6.5.5 Keep your eye on the horizon

In the heat, the excitement, the stress of an extinction event it can be easy to throw away existing processes that may be seen as constraining. While an element of this is permissible, it is wise to cast your mind forward to a time when the event is over. It can be critically important for an organisation or individual to be able to justify decisions made in the heat of the moment in order to support future activities, such as legal action or root cause analysis. Decisions should be documented, even if only pen and paper are available, and the rationale for these decisions captured and officially preserved.

Within the cybersecurity arena, there can be a focus on preserving forensic evidence for later legal action. While this is often well-intentioned, it can be a significant resource distraction, so consider carefully the benefits of doing so. Ask yourself – Am I really going to sue that nation state or criminal gang?

8.6.6 Concluding an extinction event

As the recovery gains momentum and success seems within grasp, it is important to manage those closing elements.

8.6.6.1 Celebrating success

Throughout the event, it is the small things that matter most to people. Share those little successes, while keeping them in the context of the overall picture.

Motivation vignettes that show the organisation working together can be powerful, and a small video clip from an employee's smart phone is likely to be more effective than a corporately presented missive. While individual acts can be acknowledged, be wary of creating a hero culture. In a complex event many people may make significant, but unseen, contributions to the recovery.

8.6.6.2 Closure

It is important that the event has a clear closure. This may not actually be the business as normal state, but a state where the recovery has gained sufficient momentum to note that there is a return to business, albeit short of a normal state. This point should be clearly communicated as it can have an important motivational effect.

8.6.6.3 Post-event

During the recovery period, remarkable things will have happened. Projects that ordinarily would take a year to complete will have been achieved in a few short weeks. People will have worked and excelled in different roles to their everyday roles; they may have been entrusted with making 'once in a career decisions'; they will have been empowered. In the vernacular, they will be 'pumped'.

Bringing staff back to normal must be planned and managed for those who have performed above their station, but equally importantly for those who did not cope well with the event. Normality must be restored. While this is not typically a task for the cybersecurity function, it is important to acknowledge that this back-to-normal transition may be prolonged and may be reflected in a changing and normalising attitude to risk and a reluctance to adopt procedures.

8.6.7 Learning from the event

The temptation is to run a root cause analysis of the extinction event and then to spend the (suddenly surprisingly accessible) budget on a series of shiny new technical controls that would have stopped the event dead in its tracks. Although obvious gaps in overall cyber-hygiene should be addressed, this strategy will ultimately fail to meet expectations. The sophistication, innovation and sheer brutality of the attack is what made it effective, and the next attack will be even more sophisticated, innovative and brutal. To try and second guess its nature is a very tall order.

The working assumption should be that the next extinction event attack will also be able to circumvent technical defences, and investment should be instead focused on resilience, recovery and continuity.

8.6.7.1 Resilience

The extinction event is infrequent (low probability) but high impact. Day-to-day nuisance attacks should not be neglected, and a comprehensive approach to good cyber-hygiene should always form part of the defence strategy. In addition, core systems to the business operations should be reviewed and their resilience to a high impact event improved, such that they are able to continue to operate at some reduced level.

8.6.7.2 Recovery

Being able to recover faster, quicker and with less effort should be a primary focus. This will involve simulations, training, documentation and gaining a better understanding of how the business works in reality. Short-cuts back to normality should be identified and documented (on paper), and made readily available to the requisite teams.

8.6.7.3 Continuity

In the race to digitise businesses, a lot of faith is being placed in technology, which is generally not justified by reality. Technology can (and will) fail and, as organisations become more connected, they will increasingly be vulnerable to cyberattacks and the extinction event.

Jim Snabe, the Chair of A. P. Moller-Maersk, speaking about the NotPetya attack, noted that '...in the near future, as automation creates near-total reliance on digital systems, human effort won't be able to help such crises'[‡‡]. These words reflect the impact of Industry 4.0, as the business knowledge is absorbed into the IT systems. That deep and practical knowledge, so vital in an extinction event, will become lost – or at least off-line. A renewed focus on business continuity should therefore form part of the response to the extinction event.

8.6.8 Conclusion

An extinction event – something that can defeat the best technical controls and take a business back to pen and paper – is still relatively rare; however, there is now sufficient evidence to affirm the need for organisations to plan for it as a realistic scenario. Digitisation, Industry 4.0 and the Internet of Things will all amplify the attack surface, as will normal business activities such as mergers and acquisitions.

Growing government focus on the resilience of critical national assets, encompassing much of the transport sector, will result in more regulation that may force a compliance-focused approach to cybersecurity. Such an approach can focus on technical control capability and therefore may not be helpful in addressing the risk of an extinction event.

While acknowledging the importance of a high level of cyber-hygiene, a well-prepared organisational approach to dealing with an extinction event forms a relatively low-cost, pragmatic response to an emerging, but real, danger.

References

[1] Cichonski P., Millar T., Grance T., Scarfone K. *Computer Security Incident Handling Guide*. Gaithersburg: National Institute of Standards and Technology; 2012.

[2] Luttgens J., Pepe M., Mandia K. *Incident Response and Computer Forensics*. 3rd edn. New York: McGraw-Hill Education; 2014.

[3] Greenberg A. *The Untold Story of NotPetya, the Most Devastating Cyberattack in History* [online]. 2018. Available from https://www.wired.com/story/notpetya-cyberattack-ukraine-russia-code-crashed-the-world.

[4] *Advanced Persistent Threat Groups* [online]. Available from https://www.mandiant.com/resources/apt-groups.

[‡‡]https://www.theregister.co.uk/2018/01/25/after_notpetya_maersk_replaced_everything/.

[5] Reuters S. *China-Linked Hackers Attack American Airlines, Sabre Systems*. Reuters; 2015.

[6] *Cyber Attack Led to Bristol Airport Blank Screens* [online]. 2018. Available from https://www.bbc.com/news/uk-england-bristol-45539841.

[7] Stoneburner G., Goguen A., Feringa A. *Risk Management Guide for Information Technology Systems*. Gaithersburg: National Institute of Standards and Technology; 2012.

[8] Cichonski P., Millar T., Grance T., Scarfone K. *Computer Security Incident Handling Guide*. Gaithersburg: National Institute of Standards and Technology; 2012.

[9] *CISA National Cyber Incident Scoring System* [online]. Available from https://www.cisa.gov/uscert/CISA-National-Cyber-Incident-Scoring-System.

[10] *Security Update* [online]. Available from https://about.fb.com/news/2018/09/security-update/.

[11] *Exercise in a Box* [online]. Available from https://www.ncsc.gov.uk/information/exercise-in-a-box.

[12] *Live Response Using PowerShell* [online]. Available from https://www.sans.org/white-papers/34302/.

[13] *Active Directory* [online]. Available from https://en.wikipedia.org/wiki/Active_Directory [Accessed 10 Jan 2020].

Chapter 9

Autonomous vehicles – cybersecurity and privacy challenges and opportunities

Hongmei He[1], Amy Ertan[2], Rory Hopcraft[2], Raja Naeem Akram[3], and Hafizah Mansor[4]

With technical innovation accelerating through the beginning of the twenty-first century, it is clearly feasible that truly autonomous vehicles without human intervention will be realised in this generation. It has been shown that all fields related to autonomous vehicles, including cybersecurity and privacy, are constantly evolving, with frequent technical breakthroughs. However, cybersecurity and privacy will be critical challenges for the success of implementing autonomous vehicles. This chapter provides an overview of the current state of the art on the cybersecurity and privacy aspects of autonomous vehicles and focuses on crucial concepts and principles in cybersecurity and privacy of autonomous vehicles rather than the current technical status of each research strand. It is expected that this chapter could provide a high-level resource, through which, readers can understand the challenges and the landscape of cybersecurity and privacy of autonomous vehicles in the incredibly connected and technology-orientated world. The chapter also addresses the economics of autonomous vehicle security – a critical element in the cybersecurity domain, as well as the infrastructure investments for improving the security and privacy of autonomous vehicles. Finally, a case study of maritime vehicles further demonstrates the critical challenges of cybersecurity and privacy, as well as the need of technology advance for the safety of autonomous vehicle in maritime, which might be threatened by the cyberattacks or cybercrimes.

9.1 Introduction

A vehicle is a machine that transports people or cargo. Modern vehicles could include motor vehicles (motorcycles, cars, trucks, buses and vans), railed vehicles (trains and

[1]School of Computer Science and Informatics, De Montfort University, Leicester, UK
[2]Information Security Group, Royal Holloway, University of London, Egham, UK
[3]Department of Computer Science, University of Aberdeen, Aberdeen, UK
[4]International Islamic University Malaysia, Selangor, Malaysia

trams), watercraft (ships and boats), amphibious vehicles (screw-propelled vehicle and hovercraft), aircraft (airplanes and helicopters) and spacecraft. Autonomous vehicles, especially autonomous cars, have captured the imagination of consumers, corporations and governments. Updates and discussions of autonomous vehicular technologies frequently appear on major media outlets, and these updates keep attracting public interest as the technology advances. There are already some automated vehicles currently in the market, with certain tasks no longer requiring human intervention. Such tasks include lane management, speed control and emergency braking mechanisms. However, strictly these vehicles cannot be termed to be 'autonomous vehicles' but *automated* vehicles.

Similarly, remote piloting of such automated vehicles often referred to as 'cooperative vehicles' represents a partially distinct but equally the relevant field of autonomous vehicles. An autonomous vehicle is 'a driving automation system that performs all of its tasks independently and self-sufficiently – without requiring any input from a human'. This is the working definition that we will base for the discussions in this chapter.

An autonomous vehicle is capable of sensing its environment and moving safely without human intervention. Hence, an autonomous vehicle system is a collection of sensors, actuators and an advanced control system, connected over a communication network. The advanced control system is the brain of the vehicle, powered by artificial intelligence (AI) technology. A variety of sensors are used to perceive the surrounding environments of the vehicle, such as radar, laser-based detection (LiDAR), sonar, Global Positioning System (GPS), odometry and IMUs. Advanced control systems interpret and process sensory information about the surrounding environment with AI technology, thus instructing relevant actuators to take the necessary actions required for a safe journey. The whole system is based on digital technologies that are, in most cases, connected to the Internet.

Many factors could influence the acceptance of autonomous vehicles. Kaur and Rampersad [1] provided an acceptance model (Figure 9.1), where the trustworthiness of users is decided by reliability, security and privacy, with the adoption of connected and autonomous vehicles (CAVs) depending upon the trustworthiness and performance expectancy. It is of paramount importance that autonomous vehicles are highly reliable systems. The reliability of autonomous vehicles is directly

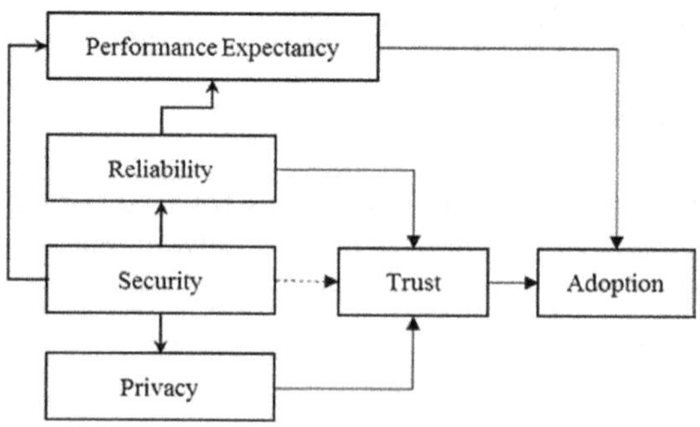

Figure 9.1 An acceptance model for CAVs

related to their safety, depending on the health of the systems and the responses to the surrounding environments, which requires the system to be able to correctly sense the surrounding environments. In the instance that a sensor reports incorrect information, the advanced control system receives an incorrect input, which could compromise the data-processing action and potentially lead to an incorrect instruction sent to actuators. These incorrect actions may be detrimental to the safety of the passengers in the vehicle, property and people in the surrounding of the vehicle. The navigation techniques well validated on robots could be applicable for improving the navigation of autonomous vehicles.

Zhang *et al.* [2] did not contend that security is an important factor affecting the acceptance of CAVs hence, the dependent relation from security to trust is depicted with dashed arrow. If we consider a denial-of-service (DoS) attack, then cybersecurity could directly affect the reliability and privacy of the system. For example, a DoS attack, using a flood of arbitrary packets on an electronic throttle control (ETC) system, may feasibly cause a system malfunction, leading to a stuck or inoperable throttle [3] and raising potential safety concerns. It can also affect the Controller Area Network (CAN) bus of a CAV; thus, it can subsequently impact the critical control systems including brakes, throttle and steering. These systems could be disabled by an overload of processes or an overhead on them, resulting in a slowing down or a malfunction in the response of those systems [4]. The slowing down behaviour or malfunction of a CAV, due to a DoS attack, could lead to an unexpected incident of the CAV. Hence, a cyberattack (e.g., a DoS attack) can directly affect the performance and the reliability of the vehicle. As another example, access control is important in cybersecurity. If a malicious actor breaks the access control mechanism of a CAV, the information stored in the database recording the history of CAV is revealed, and thus the privacy may be offended. This shows that cybersecurity has the same importance as the reliability and the privacy of CAVs, but the importance is hidden or indirect for the acceptance of users to autonomous vehicle. Therefore, the acceptance model of CAVs needs to be updated to include the relationship of security to three properties: performance, reliability and privacy. When a new technology emerges as a result of rapid innovation, there is an expected level of suspicion and scepticism, especially for products or services that may be linked with critical safety concerns. This remains the case even where autonomous vehicles may appear to present efficiency and safety advantages compared with manually operated vehicles. But cybersecurity and privacy features may hold significant implications for the implementation of autonomous vehicles. As a dynamic and evolving field, a high quality cybersecurity defence approach requires constantly updating threat models with new techniques and controls to protect a specific asset. This requirement applies to the cybersecurity of autonomous vehicles, as autonomous systems may represent a target for malicious actors as individuals or groups. The cybersecurity of autonomous vehicles depends on how robustly they deal with anomalies from cyberspace. For example, malicious entities, attacking autonomous vehicles, may deliberately compromise the security mechanisms of autonomous vehicles or directly replace a part of control system software, take the control of the autonomous vehicle and instruct the control system in an unsafe way. This greatly raises the risk of affecting the integrity and safety of the autonomous transport system. Therefore, all potential risks in the safety and security of autonomous vehicles should be eliminated or greatly mitigated before autonomous vehicles go to reality on roads, and they should be robust enough to overcome the potential future challenges.

In this chapter, the increasing challenges in cybersecurity and privacy of autonomous vehicles will be addressed thoroughly, and thus helping designers, engineers and manufacturers to achieve a forward-looking approach to cybersecurity of autonomous vehicles, with a priority of mitigating potential cyber threats.

The chapter is organised as follows. Section 9.2 briefly introduces the broad cybersecurity of the ecosystem for autonomous vehicles, which are connected to form a wireless ad-hoc vehicular networks and AI for the automation of cybersecurity. Section 9.3 identifies the challenges in privacy and explores the privacy-preserving mechanisms of autonomous vehicle systems, regarding data ownership and stewardship, with reference to the influential General Data Protection Regulation (GDPR). In Section 9.4, economic factors related to the security and privacy of autonomous vehicles are discussed within a wider smart-city transport landscape. This discussion provides an insight into the market justification for security and privacy. Finally, in Section 9.5, a case study in maritime vehicles is done to demonstrate the significance of cybersecurity and privacy in maritime vehicle systems, and the state of the technology advance for the safety of autonomous vehicle in maritime, which could be threatened by cyberattacks and cybercriminal activities.

9.2 Cybersecurity of autonomous vehicles

This section first addresses the vehicular network and communication, followed by the predictive security paradigm for autonomous vehicles, a discussion on AI as an automatic mechanism for the cybersecurity of CAVs, and the open challenges to implement automated cybersecurity practices.

9.2.1 Vehicle networks and communications

Reliable and secure networks are required for fast information exchanging to ensure the functionality, safety and efficiency of autonomous vehicles as well as the comfort of vehicle users. Vehicular networks could be either wired or wireless. The communication of an on-road motor vehicle could take between on-board units within a vehicle and between the vehicle and other surrounding components, including other vehicles, other road users, such as pedestrians and cyclists, as well as wider road infrastructure (e.g., roadside units [RSU]). Simply, vehicular communications can be divided into two categories: in-vehicle and extra-vehicles (i.e., vehicle-to-everything (V2X)). The communications between various entities for information exchanging use different communication protocols [5]. For example, the IEEE 802.11 p [6, 7] was designed to facilitate vehicle-to-vehicle (V2V) and vehicle-to-infrastructure (V2I) communications, and LTE-V standards (Bazzi2018b, Molina2017) were designed for vehicular connectivity over the existing operator mobile network infrastructure [8, 9]. 5G V2X is the future trend for autonomous vehicles, and the 5G Automotive Association (5GAA) was established to promote communications solutions to support connected mobility, road safety applications, ubiquitous access and integration into smart cities and intelligent transportation [10]. Table F.1 in the Appendix (IoT communication protocols) shows the properties (frequency, data rate, distance etc.) of IoT communication protocols at communication/transport layer and their application domains.

9.2.1.1 In-vehicle communications

Electronic control units (ECUs), sensors and actuators communicate within in-vehicle wired networks, called on-board or in-vehicle communications. Wired vehicular networks include CAN, Local Interconnect Network (LIN), Media Oriented Systems (MOST), FlexRay and Ethernet. Table 9.1 shows the different in-vehicle networks, their general applications and their data rates.

9.2.1.2 Extra-vehicle communications

Extra-vehicle (i.e., V2X) communications refer to the communications between a vehicle and the outside world, including communications of V2V, vehicle-to-infrastructure/RSU (V2I or V2R), vehicle-to-pedestrians (V2P), vehicle-to-cloud (V2C) etc. V2X communications use wireless techniques for information exchange, including cellular-based networks, Global System for Mobile communications, long-term evolution (LTE), radio frequency (RF), 3G, 4G, 5G, dedicated short-range communication (DSRC) and the [11] wireless access in vehicular environments (WAVE) (based on IEEE 802.11 p), GPS, LiDAR and radar, all of which form a vehicle ad-hoc network (VANET). Moreover, V2X communications can provide data to intelligent transport systems (ITSs) to enhance traffic management. These technologies could become mandatory to build more reliable autonomous vehicles on roads in future [5].

With V2X communications, data (e.g., position, speed and other state parameters) can be transmitted from a vehicle to another vehicle, road infrastructure and pedestrians, etc., within an ad-hoc mesh network, thus preventing or reporting possible accidents. A vehicle is a safety critical system, and its safety is entirely dependent on the functionality of onboard sensors, cameras and radars. Hence, V2X communications should be far more effective than general embedded systems in

Table 9.1 In-vehicle networks, data rate and general applications

Network	Data rate (bps)	Applications
CAN	125 k to 1 M	Powertrain and engine, chassis and safety
LIN	10 k to 20 k	Body control
MOST	25 M to 150 M	Multimedia and entertainment
FlexRay	10 M	Safety critical, Drive-by-X
Ethernet	100 M	Diagnostic

other non-safety-critical application domains. The current autonomous security systems from original equipment manufacturers in a vehicle include blind spot monitoring [12], anti-schlupf regielung/traction control system/automatic stability control [13], electronic stability program [14], forward collision warning [15], automatic emergency braking [16], brake assist system [17] and lane departure warning system [18]. The cooperation with these existing security systems through V2X techniques allows efficient management of possible hazards for autonomous vehicles on roads. The critical challenge of V2X communications in an ITS is the reinforcement of system security, privacy and confidentiality of data sent in broadcasts and multicasts, agreed by vehicle manufacturers and the stakeholders on vehicle ecosystem supply chains.

9.2.2 Cyber threats to CAVs

A key characteristic of cybersecurity is the constantly evolving threat landscape, a challenge exacerbated further within safety-critical systems such as autonomous transport. Autonomous transport systems have several key features that are of note when considering cybersecurity.

1. Autonomous vehicles operate within a physical environment, safely navigating to ensure collision free with obstacles and effectively interacting with human agents;
2. Autonomous vehicles have the potential to affect the environment;
3. It is technically and operationally possible for adversaries to access both the environment and the vehicles' systems remotely or physically.

The compromise of autonomous vehicles' functionality due to cyberattacks represents a number of significant risks against transport safety, resulting in serious injuries or even death to road users. The issue of cybersecurity is therefore a critical challenge for CAVs.

The robustness of autonomous vehicle software requires automobile manufacturers and their suppliers to develop software that can perceive and comprehend the environment, predict the behaviour of dynamic agents within the surrounding environment and execute manoeuvres in a way that does not contribute to an unsafe case and does not cause the user any discomfort. Therefore, the on-board software of an autonomous vehicle is complex and critically significant. Modern CAVs have approximately 100 million lines of code [19], while Microsoft Windows code is similarly based on 70+ million lines of code [20]. Such a large volume of code is inherently insecure as highlighted by the discovery of vulnerabilities and the constant release of security alerts from all operating systems (e.g., Windows). This presents a cybersecurity dilemma for CAVs: should we keep on following the same software security path as the operating systems with the traditions of potentially overwhelming software update management or may a different security model be considered?

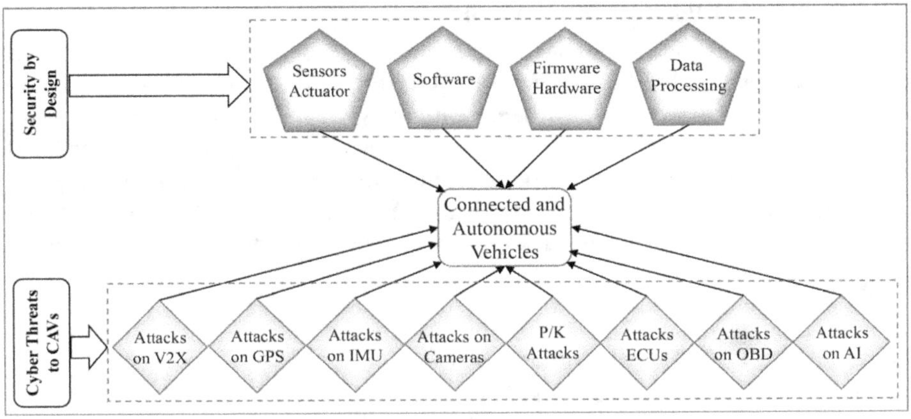

Figure 9.2 Security by design against cyber threats to CAVs

To understand the cyber risk posed by cyber-physical systems (such as CAVs), the asset should be considered as the integration of its hardware and software based sub-components [21] – opting for a collective and holistic risk management and cybersecurity countermeasures rather than a solo approach of protecting components individually. Figure 9.2 provides the framework of security by design, concerning four components, sensors/actuators, software, firmware/hardware and data processing for privacy, and mitigating the identified cyber threats to CAVs.

9.2.3 Attacks on CAVs

CAV security is a broad topic, as displayed through the huge variation in deployed applications across diverse operational environments. Different types of vehicle systems have different structures in hardware, software and communication modes. Hence, they may have different attack surfaces, facing different types of cyber threats for different motivations or purposes. We intend to cover general aspects of existing research on the cybersecurity of CAVs and provide a reference for the implementation of security by design in the development of CAVs. With a constantly evolving threat landscape, strong research and development investments in cyber defence should be considered a continuous task. For more in-depth coverage on VANET security challenges, one can consult [22–27]. ITS cybersecurity and cryptographic challenges are addressed in Hamida *et al.* [28]. In Petit and Shladover [29], various cyberattacks on CAVs are compared in terms of feasibility, severity and preventable measures, and the security vulnerabilities of CAVs and mitigation efforts are overviewed. Furthermore, a good resource about knowledge gaps, which may be used as a road map, addressing the cybersecurity challenges in the CAV sector is provided in Parkinson *et al.* [30]. Here, a very brief study of important attacks that could produce severe impact on CAV systems is provided as follows.

9.2.3.1 Global Positioning System

The vehicle navigation is a crucial component of CAV operations and heavily relies on the data from Global Positioning System (GPS). GPS signals are vulnerable to in-band interferences because of being extremely weak broadcasted signals over wireless channels. Therefore, even a low-power interference can easily jam or spoof GPS receivers within a radius of several kilometres [31]. Such GPS jamming and spoofing could compromise CAVs or existing vehicles [29, 32–35]. The exploitation of GPS vulnerabilities can have a severe impact on the reliability and safety of the CAV operations.

9.2.3.2 Inertial Measurement Unit

Inertial Measurement Units (IMUs) refer to the combination of the Gyroscope and accelerometers, which provide data on the velocity, acceleration and orientation of a vehicle. Such sensors can be exploited by an advanced persistent threat with the interference, subsequently impacting on the performance and reliability of CAV operations [30].

A crucial requirement is that a vehicular system should be able to understand the environment, handle motion around obstacles, avoid any potential collisions with obstacles in the surrounding environment (including other moving vehicles). At the current time, most of these operations rely on LiDAR systems, and an adversarial attack on these systems could feasibly influence CAV operations, with negative consequences for the safe navigation of CAV [29, 36, 37].

9.2.3.3 Monoscopic and stereoscopic cameras

CAVs use different types of visual-recognition systems (cameras) for lane detection, traffic sign recognition, headlight detection, obstacle detection and so on. Accurate functioning and visual processing may be partially disabled by the use of high-beam headlights or headlights of the opposite vehicles – a very simple way to currently subvert sensor measurements [29]. It is also possible to disable the visual-recognition system with a malicious intrusion software.

9.2.3.4 Passcode and key attacks

Access to a vehicle (in particular, an on-road vehicle) is usually carried out by using a key, passcode or similar security authentication feature. In recent years, such security features are continuously compromised. This requires significant technical improvements [38, 39, 40].

9.2.3.5 V2X network attacks

CAVs are enabled to communicate with other vehicles in the surrounding environment and infrastructure in VANETs to improve the safety of vehicle navigation. However, the channels for V2X communications represent potential vulnerabilities that increase the possibility of attacks by malicious actors [41–44]. A malicious actor can remotely connect to a CAV via an interface and intends to influence the operations of the vehicle, once finding an appropriate vulnerability in the vehicle.

9.2.3.6 On-board diagnostics: port-based attacks

On-board diagnostics (OBD) and OBD ports are presented in almost all the vehicles [19]. These ports are used for collecting any vehicle faults and performance, and a user can communicate with the Engine Control Unit (ECU) through OBD ports and CAN bus. This could provide a way for a malicious actor to modify the ECU via these OBD ports [12, 45, 46].

9.2.3.7 ECU firmware tampering attacks

A malicious entity can modify the firmware of an ECU using the external interfaces (like OBD ports), thus changing the behaviour of the ECU [47–50].

9.2.3.8 Attacking machine learning models

The autonomy of CAVs greatly relies on AI technology. Machine learning (ML) technology, as a branch of AI, can be used to pre-process the data from sensors, fuse the data and make the decisions/prediction for CAVs. In recent years, it has been shown that the training of a ML algorithm could be subverted due to adversarial sampling, thus leading to incorrect and unreliable predictions and decisions [39, 51–53]. Hence, the security of ML models should be considered in the design of an ML decision system, thus to minimise the errors brought by adversarial sampling attacks, which use adversarial examples to create ML models to make a mistake and other possible attacks.

9.2.4 AI as a cybersecurity mechanism

In the cybersecurity domain, a protection system needs to adapt the constant evolution of adversarial behaviours and cyberattack techniques. According to the Symantec Cybercrime Report [54], the overall number of vulnerabilities has increased by 13% in 2018. Similarly, according to cybersecurity ventures [55], zero-day exploits will grow from one per week in 2015 to one per day by 2021. He *et al.* [56] explored the challenges of cybersecurity in IoT-enabled cyber-physical systems and the opportunities of computational intelligence for cybersecurity. The value of AI technology has been demonstrated in cyberattack detection or anomaly monitoring in cyber-physical systems. Hence, AI could be increasingly employed for cognitive cybersecurity, as an addition to cybersecurity countermeasures of CAVs.

On the other hand, it is impossible for a human to keep pace with the large number of cybersecurity events and related activities on a daily basis on top of an already daunting threat landscape [57]. Also, there is a crisis of lack of skilled cybersecurity practitioners. According to Ciccone, the cybersecurity job market will grow by approximately 6 million roles globally by 2019, with potential shortages of trained professionals up to 25% [58].

Hence, the automation of decisions and actions, generating the alerts when an anomaly occurs in a network and system, has the potential to help overcome the challenges of cybersecurity – in both technological and business-operations (e.g., labour shortages).

Considering the increasing financial burden of cybersecurity, coupled with the increased number of attack vectors and the lack of skilled cyber-security practitioners, AI technology may be highlighted as a viable security tool to facilitate the automation of cybersecurity tasks. ML has been successfully deployed in many domains, such as image classification [59], object detection and recognition [60], language translation, and voice synthesis [61]. Recently, there has been a lot of exploration of ML for data-driven cognitive cybersecurity, for example, incremental support vector machine [62] and linguistic attribute hierarchy [63] for spam detection and various ML techniques for malware detection [64] and intrusion detection [65].

Deep learning (DL) is part of a broader family of ML techniques based on artificial neural networks. DL generally requires fewer resources for manually engineering of feature extraction and less special knowledge [66]. Signature-based threat detection works only for known threats through pattern recognition, and signature creation takes time, once a new malicious intrusion or virus is detected. Sandbox-based threat detection that performs dynamic analysis on files in a virtual environment also has limitations. Certain file types (e.g., Deep Learning Library) and large file sizes cannot be analysed in malware sandboxes. DL is not the panacea for all security problems, but it is ideal for detecting known and unknown cyber threats, and can do so in a fraction of a second to keep pace with the onslaught of attacks. As DL can transform raw data into abstracted representation, it is good for zero-day vulnerability and zero-day exploitation, where the term 'zero-day' refers to a newly discovered software vulnerability, which has not been fixed. There has been some research on DL for detecting malware [67, 68], cyberattacks [69] and intrusions [70].

AI as a technology for the automation of cybersecurity is expected to continue to capture the attention from researchers and professionals, with clear impact already on cybersecurity [71]. Hence, there is sufficient market interest in both research community and commercial environments. However, as discussed in [72, 73], ML models could be vulnerable to adversarial examples. When a ML model is trained with adversarial examples, it could overfit to the adversarial examples, thus leading to wrong results at test. Therefore, one of challenges of deploying AI-based techniques (ML/DL) to security domains is to solve the overfitting problem. In real-world applications of cybersecurity, it is difficult to check if the collected data are wrong or not. Namely, ML/DL techniques themselves do represent a grand solution for the automation of cybersecurity but requiring correct (or trustworthy) features or data to be available. In addition, an AI model, as a program in a system, could be hacked, like other programs in the system. This could produce unexpected results. Therefore, the robustness and security of data collection systems and ML models need to be investigated in the system design.

9.2.4.1 ML/DL in CAVs

Intelligent transportation systems (ITS) refer to the entire range of components within vehicular systems and are deployed in a variety of CAVs [74–76]. CAV's autonomous operations require processing of a large volume of ITS data, often collected from a large number of sensors and communication links. Designing a reliable and safe ITS requires demonstrable robustness to attacks on the vehicle as a whole as well as the system's sub-components. The existing research related to CAV security often only considered a static attack model. However, in the real-world applications, malicious actors actively update attack strategies to increase the likelihood of a successful attack on an ITS. To evolve security defence mechanisms to match the attack techniques, adaptive ML/DL could represent a valued mechanism for developing robust, reliable and secure CAV systems.

ML models can be evaluated with different performance indicators in terms of the tasks within a CAV environment. Time sensitivity and computational resource restriction are part of rigorous performance criteria on ML/DL usage for cybersecurity of CAVs. There has been much research on ML/DL for safety issues of CAVs – covering the data fusion of various sensor inputs (e.g., camera and radar data) for obstacle detection, trajectory planning, decision-making and eventual vehicle control (braking, acceleration and steering) [77]. As a CAV is a safety-critical system, cybersecurity must constantly evolve to incorporate with the original safety requirements under the absence of human intervention such that a CAV is able to respond to not only the safety threats from surrounding environments but also the security threats from the cyberspace. The dynamic landscape of cybersecurity makes it difficult to take a meaningful abstract and uniformly apply a performance measurement framework [78].

Burton *et al.* [79] provided a comprehensive checklist to assess the scope of ML/DL applications to CAV safety issues. With minor modifications, the checklist could be applicable to consider the scope of ML/DL applications for the security of CAVs. The checklist includes the following requirements:

- The application context is well defined and reflected in training data.
- The function is robust against distributional shift in the application context.
- The function exhibits a uniform behaviour over critical classes of situations.
- The function is robust against differences between its training and execution platforms.
- The function is robust against changes in its system context.
- The residual risk associated with functional insufficiency in the detection function is acceptable.
- The functionality and performance of the detection system are articulated and acceptable to the overall system stability and reliability.

An appropriate ML/DL algorithm for cybersecurity or safety of a CAV should rightly answer the checklist above, and it can be designed as a component/function of a specific or holistic system. The design of high performance ML/DL algorithms for CAV cybersecurity is a challenge. There have been some interest results for

CAV cybersecurity, using ML/DL techniques [11, 80–84], but implementing a comprehensive and secured ML/DL framework for holistic cybersecurity and safety of CAVs is still a long way to go.

9.2.5 Open challenges

There are a number of open questions in the field of ML applicability to cybersecurity. They should be applicable for the cybersecurity of CAVs too.

1. Impact analysis of policy change: It is expected to check and update policies and procedures regularly, regarding the security and privacy in an corporate environment. The impact assessment of such policies on the enterprise environment is usually based on the knowledge of experts in cybersecurity. If ML/DL is used as an approach to solving the issues of security and privacy, the policy should be set to reflect the learning and execution of the ML/DL method. Potential security metrics cover a broad range of measurable features from security audit logs of individual systems to the number of systems within an organization that were tested over the course of a year [85]. An operational cybersecurity metrics is very crucial tool for specifying the measurement of security states and susceptibilities in a corporate environment, thus to improve security assurance [86]. Predictive impact analysis of policy changes on ML/DL-based security and privacy should be part of the cybersecurity metrics.
2. Policy-oriented ML/DL: Evolving security and privacy objectives reflect the policy changing. The implementation of ML/DL should adapt to the policy changing. Hence, the training data of ML/DL should represent the change of security policy. Updating the training data and retraining ML/DL models require additional performance, time and financial resources. The challenge lies in finding a good trade-off between continuous update of ML/DL for improving the cybersecurity of a system and the updating cost.
3. Preparing ML/DL to cope with the 'future': The evolving cybersecurity landscape requires ML model adaptive to the continuous change. A semi-supervised learning with generative models could allow for effective generalisation from small labelled data sets to large unlabelled ones. Hence, a semi-supervised learning has the capability to recognise new patterns, which will be added to the training set to further improve the ML/DL model. Creating such an effective mechanism for semi-supervised ML/DL is a challenge.
4. Isolated or collaborative learning: Isolated learning has both advantages. The positive of isolated learning lies in that the training set includes the behaviours specific to the organisation, and the learned ML model well performs for the specific organisation. However, due to the limited data, the learned model is not generic for more actors. The ML model through isolated learning may not recognise a new cyberattack. In contrast, a collaborative learning shares data (more instances) from multiple actors, and each actor of which benefits from the learned model. Collaborative learning has the potentials to improve security

defences against developing, previously unknown attacks. However, collaborative learning also introduces some additional challenges:

   (a) Knowledge-based collaboration: In collaborative learning, the learning algorithms share their learned knowledge or simply share the raw records of the out-of-profile observations. The challenge is how to find an optimal approach to sharing prior knowledge between multiple ML/DL instances.
   (b) Raw record-based collaboration: Raw record sharing could lead to the leak of sensitive data, and subsequently violate privacy requirements. For raw record-based collaboration, efficient and strong anonymisation techniques must be developed to protect privacy and secure sensitive data, but retain sufficient features to train ML/DL models.

5. Ignoring some data: Some specific data samples from the knowledge base may need to be updated/removed in some situations.
   (a) Mislabelled data in the training data set should be revised, once they are discovered;
   (b) Adversarial samples need to be removed from ML/DL knowledge;
   (c) The information related to the privacy of a consumer/user, in the exercises of RtR (Right-to-Rectification) or RtF (Right-to-Forget) under GDPR, should be removed from the training set for ML/DL training. This remains an open question as to how this requirement can be best achieved, with increasing public awareness on privacy issues in future, and adversaries are becoming better-equipped to implement adversarial attacks on ML/DL systems.

9.3 Privacy in CAVs

Privacy may be viewed as a branch of security. A data breach has the potential to severely damage an organisation's bottom-line, including but not limited to financial, regulatory or reputational damage. The global cost of data breaches has increased by 6.4% [87], even without taking the potential penalties imposed by the GDPR into account [88]. As per the GDPR, an organisation can be fined up to AC10 million or 2% of the firm's global turnover for a small offence (whichever is greater). For a serious offence, an organisation can be fined up to AC20 million or 4% of a firm's global turnover (whichever is greater) [88]. Obviously, privacy issues exist on CAVs and in CAV supply chains as well.

9.3.1 Privacy issues of CAVs

In the context of continued proliferation of information, the explosion of devices connecting to the expanding communication infrastructure, and a rapidly evolving threat environment, a greater number of vehicles are increasingly connected to the outside world through different communication networks. The automotive industry faces various cyber threats in the attempt to move towards the successful implementation of CAVs. In 2017, for the first time, there were more cars added to the cellular networks in the United States than phones. At the same time, 48% US customers were not aware that connected cars could store their personal data. The huge amount of data is produced by connected vehicles, including GPS tracking, use of systems or wider sensor data, and these data could be shared with third parties. A Deloitte survey [89] from the United Kingdom shows 68% of respondents would share their personal information with automotive manufacturers or dealers, and 51% of respondents would share their personal information with commercial third parties. These results highlight the privacy implications of collecting and sharing these data.

New technology continues to constantly emerge, often with benefits that can (or at least, appear to) be obviously observable and explained. However, there is a risk in ignoring some of the effects and consequences in implementing these new technologies. For example, consumers may relinquish their privacy to use many of the convenience applications on autonomous cars, similar to behaviour observed in existing human–computer interaction with smart phones or social media usage. The automotive industry may record any, and all, relevant data with intentions to improve the performance, safety and security of a vehicle, as well as for cost-saving purposes. Data collected and shared by autonomous vehicles may include the present location of an autonomous vehicle user, the user's past travel patterns as well as their future travel plans, thus making the user susceptible to targeted marketing, law enforcement or surveillance [90]. Autonomous vehicles that select routes on their own may limit human autonomy and compromise privacy in terms of the route they undertake [91]. The V2V technology allows the communication between vehicles to exchange information dynamically [92]. Such instant information exchange between vehicles enables them to have the capacity of sensing threats and hazards, assessing risks and estimating an anticipated situation with AI technology, thus to avoid and mitigate accidents. However, V2V communications for safety can compromise privacy by controlling communication with other cars [93]. The risks and awards of autonomous vehicles are summarised in Table 9.2.

In November 2015, the German motorist organisation, Allgemeiner Deutscher Automobile Club, discovered that large amounts of data were being captured by the OBD system of a BMW320d. Data included driving destinations and phone contacts, captured without the permission of users [95]. Other models of vehicles exhibit the same concerns. Previously, these data could only be accessed by directly connecting to the OBD, which means that only relevant persons in manufacturing process could access the data. However, now the data can be transmitted through

Table 9.2 Risks and awards of car data sharing [94]

Risks	Awards
Location tracking: Personal habits and schedules could be taken advantage of by unwanted third parties	Enhanced safety: Improved road safety; enhanced security through criminal tracking
Profiling: Data could be mined on a grand scale for social engineering	Personalisation: Entertainment, content and settings all aligned to individual user-preferences
Private sphere: Individuals could lose control over their public persona	Cost reduction: Lower insurance premiums based on real, individual driving patterns

wireless communication, which highlights the potential for data loss through a targeted cyberattack.

Sensors, radars and cameras work together to process information that a human driver may not see, and as systems are connected outside world through communication networks, any accident or possible threat record can be reported immediately by the CAV, thus reducing the risk of accidents. However, while advanced IT technology is helpful for data collection, this does not negate the need for privacy as a priority issue in responsible technological development. During the development of CAVs, privacy has been raised as a central concern, leading to a developing awareness of privacy implications in the autonomous vehicles' space. This raises higher privacy requirements to CAVs and various ecosystem services. To avoid privacy invasion, privacy by design should be implemented in automotive manufacturing. This represents some critical challenges to the automotive industry.

9.3.2 Data generated by autonomous vehicles

The amount of data being captured from CAVs has grown significantly in recent years, and this trend is expected to continue, as CAVs are widely considered as 'mobile data centres on wheels'. These changes in the scope of data-collection processes may not always be obvious to the user. It is predicted that autonomous cars will generate more than 300 Terabytes of data per year [96]. There were 38.2 million licensed vehicles in the United Kingdom alone at the end of 2018 [97]. Such a huge amount of vehicle-generated data, available to share in the age of connected cars, raise obvious questions about data privacy. Figure 9.3 shows the data generated on connected cars and the data usages of online activities per hour. [98]

The United States defines six levels of autonomy of cars, from level 0 (representing zero automation) to level 5 (fully automated). At present, the most advanced cars on the market are only at level 2 autonomy [96]. Audi calls its new A8 a

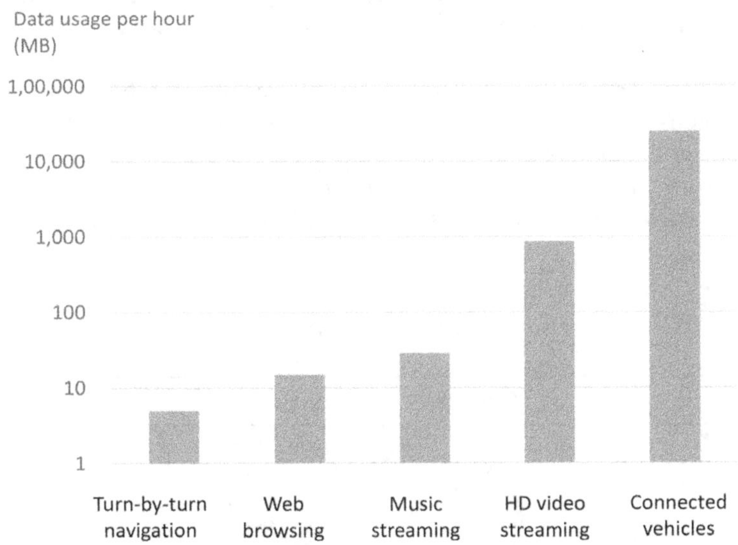

Figure 9.3 Big Data on wheels

level 3 ready autonomous car – meaning the car has the potential to drive itself in certain circumstances, where it will assume control of all safety-critical functions, and twenty car makers say they'll sell autonomous cars in the United States by 2022 [99].

The level of a car's autonomy is determined by the number of installed sensors, which are used to ensure an autonomous system operates as designed and the inherent complexity to exploit the sensor data. Dmitriev [96] estimates some sensor-generated data, as shown in Table 9.3.

Besides sensor-generated data, other data may be collected for different purposes. The infotainment centre of an autonomous vehicle may have additional capabilities compared with today's technology: (1) the data from play-lists from customers', the location of preferred sites (e.g., usual coffee shops) may be sent to the vehicles manufacturer, (2) the robust navigation system records the journeys of users and (3) mobile payment may be shifted to mobile in-car payment; hence, personal bank information and shopping will have be captured. These represent just some of the possibilities that autonomous vehicles may have the data to be 'marketised' at the request of various non-safety related business cases.

CAVs may also have a 'live box' to record the inside of vehicles, capturing data if an accident occurs. These data may plausibly be used in court cases as a reliable source of evidence. This 'live box' is different to a black box (i.e., used to record any accident on aeroplanes, which is only opened after an accident occurs). With advanced communication techniques, the box on CAVs may be connected to the outside world ('live') and hold the intelligent capacity to report any anomalous feature of the vehicle or wider environment.

Some customers may put data-collecting devices in their cars. Transponders are a convenient way to pay tolls without having to stop. Also, vehicle monitoring devices provided by insurance companies may help good drivers save money on their vehicle insurance premiums. When considering a future with fully autonomous cars, the insurance company may insure the automakers or service providers rather than customers.

The Federal Trade Commission categorises three different types of CAV data: aggregate, non-sensitive and sensitive [100]. Aggregate data are generated from raw data from a group of cars through statistics or AI techniques to discover solutions for traffic management and journey patterns, amongst other purposes. Non-sensitive data are non-identifiable information about a specific car, for example, measuring

Table 9.3 Sensor-generated data in an autonomous vehicle [96]

Sensors	Quantity	Data generated per sensor
Radar	4–6	0.1–15 Mbits/s
LiDAR	1–5	20–100 Mbits/s
Camera	6–12	500–3500 Mbits/s
Ultrasonic	8–16	<0.01 Mbits/s
Vehicle motion, GNSS, IMU	1	<0.1 Mbits/s

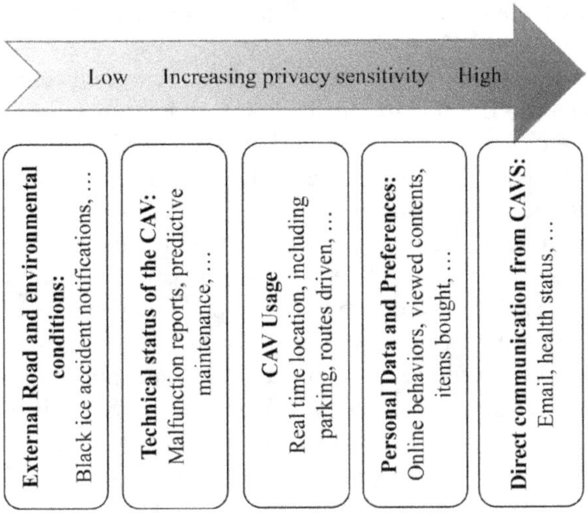

Figure 9.4 Increasing privacy sensitivity of CAV data [112]

vehicle's health and various weather conditions. Sensitive data are personal data specific to customers, including addresses or frequently visited places. Figure 9.4 depicts the increasing privacy sensitivity of data from CAVs. This framework offers a simple yet effective method to understand the data that CAVs collect and share, and how this impacts data privacy decisions.

9.3.3 Who wants these data?

Now that data have become a valued industry asset, data from CAVs may be used by a number of different stakeholders.

Automotive manufacturers may collect data for many reasons, though it is expected the main reason would be to facilitate service provision for vehicle owners. The data, collected from sensors and cameras that track performance and surroundings of a CAV, are likely to be highly valuable. It is speculated that by 2020 automotive manufacturers will be able to make more revenue through selling vehicle data as opposed the actual vehicles themselves [101]. The main goal of collecting vehicle data is to 'teach' driverless cars how to operate, therefore building an autonomous vehicle that may safely operate on roads with the capability to deal with unexpected situations effectively. In the case of the automotive industry, large data pools may produce insight into how vehicles are performing once they have left the manufacturer. The next step is for manufacturers to extract the value from this insight, creating actionable plans for innovation, in order to create the best possible vehicles for their customers. For an automobile manufacturer, the most precious data that a connected car might provide relates to driver behaviour and usage information. This indicates how the user interacts with the components that make up the car – from how often they engage the windshield wipers to the use of their hazard lights. In isolation, data from a single car are not inherently valuable to a manufacturer. However, when these data points are aggregated and analysed from potentially millions of vehicles, manufacturers may properly evaluate performance and identify improvements.

Manufacturers of infotainment centres may also scrape customers' data through apps, for data such as GPS record, music, home address, emails, contact numbers, online footprints etc. From these types of data, they may discover new patterns or the preferences of customers. Thus, data analytics could help shape the nature and performance of infotainment systems.

Insurance companies would likely be willing to invest significant financial resources to efforts identifying and deriving the driving habits of users, as driving records may prove a valuable indicator of a driver's competency. With the increase of autonomous cars, there is a chance that the number of drivers will decrease. However, this would nonetheless involve a significant transitional period, during which cars have an option to operate with human agents as drivers, and there are likely to be users who prefer aspects of manual operation even in the age of fully automated vehicles. As an example of current user choices in autonomous functions, Volvo's IntelliSafe Auto Pilot interface allows drivers to switch in and out of autonomous mode [102].

Third parties may collect data through dongles that connect to a port in a car, using data analysis to discover new information from the raw data or any pattern to improve performance of traffic management or performance of the CAV. Analysing the driving habits of road users might help influence the future design of autonomous vehicles and wider infrastructural structures (e.g., charging stations).

Hackers are expected also want to have the data from CAVs: by tapping into a CAV, a hacker could control the vehicle remotely. The personal information taken from CAVs may then be sold or leaked to the public domain. Therefore, manufacturers need to work closely with cybersecurity experts to ensure connected cars are being designed securely, implementing both security by design and privacy by design principles.

9.3.4 Compliance with GDPR

The EU's GDPR is a data protection and privacy regulation that every company operating in the EU or handling the personal data of EU citizens has to comply with. It has taken effect as of 25 May 2018. The principles of and rules on the protection of natural persons with regard to the processing of their personal data should, whatever their nationality or residence, respect their fundamental rights and freedoms, and in particular their rights to the protection of personal data. Data play an increasingly important role for vehicle manufacturers and manufacturer of infotainment systems, so ensuring that they comply with the GDPR is essential.

GDPR aims to protect consumers from actors who would try to misuse or steal user information, including those cases, which could potentially compromise consumers' digital identities. To remain compliant with GDPR, it is essential that manufacturers ensure they can manage and store all required data without violating the data rights of customers. Explicit customer permission is a core principle in GDPR. Manufacturers must ensure that connected car data are not captured, processed and shared without a customer's proactive consent. This is one of the privacy challenges both for vehicle manufacturers and manufacturers of infotainment systems of CAVs.

Privacy regulations distinguish between the data that can be used for customer intelligence and the data that constitute personally identifiable information (PII). The US government agency, the National Institute of Standards and Technology (NIST), provides a formal definition of PII [103]: PII is any information about an individual maintained by an agency, including (1) any information that can be used to distinguish or trace an individual's identity, such as name, social security number, date and place of birth, mother's maiden name or biometric records, and (2) any other information that is linked or linkable to an individual, such as medical, educational, financial and employment information. Under GDPR, much of the collected data could be defined as PII broadly, as even an anonymised vehicle identity number may be used to find out who the owner is with access to the right databases.

Connected automobile companies have been collecting data from connected cars for years. GDPR does not prevent car companies from collecting these data, but it does put heavy restrictions on how this data is used and who may access to the data.

Manufacturers must have customers' consent in order to collect or use any customer data, and customers retain the right to request data deletion at will. Connected car initiatives have driven manufacturers to ask customers to sign a Connected Car Privacy Policy as part of an account set up. These policies must be reviewed to ensure they are in line with GDPR requirements with particular attention paid to the rights which individuals have [95]. A negotiation between privacy and CAVs could run in a certain period.

9.3.5 Privacy by design for CAVs

The concept of 'Privacy by Design' (PbD) has been widely accepted by regulators around the world as a fundamental aspect of privacy protection and is a key concept in GDPR. Autofacets [104] illustrate the stakeholders involved in the CAV life cycle and wider ecosystems, as shown in Figure 9.5. Under GDPR, it is good practice for all parties involved in CAV-related data to document their data processing activities. First, all engineers in the life cycle of CAVs should understand the meaning of PII and the implications for handling such data. This should be a reflection of a wider awareness of privacy issues by developers and designers of autonomous vehicles, encouraging greater PbD in the develop awareness would likely filter to a stronger awareness of wider privacy issues at early stage of their development and make good design solution for privacy in the system.

PbD implies that privacy should be taken into account throughout the whole engineering process, implementing appropriate technical and organisational measures at each stage. This ensures automakers or other stakeholders of CAVs or CAV ecosystems to secure data at every stage of the value chain in Figure 9.5. At the design stage of a CAV related project, the following aspects, but not the last, need to be considered:

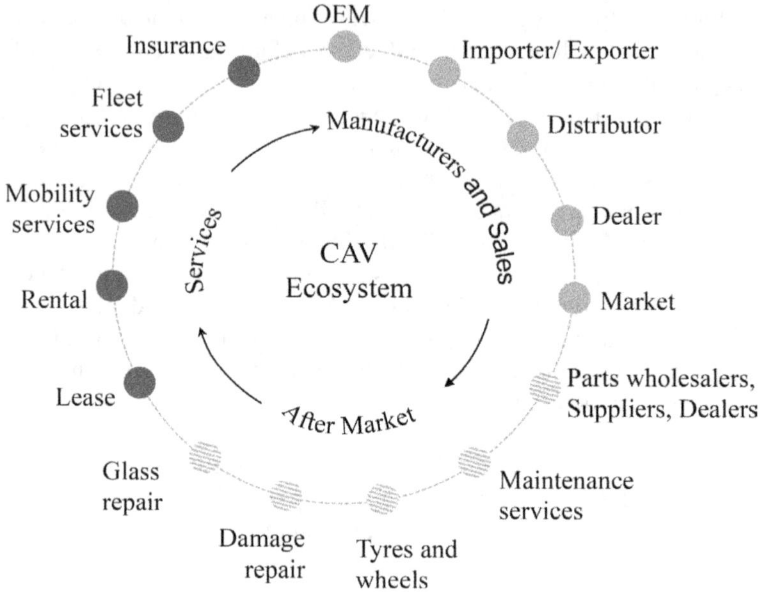

Figure 9.5 Stakeholders in the life cycle of CAVs and Ecosystems [103]

- Which kinds of data from a CAV or a CAV service count to PII?
- How much data should be collected? Data collection should be minimised.
- How should data be stored? Should personal data be stored separately to non-personal data?
- How long the data should be stored? Data storage must not be stored longer than necessary and must satisfy GDPR requirements.
- Are data access controls compliant with regulations (e.g., GDPR)?
- What internal policies or strategies need to be developed to reflect privacy considerations?
- How should data be shared with external parties, remaining mindful of privacy considerations? For example, should a third party wish to use CAV data, a data retrieving system should be capable to only distribute non-sensitive information, with mechanisms to prevent the sharing of hidden, sensitive or personal data.
- How should customers be made aware of the use of their data? When customers use a CAV or a CAV ecosystem, a warning should be provided to the customers, indicating what personal data will be collected and how the data will be used. Consent should be collected from the customer before and data collection or analysis.
- A Connected Car Privacy Policy, aligning with GDPR or other regulation, should be automatically produced, and allow customers to sign as part of their account set up, as same as all IT services doing.

- Creating a dynamic data flow to show the data process activities, where the data are held, where and how they are transferred and how the data are processed, etc.
- An appropriate and structured review process should be in place that involves the relevant stakeholders, including product development teams, connected-car teams and legal departments.
- How do customers delete data on themselves? There should be a mechanism to allow customers to delete their personal data or allow customers to amend their consent to data collection and processing. This mechanism should enable the automatic removal of personal data as required by the customers.

As a real-world example, BMW recently launched CarData, positioning itself at the forefront of GDPR compliance. This new service provides a high level of transparency, security and puts their customers in control of their telematics data [95].

In the journey towards complete connected and autonomous vehicles, the interaction between privacy and CAVs is likely to be an ongoing negotiation between manufacturers, regulators and other stakeholders. Regulations considering aspects of privacy may require updates or subsequent additions as the development of CAVs continues. All stakeholders with experiences of data protection should keep their eyes on both technical innovation and the regulatory landscape, and update systems and services to realigning with the development of CAVs (and CAV) ecosystems based on regulatory compliance.

9.4 Autonomous vehicle security: economics and wider landscape

Due to common vulnerabilities, instances of security breaches occur across firms and in patterns that are difficult to anticipate. It was estimated that malicious cyber activity cost the US economy between \$57 billion and \$109 billion in 2016 [105]. In FY2018, DARPA allocated about 10% (\$41.2 million) of its research budget to cyber sciences. There are several significant hurdles to turning prototype autonomous vehicles into mass market models. These include the challenges of ensuring safety levels to satisfy regulatory requirements across transport domains of ensuring the economics of data, asset (vehicular) and supporting infrastructure ownership models promote the development of a mass market; and of managing the manufacturer's incentive to decrease per-unit production costs. That said, the area energising engineers most concerns the challenges relating to functional safety and cybersecurity, according to the survey of 54 automotive industry experts [106].

9.4.1 Investment in automotive vehicles

There are many parts to the autonomous vehicle landscape, from investment in ML (including navigation and sensor technology), to the assistance and system communication mechanisms involved in braking, steering or navigation. Beyond obvious safety and efficiency considerations, in a rapidly growing market, many major investors could have commercial motivations to deploy ahead of competitors, and organisations have an economic motivation to offer a superior product, thus establishing their priority in the market, benefiting from 'first mover' competitive advantage, and the improved reputation as market-leader. This could strengthen the organisation's ability to dominate and influence both the maturing consumer market as well as the developing infrastructure. Late entrants to the market may find higher barriers to entry as the market usually tends to consolidate towards the largest providers. Autonomous vehicle manufacturers must therefore address the innovation of cybersecurity to make CAVs widely acceptable by customer market.

The exact expenditure on the R&D of autonomous vehicles is difficult to determine, as the competitive nature of the market dissuades actors from making investments open-source, but estimates have valued the market tripling from 2019 to reach US556bn in 2026 [107]. A broad range of players are investing heavily in automotive technology. Not only some of the larger corporations, such as Google, Uber and Tesla, are working on autonomous vehicle technology but also a large range of more specialised start-ups are investing in this field. An abundance of smaller organisations are highly active in more specialised research, including non-profit organisations and governments and engaging the exploration of the future of connected transport and automated infrastructure. Investments in autonomous vehicles are targeting progress across a range of objectives, from technical excellence (including performance and safety) to broader societal goals of connecting communities, thus developing real-world connected cities and providing optimal transport options, instead of traditional private vehicles.

9.4.2 Innovation in security and safety

International standards, International Organization for Standardization (ISO 26262) and Society of Automotive Engineers (SAE J3061) have been proposed for vehicle safety and security, respectively, but mainly for manual vehicles rather than autonomous vehicles. The international standard SAE J3016 is for level 3+ automation vehicles. Nevertheless, the security challenges in autonomous vehicles need to be invested along with their critical safety, and different driving automation levels need to be taken into consideration when designing the safety and security of autonomous vehicles. The co-design of safety and security of CAVs is demanded. A level-based design of safety and security for autonomous vehicles have been proposed in Cui and Sabaliauskaite [108].

Also, a level of risk acceptance must be agreed for deployment, at which each autonomous vehicle technological breakthrough is deemed to meet the threshold for safety. A number of viewpoints feed into the risk acceptance process, by the manufacturers and all stakeholder members, such as security testers, researchers, technical auditors, as well as government and independent regulatory and standard inputs. From an engineering, performance, design, safety and security angle, it is reasonable to expect rigorous review by various agencies before approval is given for deployment.

Innovation should keep pace with the rapid change in the CAV domain. To implement fully autonomous vehicles, for example, the capacity of vehicles in sensing surrounding environment and fast response to the dynamic environments need to be improved to ensure the safety of CAVs. Safety challenges are expected to remain in outlier cases (perhaps including extreme weather conditions, unique terrain and cases of malicious subversion). A greater investment will be required to address specialised cases.

9.4.3 The autonomous vehicles landscape

Self-driving vehicles have been on public roads since 2012 (in the United States), but it is widely accepted that a mass deployment requires vast updates to urban infrastructure. The development of autonomous vehicles further integrates digital communication within urban service provision will facilitate the adoption of 'smart city' infrastructure. Long-term deployment is likely to involve the changing the mobility, with automated shuttles or interconnected modes of autonomous vehicles (covering earth, land and air) to door-to-door automated transport solutions. Outside major cities and urban areas certain modes of autonomous vehicles may require overcoming additional challenges such as diverse road conditions and a relative lack of informatics. Government policy, infrastructure investment and urban development are just some of the key variables that will determine the rate and success of widespread vehicular autonomy over the proceeding decades. In particular, government subsidies over private CAV ownership may be deployed as measures to encourage market adoption, as would policies designed to facilitate local authority procurement of CAVs as the option of choice. Regulation may also assist in driving adoption, whether light-touch regulation or stricter requirements that encourage potential users to benefit from connected aspects of autonomous transport. Lastly, public investment in open areas of research, relating to innovation on sensors to further promote collision-avoidance technology, or on urban planning and smart-city design, would prove beneficial in creating an environment in which autonomous vehicles may be deployed at scale.

As a whole, the competitive nature of the automotive industry suffers a difficulty for the large-scale data-sharing activities to improve autonomous processes. The benefits and risks of sharing vehicular data have been listed in Table 9.2 [113]. It is desirable that automated transport solutions would optimise information exchange to provide continuous updates on traffic congestion, road conditions and navigation. Currently, the industry is short of the collaborative data sharing, this is partially because there lacks an incentive when sharing sensitive data (which represents potential risk exposure). The requirements and consideration for data sharing could become more pertinent as the autonomous vehicles industry continues growing. Data sharing may be encouraged in response of the requirement of a change in the current market structure, by either a new incentive of regulatory requirements or the consolidation of industrial companies.

9.5 Maritime case study

The maritime industry is no exception to the trend towards the automation process over the last decade, and it is increasingly reliant on cyber-enabled technology to assist in everyday practices. The increased technology advances have led to a rise in both digitisation and automation, both of which are reliant upon data and its security.

In 2013, a group of researchers from the University of Texas showed that it is possible to spoof the GPS signal of a superyacht, thus changing the course of the yacht [109]. A hacker then is now able to disrupt a vessel without physical proximity to the vessel. As demonstrated at the InfoSec 2018 conference, it is relatively easy to gain access to a vessel's electronic charts (ECDIS), via the target's satellite communication terminal [110]. By changing the program code or parameters within an ECDIS system, an attacker is able to change the digital position and size of the vessel, triggering numerous alarms and warnings. These newsworthy examples highlight the risks to this increasing reliance on data-driven processes and the need for data protection.

In a commercially driven industry, like the maritime, automation is motivated by the desire for greater efficiency, through the reduction in the crew size or the time spent in port. Over the past few years, there have been numerous examples of remote or semi-autonomous processes being implemented across the world's oceans. While many of the fully autonomous ventures remain within the testing phase, the amount of data that vessel sensors collect and send back to shores is on a rapid rise, raising the criticality of securing the data.

The United Nations International Maritime Organization (IMO), the primary regulator of the maritime space, has been considering its role within the realm of autonomous vessels since 2017, and IMO has realised and accepted that autonomy is the future trend in maritime industry.

This case study will raise the awareness of the autonomy of maritime vehicles and the importance of data to the maritime industry and then show the effort that the international community through IMO has made to implement the autonomy. Finally, the challenges faced by the community are addressed in the process that maritime industry continues to move from low automation to full autonomy.

9.5.1 Autonomy and data: what it means to the maritime industry

For the maritime industry, autonomy means unmanned engine rooms, automatic route planning/correction or increased sensor data collection and processing. IMO divided autonomous ships into four categories, depending on the degree of autonomy found onboard (Table 9.4). These degrees highlight the range of possibilities that automation offers to this sector.

Table 9.4 Maritime degree of automation [111]

Degree of automation	Description
1	Ship with automated processes and decision support
2	Remotely controlled ship with seafarers on board
3	Remotely controlled ship without seafarers on board
4	Fully autonomous ship

The maritime industry has grown with the development of port infrastructure over the past decade. Appropriately used automation can improve efficiency and safety, and thus gain profits in a low margin industry. A vast majority of ports now boast a range of autonomous features, including unmanned container cranes, automatic guided vehicles and digital business systems that handle the multitude of documents required for the loading and off-loading of cargoes.

Initially, operators introduced basic task automation onto vessels for both safety and efficiency gains. Now, it is expected that the maritime industry will have fully autonomous ships sailing the high seas in the next 3–20+ years.

Regardless of the degree of automation or the types of on-board systems, data carry significant importance throughout the maritime sector. Without data, autonomy could not occur, and without secure data, the wrong actions could be carried out unknowingly by the vessels systems. It is significantly important to protect data in the age of the container.

The largest container ships can carry over 21,000 TEUs (containers), all of which look the same from the outside. Extensive data are therefore required to support the successful shipping of a container. It is vital that data pertaining to: the contents of each container, where the goods need to go, where they should be unloaded, and to whom they belong to, is stored securely.

If the data that drive the modern maritime industry is lost, it can have a severe impact on the wide economy. In 2011, a Dutch-based drug smuggling group employed a group of hackers to infiltrate the computer networks of two companies operating in the Port of Antwerp. Once getting the access to their systems, the hackers manipulated the companies' databases to discover the location of target containers, thus allowing the collection of those containers by the smugglers [112]. This demonstrates how malicious actors can, and will, exploit weaknesses in the security of the data used within maritime systems.

Ports are not the only targets of criminals through utilising data to monetise their actions within the maritime domain. In 2017, Malaysian authorities arrested and jailed a group of pirates who were believed to target specific vessels by using the Automatic Identification System (AIS) [113]. AIS is used by hackers for two reasons: (1) a simple receiver can be built from readily available parts online, following basic instructions and (2) as AIS data packets contain the location, speed and heading of a vessel. Hackers can use the data received via the AIS to target high-value cargoes and crews.

It is important to note that AIS data is utilised by a ships Electronic Chart Display and Information System (ECDIS). ECDIS is designed to assist seafarers in daily operations, by allowing them to see a comprehensive visualisation of their own location and route in relation to other vessels and obstacles. These data are also fed into collision avoidance systems, which help to reduce the likelihood of a maritime accident.

This human-centric approach to data facilitates improvements in seafarer situational awareness as it provides information and data in a user-friendly way. However, if these data were manipulated, the ships ECDIS and other interconnected systems could start displaying inaccurate information. Seafarers on board may be able to identify these discrepancies and manually override them. However, in the case of a fully autonomous vessel this last line of defence is no longer present on board. Therefore, the security of this data is critical to ensure that these vessels are able to operate in the intended way, without increasing the risk of accidents.

9.5.2 IMO approaching to autonomy

In 2017, the Maritime Safety Committee (MSC) of IMO established the Maritime Autonomous Surface Ships Working Group (MASS), outlining a two-phase work process. Stage one, a regulatory scoping exercise, is to identify current provisions within IMO instruments and assess their applicability to various levels of autonomy, ending by mid-2019, with the submission of Interim Guidelines for MASS Trials at MSC 101 [114]. Phase Two will then start to analyse the technical and operational factors that will determine the safe conduct of MASS trials on the world's oceans. Therefore, at an international level, these regulatory discussions of maritime autonomy are still in their infancy.

'MASS' has become a buzzword and has led to a significant discussion in both the Working Group and the Plenary of the major committees within IMO. This discussion has highlighted how contentious the issue of automation and autonomy is within the sector. At MSC 101 (June 2019), submissions under the MASS agenda included one debating the terminology being used to describe and define autonomous systems [115]. This document suggests that the acronym MASS should be used to mean 'Maritime Automated Smart Ship' broadening the types of automation encompassed by the term. This highlights not only the contentious nature of the discussion but also the infancy of the discussions. It highlights that the progress of implementation towards autonomy is outpacing the relevant regulation, which in turn will impact on the security of the data being used.

While the IMO is the international framework charged with overseeing the creations of MASS regulations, it lacks any enforcement capability. The IMO relies upon member states, ratifying regulation into their national laws and enforcing it through their authorities. Alongside the national authorities, the Classification Societies have significant power within the enforcement of IMO instruments. Classification Societies issue and carry out inspections that proved a vessel with its class certificate. Without this certificate, the ship would not be able to enter foreign ports, or gain insurance. Therefore, it is against this certificate that a ship could be detained if its condition is seen to be substandard.

Through the International Association of Classification Societies (IACS), a set of unified interpretations and rules are created, ensuring the fair application of regulations across all classification societies. Therefore, IACS and the classification societies will play a pivotal role in the creation and amendments to the international framework that focuses on MASS. Their involvement will ensure that the technical requirements of the regulation are developed to a standard, which ensures the safety of the crew, ship, environment and data, in the absence of humans within those practices.

9.5.3 Challenges of autonomy in maritime

For autonomous vessels of any degree to be successful, the maritime industry needs to overcome several challenges in cybersecurity; otherwise, vessels could be in an insecure situation, making them vulnerable to attackers.

One of the major challenges that autonomous vehicles have faced is the limitations in legislation. As highlighted by IMO and IACS, working on MASS, there is a significant reliance on humans within international regulations. Therefore, for autonomous systems to be tested and implemented, the regulation should be created based on the absence of human intervention. Finland, one of the most forward-thinking state in terms of autonomous shipping, has already changed its national legislation.

Amendments to The Finland Pilotage Act allows authorised pilotage companies to utilise remote pilotage in public channels [116]. These amendments allow a pilot to forgo the boarding of a vessel, instead, allow the pilot to remotely guide the vessel through the channel. However, this regulatory change is only applicable to Finnish territorial waters. Namely, these practices are not permitted outside of this area.

Autonomous shipping is facing not only political challenges but also technological issues. The research in [117] highlights that high quality of communication is required for fully autonomous vessels. It is argued that the current satellite communication infrastructure is not sufficient to support fully autonomous ships on the high seas. In land and coastal areas, 4G/5G cellular networks could provide enough bandwidth to support fully autonomous vessels. Thus, making the cybersecurity of 4G/5G networks a critical challenge for the industry. Maritime stakeholders need to consider what data is collected, used and analysed both on board or ashore, as well as which types of data needs to be protected and how they should be protected. On the path towards the implementation of autonomy, the data protection is, as always, a critical challenge in the maritime sector.

If the data being used by autonomous systems at any degree is corrupt, inaccurate or missing, it could have a severe impact on safety, security, efficiency and the environment. The complexity of maritime systems and the inter-dependencies with shore-based systems makes securing these systems especially challenging. IMO has issued MSC-FAL.1/Circ.3 Guidelines on maritime cyber risk management [118]. These guidelines offer guidance on how to assess and mitigate cyber risks within these networks. However, this is still an ongoing challenge and becomes harder, as more data are being collected and used for autonomy.

9.5.4 The future of autonomous shipping

Autonomous shipping is attractive as its success could lead to a safer, more secure, and highly efficient form of shipping. However, there is still a long journey both technologically and politically, before it becomes a reality. It is notable that at its 101st session the IMO Maritime Safety Committee commended the third version of the Industry Guidelines on cyber-security on Board Ships [119]. These guidelines produced by an agglomeration of industry stakeholders helps to enhance the understanding and awareness of cyber risk management, both technically and regulatory.

The document covers information on the response to and recovery from cyber incidents, as well as the information for ship owners and operators on how to comply with resolution MSC.428(98) *Maritime Cyber Risk Management in Safety Management Systems (SMS)* [120]. However, there is still much that regulators and operators must do to ensure that the data within maritime systems remains secure. Highly secured data are essential for the success of implementing autonomy safely and efficiently.

It is a challenge to implement fully autonomous shipping in maritime sector. Although many examples of lower degrees of autonomy have already been tested, fully autonomous ships are still in their infancy. Nevertheless, industry and regulators must work together in the interim to ensure that while innovation occurs it does not negatively impact data security.

9.6 Conclusions

Examining the opportunities and challenges in the development of autonomous vehicles depicts a complicated prospect. Autonomous vehicles promise opportunities in terms of user and passenger safety, as well as wide efficiencies across the transport and city infrastructure landscapes. Nonetheless, there are a number of key considerations and open issues on the way towards full vehicular autonomy. The acceptance of vehicle autonomy directly depends on vehicle reliability and data privacy, as well as indirectly on cybersecurity, which directly affects the safety and the data privacy in the vehicular ad-hoc networks. The connected nature of autonomous vehicles raises cyber risks, where malicious behaviour or technical failure may have serious implications on vehicle safety and represent a threat to human lives. Hence, cybersecurity is critical in this context.

AI is playing an important role in implementing 'Security by Design' for autonomous vehicles, and it provides an approach to the automation of cybersecurity. However, the security of ML models needs to be investigated for implementing high performance of AI models, of which, the data should be secured in real applications.

Moreover, there are a number of highlighted privacy concerns, as each autonomous vehicle represents a data-rich source of information. Information sharing can help not only improve the performance of autonomous vehicles, thus to speed up the innovation, but also improve marketing and thus maximise the commercial gain. However, consumer privacy should be seriously considered. Hence, 'Privacy by Design' is demanded, resilient to the updating of regulations (GDPR), on CAVs and on their supply chain is required as well.

The case study in maritime sector demonstrates the awareness of future trend towards autonomous ships, but the difficult journey towards full autonomy of maritime vehicles (ships), and the understanding that there is still much work to do in both techniques and regulations, especially data protection is a critical challenge in the autonomy of maritime industry.

Different types of vehicles could face different cyber threats due to their hardware, software and communication modes. Further analysis on different types of vehicles in respects of the characteristics, vulnerabilities, threats, attack surfaces is needed to guide the development of secure autonomous vehicles in different automotive industries.

Finally, various autonomous vehicles, such as cars, aerial vehicles, and maritime vessels, require changes to wide transport infrastructure. For example, the automated transport should be incorporated into the development of 'smart cities'. Though now we cannot clearly see how the future might work, with variables on private industry, integration between sectors or services geographically, privacy developments, state policy and regulations, as well as industry research and development, each of them represents part of the aspects that may shape future development. With the technologies moving so rapidly in this field, it is important to understand, predict, plan and make effective actions to promote the deployment of autonomous vehicles with the assurance of safety and security.

Appendix: IoT communication protocols

Table F.1 Communication/transport layer, most of protocols, frequency, data rate and range are extracted from [42]

Protocols	Frequency	Data rate	Range	Description
2G/3G (UMTS)	Cellular Bands	64 kbps, 1.44–4 Mbps	n miles	the 2nd/3rd generation of wireless mobile telecommunications technology
Bluetooth (BLE)	2.4 Ghz	1,2,3 Mbps	300 ft	wireless personal area network (WPAN) technology
802.15.4	subGhz, 2.4 Ghz	40,250 kbps	>100 miles	a technical standard defines the operation of low-rate WPANs
802.11 p	5/10/20 Ghz	max 54 Mbps	3,300 ft	an approved amendment to the IEEE 802.11 standard to add WAVE, a vehicular communication system
LoRa (Long Range)	subGhz	<50 kbps	1–3 miles	the first low-cost implementation of chirp spread spectrum for commercial usage
LTE Cat 1	Cellular Bands	1–10 Mbps	n miles	Long-Term Evolution, a type of low power WAN radio technology standard
NB-IoT	Cellular Bands	0.1–1 Mbps	n miles	Narrowband IoT is a Low Power Wide Area Network (LPWAN) radio technology standard

Protocols	Frequency	Data rate	Range	Description
SigFox (French)	subGhz	<1 kbps	n miles	wireless networks to connect low-power objects such as smart meters and watches
Weightless	subGhz	0.1–24 Mbps	n miles	a set of LPWAN open wireless technology standards for exchanging data between a base station and thousands of machines around it
Wi-Fi	subGhz, 2.4/5 Ghz	0.1–54 Mbps	<300 ft	a family of radio technologies commonly used for wireless local area networking (WLAN) of devices
WirelessHART	2.4 Ghz	250 kbps	300 ft	a wireless sensor networking technology based on the Highway Addressable Remote Transducer Protocol (HART)
ZigBee	2.4 Ghz	250 kbps	300 ft	an IEEE 802.15.4-based specification for a suite of high-level communication protocols used to create personal area networks
Z-wave	subGhz	40 kbps	100 ft	a wireless communication protocol used primarily for home

References

[1] Kaur K., Rampersad G. 'Trust in driverless cars: Investigating key factors influencing the adoption of driverless cars'. *Journal of Engineering and Technology Management*. 2018;48:87–96.

[2] Zhang T., Tao D., Qu X., Zhang X., Lin R., Zhang W. 'The roles of initial trust and perceived risk in public's acceptance of automated vehicles'. *Transportation Research Part C: Emerging Technologies*. 2019;98(9):207–20.

[3] Koscher K., Czeskis A., Roesner F., *et al.* 'Experimental security analysis of a modern automobile'. *IEEE Symposium on Security and Privacy*; Claremont Resort, Berkeley, CA; 2011. pp. 447–62.

[4] Rizvi S., Willet J., Perino D., Marasco S., Condo C. 'A threat to vehicular cyber security and the urgency for correction'. *Complex Adaptive Systems*

Conference with Theme: Engineering Cyber Physical Systems, CAS, Procedia Computer Science; Chicago, IL; 2017. pp. 100–5.

[5] Arena F., Pau G. 'An overview of vehicular communications'. *Future Internet*. 2019;11(27):27.

[6] Bazzi A., Campolo C., Masini B.M., Molinaro A., Zanella A., Berthet A.O. 'Enhancing cooperative driving in IEEE 802.11 vehicular networks through full-duplex radios'. *IEEE Transactions on Wireless Communications*. 2018;17(4):2402–16.

[7] Han C., Dianati M., Tafazolli R., Kernchen R., Shen X. 'Analytical study of the IEEE 802.11p MAC sublayer in vehicular networks'. *IEEE Transactions on Intelligent Transportation Systems*. 2012;13(2):873–86.

[8] Bazzi A., Masini B.M., Zanella A., Thibault I. 'On the performance of IEEE 802.11p and LTE-V2V for the cooperative awareness of connected vehicles'. *IEEE Transactions on Vehicular Technology*. 2017;66(11):10419–32.

[9] Park S., Kim B., Yoon H., Choi S. 'RA-eV2V: relaying systems for LTE-V2V communications'. *Journal of Communications and Networks*. 2018;20(4):396–405.

[10] Chen S., Hu J., Shi Y., *et al.* 'Vehicle-to-Everything (V2X) services supported by LTE-based systems and 5G'. *IEEE Communications Standards Magazine*. 2017;1(2):70–6.

[11] Alheeti K.M.A., Gruebler A., McDonald-Maier K.D. 'An intrusion detection system against malicious attacks on the communication network of driverless cars'. *The 12th Annual IEEE Consumer Communications and Networking Conference (CCNC)*; Las Vegas, NV; 2015. pp. 916–21.

[12] Liu J., Zhang S., Sun W., Shi Y. 'In-vehicle network attacks and countermeasures: challenges and future directions'. *IEEE Network*. 2017;31(5):50–8.

[13] Cao W., Wu Y., Zhou E., Li J., Liu J. 'Reliable integrated ASC and DYC control of All-Wheel-Independent-Drive electric vehicles over can using a Co-Design methodology'. *IEEE Access*. 2018;7:6047–59.

[14] Wang C., Song C., Li J. 'Research on key state parameters estimation of electric vehicle ESP based on multi-sensor'. *Proceedings of the 2015 Fifth International Conference on Instrumentation and Measurement, Computer, Communication and Control (IMCCC)*; Qinhuangdao, China; 2015. pp. 29–32.

[15] Lu Y., Yuan Y., Wang Q. 'Forward vehicle collision warning based on quick camera calibration'. *Proceedings of the 2018 IEEE International Conference on Acoustics, Speech and Signal Processing (ICASSP)*; Calgary, AB, Canada; 2018. pp. 2586–90.

[16] Chao C., Qin X. 'Research of vehicle automatic emergency braking system evaluation methods'. *Proceedings of the IET International Conference on Intelligent and Connected Vehicles (ICV 2016)*; Chongqing, China; 2018. pp. 1–9.

[17] Kusano K.D., Gabler H.C. 'Safety benefits of forward collision warning, brake assist, and autonomous braking systems in rear-end collisions'. *IEEE Transactions on Intelligent Transportation Systems*. 2012;13(4):1546–55.

[18] Vinuchandran A.V., Shanmughasundaram R. 'A real-time lane departure warning and vehicle detection system using monoscopic camera'. *Proceedings of the 2017 International Conference on Intelligent Computing, Instrumentation and Control Technologies (ICICICT)*; Kannur, India; 2017. pp. 1565–9.

[19] Klinedinst D., King C. *On Board Diagnostics: Risks and Vulnerabilities of the Connected Vehicle*. Tech. Rep. CERT Coordination Center; 2016. Available from https://resources.sei.cmu.edu/asset_files/whitepaper/2016_019_001_453877.pdf [Accessed 11 May 2022].

[20] Morrison P., Herzig K., Murphy B., Williams L. 'Challenges with applying vulnerability prediction models'. *Proceedings of the 2015 Symposium and Bootcamp on the Science of Security (HotSoS'15)*; Urbana Illinois, USA, 21-22 Apr; 2015.

[21] Keller T.M., Benjamin J.S., Wright V.L., Gold B.H. 'Vulnerabilities under the surface: tracking potential vulnerabilities by voiding warranties'. *CyberSec '18 Proceedings of the Fifth Cybersecurity Symposium*; Coeur d' Alene, Idaho, USA, 9-11 Apr; 2018.

[22] Hasrouny H., Samhat A.E., Bassil C., Laouiti A. 'Vanet security challenges and solutions: a survey'. *Vehicular Communications*. 2017;7:7–20.

[23] Liang W., Li Z., Zhang H., Wang S., Bie R. 'Vehicular AD hoc networks: architectures, research issues, methodologies, challenges, and trends'. *International Journal of Distributed Sensor Networks*. 2015;11(8):745303.

[24] Lu N., Cheng N., Zhang N., Shen X., Mark J.W. 'Connected vehicles: solutions and challenges'. *IEEE Internet of Things Journal*. 2014;1(4):289–99.

[25] Mishra R., Singh A., Kumar R. 'Vanet security: Issues, challenges and solutions'. *IEEE International Conference on Electrical, Electronics, and Optimization Techniques (ICEEOT)*; Chennai, India, 3–5 Mar; 2016. pp. 1050–5.

[26] Raw R.S., Kumar M., Singh M. 'Security challenges, issues and their solutions for Vanet'. *International Journal of Network Security & Its Applications*. 2013;5(5):95–105.

[27] Vaibhav A., Shukla D., Das S., Sahana S., Johri P. 'Security challenges, authentication, application and trust models for vehicular AD hoc Network-a survey'. *International Journal of Wireless and Microwave Technologies*. 2017;7(3):36–48.

[28] Hamida E., Noura H., Znaidi W. 'Security of cooperative intelligent transport systems: Standards, threats analysis and cryptographic counter-measures'. *Electronics*. 2015;4(3):380–423.

[29] Petit J., Shladover S.E. 'Potential cyberattacks on automated vehicles'. *IEEE Transactions on Intelligent Transportation Systems*. 2015;16(2):546–56.

[30] Parkinson S., Ward P., Wilson K., Miller J. 'Cyber threats facing autonomous and connected vehicles: future challenges'. *IEEE Transactions on Intelligent Transportation Systems*. 2017;18(11):2898–915.

[31] Jafarnia-Jahromi A., Broumandan A., Nielsen J., Lachapelle G. 'Gps vulnerability to spoofing threats and a review of antispoofing techniques'. *International Journal of Navigation and Observation*. 2012;2012(9):1–16.

[32] C4ADS Above us only stars: Exposing gps spoofing in Russia and Syria [online]. 2019. Available from https://www.crows.org/news/443783/Above-Us-Only-Stars-Exposing-G PS-Spoofing-in-Russia-and-Syria.htm [Accessed May 2022].

[33] Dawam E.S., Feng X., Li D. 'Autonomous arial vehicles in smart cities: Potential cyber-physical threats'. *2018 IEEE 20th International Conference on High Performance Computing and Communications; IEEE 16th International Conference on Smart City; IEEE 4th International Conference on Data Science and Systems (HPCC/SmartCity/DSS)*; Exeter, UK; 2018. pp. 1497–505.

[34] Giray S.M. 'Anatomy of unmanned aerial vehicle hijacking with signal spoofing'. *The 6th International Conference on Recent Advances in Space Technologies*; Istanbul, Turkey; 2013. pp. 795–800.

[35] Kerns A.J., Shepard D.P., Bhatti J.A., Humphreys T.E. 'Unmanned aircraft capture and control via GPs spoofing'. *Journal of Field Robotics*. 2014;31(4):617–36.

[36] Popko G.B., Gaylord T.K., Valenta C.R. 'Signal interactions between lidar scanners'. *Laser Radar Technology and Applications XXIV*; Volume 11005, ID 110050J 14 SPIE; 2019.

[37] Yan C., Xu W., Liu J. 'Can you trust autonomous vehicles: Contactless attacks against sensors of self-driving vehicle'. *DEF CON 24*; Las Vegas, NV, USA; 2016.

[38] Bian C., Yin G., Xu L., Zhang N. 'Active collision algorithm for autonomous electric vehicles at intersections'. *IET Intelligent Transport Systems*. 2019;13(1):90–7.

[39] Li Z., Xiao Z., Zhu Y., Pattarachanyakul I., Zhao B.Y., Zheng H. 'Adversarial localization against wireless cameras'. *The 19th International Workshop on Mobile Computing Systems & Applications*; Tempe Arizona USA, 12-13 Feb; 2018. pp. 87–92.

[40] van Wyk F., Wang Y., Khojandi A., Masoud N. 'Real-Time sensor anomaly detection and identification in automated vehicles'. *IEEE Transactions on Intelligent Transportation Systems*. 2019;21(3):1264–76.

[41] Chattopadhyay A., Mitra U., Ström E.G. 'Secure estimation in v2x networks with injection and packet drop attacks'. *The 15th International Symposium on Wireless Communication Systems (ISWCS)*; Lisbon, Portugal; 2018. pp. 1–6.

[42] Wang J., Shao Y., Ge Y., Yu R. 'A survey of vehicle to everything (V2X) testing'. *Sensors*. 2019;19(2):334.

[43] Yang Y., Wei Z., Zhang Y., Lu H., Choo K.-K.R., Cai H. 'V2X security: a case study of anonymous authentication'. *Pervasive and Mobile Computing*. 2017;41(1):259–69.

[44] Alnasser A., Sun H., Jiang J. 'Cyber security challenges and solutions for V2X communications: a survey'. *Computer Networks*. 2019;151(12):52–67.

[45] Pesé M.D., Shin K.G. *Survey of Automotive Privacy Regulations and Privacy-Related Attacks*. SAE Technical Paper, 2019-01-0479; 2019.

[46] Urien P. 'Designing attacks against automotive control area network bus and electronic control units'. *The 16th IEEE Annual Consumer Communications & Networking Conference (CCNC)*; Las Vegas, NV; 2019. pp. 1–4.

[47] Dürrwang J., Rumez M., Braun J., Kriesten R. 'Security hardening with plausibility checks for automotive ECUs'. *The 6th International Conference on Advances in Vehicle Systems*, Technologies and Applications; Nice, France; 2017. pp. 38–41.

[48] Pan L., Zheng X., Chen H.X., Luan T., Bootwala H., Batten L. 'Cyber security attacks to modern vehicular systems'. *Journal of Information Security and Applications*. 2017;36(2):90–100.

[49] Salfer M., Eckert C. 'Attack surface and vulnerability assessment of automotive electronic control units'. *The 12th IEEE International Joint Conference on e-Business and Telecommunications (ICETE)*; Colmar, France; 2015. pp. 317–26.

[50] Shanmugam K. Securing inter-processor communication in automotive ECUs. SAE Technical Paper, 2019-26-0363; 2019.

[51] Boloor A., He X., Gill C., Vorobeychik Y., Zhang X. 'Simple physical adversarial examples against end-to-end autonomous driving models'. *IEEE International Conference on Embedded Software and Systems* (ICESS); Las Vegas, NV; 2019.

[52] Clark G., Doran M., Glisson W. 'A malicious attack on the machine learning policy of a robotic system'. *The 17th IEEE International Conference on Trust, Security and Privacy in Computing and Communications/12th IEEE International Conference on Big Data Science and Engineering (Trust-Com/BigDataSE)*; New York, NY; 2018. pp. 516–21.

[53] Khalid F., Hanif M.A., Rehman S., Shafique M. 'Security for machine learning-based systems: Attacks and challenges during training and inference'. *The 16th International Conference on Frontiers of Information Technology (FIT'18)*; Islamabad, Pakistan; 2018. pp. 327–32.

[54] Symantec. Executive Summary: 2018 Internet Security Threat Report ISTR Volume 23. 2018. Available from https://docs.broadcom.com/doc/istr-23-2018-executive-summary-en-aa [Accessed 11 May 2022].

[55] Morgan S. 2019 official annual cybercrime report, Herjavec Group. 2019. Available from https://www.herjavecgroup.com/wp-content/uploads/2018/12/CV-HG-2019-Official-Annual-Cybercrime-Report.pdf [Accessed 11 May 2022].

[56] He H., Watson T., Maple C., *et al.* 'The security challenges in the IoT enabled cyber-physical systems and opportunities for evolutionary computing & other computational intelligence'. *WCCI2016*; Vancouver, Canada; 2016.

[57] Trull J. Top 5 best practices to automate security operations [online]. Technical Report Online Blog, Microsoft Secure, Enterprise Cybersecurity Group. 3 Aug 2017. Available from https://www.microsoft.com/security/blog/2017/08/03/top-5-best-practices-to-automate-security-operations/ [Accessed 11 May 2022].

[58] Ciccone S. Cybersecurity: More threats, but also more opportunities [online]. Technical report, Paloalot Networks. 2 Jun 2016. Available from https://blog.paloaltonetworks.com/2016/06/cybersecurity-more-threats-but-also-more-opportunities/ [Accessed 11 May 2022].

[59] Krizhevsky A., Sutskever I., Hinton G.E. 'Imagenet classification with deep convolutional neural networks'. *NIPS'12 Proceedings of the 25th International Conference on Neural Information Processing Systems*; Lake Tahoe, Nevada; 2012. pp. 1097–105.

[60] Ren S., He K., Girshick R., Sun J. 'Faster R-CNN: towards real-time object detection with region proposal networks'. *IEEE Transactions on Pattern Analysis and Machine Intelligence*. 2017;39(6):1137–49.

[61] Xiong W., Wu L., Alleva F., Droppo J., Huang X., Stolcke A. 'The Microsoft 2017 conversational speech recognition system'. *IEEE International Conference on Acoustics, Speech and Signal Processing (ICASSP)*; New York, NY; 2018. pp. 5934–8.

[62] He H., Tiwari A., Mehnen J. 'Incremental information gain analysis of input attribute impact on RBF-kernel SVM spam detection'. *2016 IEEE Congress on Evolutionary Computation (CEC)*; Vancouver, BC, Canada, 24–29 July; 2016.

[63] He H., Watson T., Maple T., Mehnen J., Tiwari A. 'Semantic attribute deep learning with linguistic attribute hierarchy for spam detection'. *IJCNN2017*; Alaska, USA; 2017.

[64] Rathore H., Agarwal S., Sahay S.K., Sewak M. 'Malware detection using machine learning and deep learning' in Mondal A., Gupta H., Srivastava J., Somayajulu D., Reddy P. (eds.). *Big Data Analytics. BDA 2018. Lecture Notes in Computer Science*. Vol. 11297. Cham: Springer; 2018.

[65] Biswas S. 'Intrusion detection using machine learning: a comparison study'. *International Journal of Pure and Applied Mathematics*. 2018;118:101–14.

[66] LeCun Y., Bengio Y., Hinton G. 'Deep learning'. *Nature*. 2015;521(7553):436.

[67] Yuan Z., Lu Y., Wang Z., Xue Y. 'Droid-sec: deep learning in android malware detection'. *ACM SIGCOMM Computer Communication Review*. 2014;44(4):371–2.

[68] Saxe J., Berlin K. 'Deep neural network based malware detection using two dimensional binary program features'. *The 10th International Conference on Malicious and Unwanted Software (MALWARE)*; 2015. pp. 11–20.

[69] Lambert G.M. 2017 Security Analytics: Using Deep Learning to Detect Cyber Attacks 'Msc thesis'. School of Computing, University of North Florida.

[70] Shone N., Ngoc T.N., Phai V.D., Shi Q. 'A deep learning approach to network intrusion detection'. *IEEE Transactions on Emerging Topics in Computational Intelligence*. 2018;2(1):41–50.

[71] He H., Gray J., Cangelosi A., Meng Q., McGinnity M., Mehnen J. 'The challenges and opportunities of artificial intelligence for trustworthy robots

and autonomous systems'. *The 3rd International Conference on Intelligent Robotics and Control Engineering (IRCE2020)*; University of Oxford, UK, 10–12 Aug; 2020.

[72] Carlini N., Wagner D. 'Towards evaluating the robustness of neural networks'. *The 38th IEEE Symposium on Security and Privacy (SP)*; the Fairmont Hotel, San Jose, CA, USA, 22–24 May; 2017. pp. 39–57.

[73] Carlini N., Mishra P., Vaidya T., *et al.* 'Hidden voice commands'. *The 25th USENIX Security Symposium (USENIX Security 16)*; Austin, TX, USA, 10–12 Aug; 2016. pp. 513–30.

[74] Mozaffari M., Saad W., Bennis M., Debbah M. 'Unmanned aerial vehicle with underlaid device-to-device communications: performance and trade-offs'. *IEEE Transactions on Wireless Communications*. 2016;15(6):3949–63.

[75] Zeng T., Semiari O., Saad W., Bennis M. 'Joint communication and control for wireless autonomous vehicular platoon systems'. *IEEE transactions on communications*. 2019;67:7907–22.

[76] Ferdowsi A., Challita U., Saad W. 'Deep learning for reliable mobile edge analytics in intelligent transportation systems: an overview'. *IEEE Vehicular Technology Magazine*. 2019;14(1):62–70.

[77] Chen C., Seff A., Kornhauser A., Xiao J. 'Deepdriving: learning affordance for direct perception in autonomous driving'. *Proceedings of the* 2015 *IEEE International Conference on Computer Vision (ICCV*); Santiago, Chile, USA, 7-13 Dec.; 2015. pp. 2722–30.

[78] Shalev-Shwartz S., Shashua A. 'On the sample complexity of end-to-end training vs. semantic abstraction training'. *arXiv, Cornnell University*. 2016.arXiv:1604.06915.

[79] Burton S., Gauerhof L., Heinzemann C. 'Making the case for safety of machine learning in highly automated driving'. *International Conference on Computer Safety, Reliability, and Security*; Trento, Italy; 2017. pp. 5–16.

[80] Bezemskij A., Loukas G., Gan D., Anthony R.J. 'Detecting cyber-physical threats in an autonomous robotic vehicle using Bayesian networks'. *The IEEE International Conference on Internet of Things (iThings) and IEEE Green Computing and Communications (Green-Com) and IEEE Cyber, Physical and Social Computing (CPSCom) and IEEE Smart Data (SmartData*); Exeter, UK; 2017. pp. 98–103.

[81] Inzartsev A., Pavin A., Kleschev A., Gribova V., Eliseenko G. 'Application of artificial intelligence techniques for fault diagnostics of autonomous underwater vehicles'. *MTS/IEEE OCEANS 2016*; Monterey, CA; 2016. pp. 1–6.

[82] Kang M.-J., Kang J.-W. 'Intrusion detection system using deep neural network for in-vehicle network security'. *Plos One*. 2016;11(6):e0155781.

[83] Straub J., McMillan J., Yaniero B., *et al.* 'Cybersecurity considerations for an interconnected self-driving car system of systems'. *The 12th IEEE System of Systems Engineering Conference (SoSE)*; Waikoloa, HI; 2017. pp. 1–6.

[84] Tian Y., Pei K., Jana S., Ray B. 'Deeptest: Automated testing of deep-neural-network-driven autonomous cars'. *Proceedings of the 40th International*

Conference on Software Engineering; Gothenburg, Sweden; 2018. pp. 303–14.

[85] Black P.E., Scarfone K., Souppaya M. 'Part 2. cross-cutting themes and technologies 5. system and sector interdependencies' [online]. Technical Report. Gaithersburg, Maryland: National Institute of Standards and Technology (NIST); 2008 [Accessed 02 Sep 2019].

[86] Ani U.D., He H., Tiwari A. 'A framework for operational security metrics development for industrial control environment'. *Journal of Cyber Security Technology*. 2018;2(3-4):201–37.

[87] Attn: Research Department. 2018 cost of a data breach study: Global overview. Technical Report Online Report, Ponemon Institute - Bench-mark research sponsored by IBM Security [online]. 2018. Available from https://www.ibm.com/downloads/cas/861MNWN2 [Accessed 02 Sept 2019].

[88] The European Parliament and the Council. Regulation (EU) 2016/679 on the protection of natural persons with regard to the processing of personal data and on the free movement of such data, and repealing directive 95/46/ec (general data protection regulation) [online]. Technical report, Official Journal of the European Union. 2016. Available from https://eur-lex.europa.eu/eli/reg/2016/679/oj [Accessed 11 May 2022].

[89] Woodward M., Dixon S., Hamilton J., Goodall W. Driving connectivity global automotive consumer study: Future of automotive technologies [online]. Technical report, Deloitte, UK, United Kingdom Insights. 2017. Available from https://www2.deloitte.com/content/dam/Deloitte/uk/Documents/manufacturing/deloitte-uk-driving-connectivity.pdf [Accessed 11 May 2022].

[90] Glancy D.J. 'Privacy in autonomous vehicles'. *Santa Clara Law Review*. 2012;5(4):1171.

[91] Fagnant D.J., Kockelman K. 'Preparing a nation for autonomous vehicles: opportunities, barriers and policy recommendations'. *Transportation Research Part A: Policy and Practice*. 2015;77(1):167–81.

[92] NHTSA. *Preliminary Statement of Policy Concerning Automated Vehicles*. Technical report. Washington, DC: National Highway Traffic Safety Administration; 2013. Available from https://www.nhtsa.gov/staticfiles/rulemaking/pdf/Automated_Vehicles_Policy.pdf [Accessed 11 May 2022].

[93] Kyriakidis M., Happee R., de Winter J.C. 'Public opinion on automated driving: results of an international questionnaire among 5000 respondents'. *Transportation Research Part F: Traffic Psychology and Behaviour*. 2015;32:127–40.

[94] 2025AD. Data Privacy and Connected Cars (2025AD.COM). 2005. Available from https://en.euractiv.eu/wp-content/uploads/sites/2/infographic/Infographic_Data_Privacy_and_Connected_Cars.pdf [Accessed 3 Jan 2023].

[95] Loebich M. Connected cars and privacy: shifting gear for GDPR?. 2005. Available from https://www.bearingpoint.com/en/insights-events/insights/connected-cars-and-privacy-shifting-gear-for-gdpr/ [Accessed 3 Jan 2023].

[96] Wright S. Autonomous cars will generate more than 300 tb of data per year. 2 Jul 2021. Available from https://www.tuxera.com/blog/autonomous-cars-300-tb-of-data-per-year/ [Accessed 3 Jan 2023].

[97] Department for Transport. Vehicle licensing statistics: Annual 2018. 11 Apr 2019. Available from https://assets.publishing.service.gov.uk/government/uploads/system/uploads/attachment_data/file/800502/vehicle-licensing-statistics-2018.pdf [Accessed 11 May 2022].

[98] Richter F. Big Data on Wheels. Available from https://www.statista.com/chart/8018/connected-car-data-generation/ [Accessed 15 Jan 2023].

[99] Pollard T. What are autonomous CAR levels? levels 1 to 5 of driverless vehicle tech explained. Bauer Media Group; 18 April 2022. Available from https://www.carmagazine.co.uk/car-news/tech/autonomous-car-levels-different-driverless-technology-levels-explained/ [Accessed 11 May 2022].

[100] Federal Trade Commission. Privacy & data security, update: 2018. March 2019. Available from https://www.ftc.gov/system/files/documents/reports/privacy-data-security-update-2018/2018-privacy-data-security-report-508.pdf [Accessed 11 May 2022].

[101] Ganesan A. Data security and privacy in the connected car age [online]. 15 Aug 2018. Available from https://www.eeworldonline.com/data-security-and-privacy-in-the-connected-car-age/ [Accessed 11 May 2022].

[102] Korosec K. *Volvo CEO: We will accept all liability when our cars are in autonomous mode* [online]. 7 Oct 2015. Available from https://fortune.com/2015/10/07/volvo-liability-self-driving-cars/ [Accessed 11 May 2022].

[103] McCallister E., Grance T., Scarfone K. *Guide to Protecting the Confidentiality of Personally Identifiable Information* (*PII*). Technical Report. NIST; 2010.

[104] Autofacets. AUTOMOTIVE Sectors.https://www.autofacets.com/segment-overview/ [Accessed 11 May 2022].

[105] CEA. *The cost of malicious cyber activity to the u.s. economy. Technical report, The Council of Economic Advisers* [online]. Feb. 2018. Available from https://www.insurancejournal.com/research/app/uploads/2018/02/The-Cost-of-Malicious-Cyber-Activity-to-the-U.S.-Economy.pdf [Accessed 11 May 2022].

[106] Cambridge Network. Safety, cybersecurity and cost delaying mass deployment of autonomous cars. 17 Apr 2019. Available from https://www.cambridgenetwork.co.uk/news/safety-cybersecurity-and-cost-delaying-mass-deployment-autonomous-cars [Accessed 11 May 2022].

[107] Garsten E. Sharp growth in autonomous car market value predicted but may be stalled by rise in consumer fear [online]. 13 Aug. 2018. Available from https://www.forbes.com/sites/edgarsten/2018/08/13/sharp-growth-in-autonomous-car-market-value-predicted-but-may-be-stalled-by-rise-in-consumer-fear/?sh=5bb3d874617c [Accessed 11 May 2022].

[108] Cui J., Sabaliauskaite G. 'On the alignment of safety and security for autonomous vehicles'. *The 2nd International Conference on* Cyber-*Technologies and* Cyber-*Systems* (*Cyber2017*); Barcelona, Spain, 12-16 Nov 2017; 2017.

[109] University of Texas at Austin. *UT Austin Researchers Successfully Spoof an $80 million Yacht at Sea* [online]. 29 July 2013. Available from https://news.utexas.

edu/2013/07/29/ut-austin-researchers-successfully-spoof-an-80-million-yacht-at-sea/ [Accessed 11 May 2022].

[110] Kelion L. *Ship hack risks chaos in English Channel* [online]. *BBC*. 7 Jun 2018. Available from https://www.bbc.co.uk/news/technology-44397872 [Accessed 11 May 2022].

[111] International Maritime Organisation. *MSC 100/WP.8 - Regulatory Scoping Exercise for the Use of Maritime Autonomous Surface Ships (MASS)*. London: IMO; 2018.

[112] Bateman T. Police warning after drug traffickers' cyber-attack [online]. BBC. 16 Oct 2013. Available from https://www.bbc.co.uk/news/world-europe-24539417 [Accessed 11 May 2022].

[113] Wee V. Pirates used ais tracking app to identify ship to hijack [online]. 2017. Available from https://www.seatrade-maritime.com/asia/pirates-used-ais-tracking-app-identify-ship-hijack [Accessed 11 May 2022].

[114] International Maritime Organisation. *MSC 101/WP.8 - Interim Guidelines for Mass Trials*. London: IMO; 2019.

[115] International Maritime Organisation. *MSC 101/5/4 - Proporal for Terms to be Avoided, Recommended Terms and Draft of Glossary*. London: IMO; 2019.

[116] Finland Ministry of Transport and Communications [online]. Pilotage Act 2019. 2019. Available from https://finlex.fi/fi/laki/kaannokset/2003/en20030940.pdf [Accessed 11 May 2022].

[117] Höyhtyá M., Huusko J., Kiviranta M., Solberg K., Rokka J. 'Connectivity for autonomous ships: Architecture, use cases, and research challenges'. 2017 *International Conference on Information and Communication Technology Convergence (ICTC)*; 2017.

[118] International Maritime Organisation. *MSC-FAL.1/Circ.3 – Guidelines on Maritime Cyber Risk Management*. London: IMO; 2017.

[119] BIMCO. The Guidelines on Cyber Security Onboard Ships V4. 2020. Available from https://www.bimco.org/about-us-and-our-members/publications/the-guidelines-on-cyber-security-onboard-ships [Accessed 11 May 2022].

[120] International Maritime Organisation. *MSC.428(98) – Maritime Cyber Risk Management in Safety Management System*. London: IMO; 2017.

Chapter 10

Continued transport modernisation and the implications for security

Miriam Howe[1]

10.1 The changing environment

Modern transport systems, as illustrated in the previous chapters, are changing in several ways that are impacting their cybersecurity profile, and in a world where cybersecurity threats are becoming ever more sophisticated and targeted. New transportation systems on the ground, in the air, in water and even in space are being researched and engineered, integrated and operated, and these are generating new types of data and will be operated with innovative means of control. Moreover, some high profile incidents illustrating national exposure to new risks have raised the urgency to the most senior levels of government, evidenced by US President Biden's 100-day review of the US electric supply chains, equipment and programmable components [1]. The development of electrical-based transportation has resulted in the re-evaluation of supply networks to encompass new dependencies, such as on electric instead of hydrocarbon resources. Some key themes of this evolution are outlined below, along with some implications for cybersecurity.

The growing population means public transport is under increasing strain. Much of the challenge therefore centres on capacity. Occasionally, there are large projects that invest in new infrastructure and solutions such as HS2 and Crossrail in the United Kingdom [2], Hyperloop (Box 10.1) and Maglev in Japan and China. The HS2 line is currently under construction [3], the first part of Crossrail is set to open in the first half of 2022 [4]. The Hyperloop pod and rail is being prototyped in the United States, Europe, Dubai and India, with Virgin's Hyperloop currently being prototyped in Nevada and targeted to enter service by 2030 [5]. Additionally, China has already completed their fast electromagnetic Maglev train [6].

More commonly, transport systems of the future have to make use of existing infrastructure, often using technology solutions to optimise its usage and expand its throughput. For example, the national rail network in the United Kingdom is already dual use: for high speed (120–30 mph), low speed and slow freight trains (30 mph).

[1]BAE Systems Applied Intelligence, United Kingdom

Box 10.1 Hyperloop

Hyperloop promises to transport high volumes of people at significantly higher speeds than current and imminent rail projects. Safety failures could therefore have significantly greater impact. More people being transported through a space per time unit increases the potential shock impact of a collision, and increasingly complex and embedded systems increases the surface area for cyber-related vulnerabilities.

Hyperloop, which is essentially designed as a pod travelling within a tube in a vacuum, introduces an aviation type of safety risk to train travel. If there were a crack, a crash or an uncontrolled evacuation (such as a bomb scare), passengers would not be able to breathe. An additional safety implication is that in a crash situation, the shock waves following impact would affect everyone within the pod. These physical threats to the integrity of the tube and the pod, resulting from bomb scares, gun crime or natural disasters, need to be built into the safety and security cases for Hyperloop.

A technology solution to further optimise the usage of the rail network is to move away from box signalling, to a system where trains signal to each other rather than via the infrastructure. This could increase the capacity of the fixed track infrastructure itself, but at the same time widens the vulnerability to coverage gaps and GNSS/GPS jamming. This has been observed in military operations [7, 8], and there exist jamming tools available to low capability threat actors.

Heavy goods vehicle 'platooning' is a concept for groups of connected vehicles to drive closely together with a single driver in the lead vehicle. Sensors and vehicle-to-vehicle (V2V) communications are used for collision avoidance, and the front driver controls speed, acceleration and braking. This was first trialled in 2017 and future on-road trials are scheduled within the Helm UK [9] project. The testing aims to show that platooning is safe and increases both fuel and commercial efficiency, making more efficient use of the existing road network (Box 10.2).

Other efforts focus on minimising and optimising maintenance windows that take the track, runway, aircraft or car out of service. The drive to reduce outage times within 24/7 operations, seen particularly at airports, can result in surrendering sufficient windows of downtime, during which necessary maintenance activities can be completed – including security patching and upgrades. The danger of accumulating a patching deficit on a legacy estate is challenging, particularly where technical interdependencies between products make upgrades unfeasibly complex. This challenge is equally recognisable in operational technology, as we drive more capacity from the fixed infrastructure underpinning future transport systems.

Driving the efficiency of future transport systems in getting people from A to B results in a greater flow rate of people within a fixed space. This means that a cybersecurity breach that causes a physical effect could have a higher impact on passenger safety than currently. The faster a vehicle travels and the more people inside

Box 10.2 Helm UK platooning trials

Helm UK is a project being run by Transport Research Laboratory but no longer (TRL) that is aiming to conduct the world's first live trials of platooning technology on United Kingdom roads, in a commercial environment.

Platooning is where vehicles – in the case of Helm UK, heavy goods vehicles – drive closely together, using sensors and V2V communications to ensure that they do not collide. The driver of the front vehicle controls the speed, acceleration and braking of the entire platoon.

Platooning results in better use of space on the road network, fuel efficiency increases due to reduced wind resistance. Fuel efficiency increases will also reduce the cost of delivering goods by road. Helm UK is intended to show that these benefits can be realised in a commercial environment and demonstrate evidence of the safety of platooning drivers, their interaction with other drivers and the security of the V2V communications used.

The testing was intended to take place throughout 2020, but COVID has impacted progress.

that vehicle, the larger the risk of injury and damage in the event of a cyber-physical attack or disruption. There has been a tendency towards connecting traditional enterprise IT into operational technology to enable remote management, which reduces the overhead of an engineer visit to the roadside or track, with the ability to monitor and control multiple endpoints or locations at once. This has increased the risk of cyber-physical disruptions, as sensors and actuators can be accessed and manipulated via the IT network. The paradox is that systems designed to improve safety may be manipulated remotely to cause a hazardous state.

While the accidental ramifications are daunting, the possibilities of a deliberate attack include scenarios worthy of horror films. Autonomous vehicles could be used to ram pedestrians; traffic signals at an intersection could be turned green at the same time to induce collisions; a sensor on a train track could be manipulated so that it does not detect a stopped train and allows another train to crash into it; and the picture provided by air traffic control could be disrupted. Automatic dependent surveillance-broadcast (ADS-B) is the latest generation of air transportation surveillance and communication, but studies have shown how it could be manipulated [10]. Due to ADS-B's use of clear and open broadcast of messages, it is possible to inject spurious aircraft or remove real aircraft from the air traffic pictures [11] by using inexpensive hardware and open source software [12]. Frameworks have been created to secure ADS-Bs [13] by using cognitive radios, cryptography [14], or by using timestamp values and location data to identify spoofed messages [15]; however, none provides a fully secure protocol. Mechanisms for collision avoidance are only as good as navigational data that they collect as inputs.

As discussed in Chapter 1, operational technology is increasingly recognised as a powerful vector for cyber-physical attacks. It therefore becomes inevitable that we need to consider large foreign actors as a disruptive threat to our transport systems. Indeed, nations have for decades been investigating and exploring the use of cyberattacks as an unregulated means of attacking another sovereign state and/or commercial entity that it wishes to disrupt. At a national level, there are ongoing studies to review readiness for cyber-physical incidents that impact critical national infrastructure and to build resilience against the impacts and cascade effect [16]. Recent events such as a remote attack on the Florida water treatment plant [17] resulting in harmful adjustments to the town's water supply, and a ransomware attack on the Colonial Pipeline in the US led to significant disruption of fuel supply to the US east coast [18].

A further issue is that the increased digitisation of vehicles and connectivity has been compounded by a general proliferation of Internet-connected devices at the roadside, in order to create ITS which digitally sense and understand the physical environment.

The rise of the smart city, led by cities in such as Singapore and Dubai [19, 20] along with the introduction of autonomous vehicle systems, allows for the advent of 'Mobility as a Service'; where in the future we will no longer own vehicles for ourselves but instead make use of an intelligent transport system for all of our travel needs (i.e. all transport is shared amongst members of the public, including road vehicles). As an example, Singapore has invested SGD$2.4 billion (US$1.73 billion) [21] into achieving this and is currently thought to be amongst the smartest cities in the world [22]. Singapore has created an autonomous fleet encompassing ferries, shuttles, vehicles and buses, which will reduce over-crowding on buses by 92%. However, since open data are used to optimise and plan the Singapore transportation system [23], there have been concerns raised over its accessibility and vulnerability to attackers [24].

Inherent within the concept of smart cities is the idea of user services across all modes of transport (planning, weather and payments for multiple journey legs). Such services are provided by applications that allow users to interact with the transport system, for example, reserve seats, pay, reserve, track, report and change. Dubai is currently transitioning into a more paperless transaction system, making payment for transport services through smart cards. This paperless strategy will mean that over one billion government paper transactions will be eliminated and stored electronically [25].

An increase in user interaction within the transport system introduces a much greater and unregulated/uncoordinated ecosystem with multiple attack vectors that could either indirectly or directly impact passengers. Furthermore, the personal data being collected and disseminated around transport systems are further enriched as the human increasingly becomes an additional node/entity in the digital network. Supplemented with IoT devices, such as mobile phones and wearable technology, humans will increasingly be connected to the cars and intelligent transport systems (ITS) around them. All this means that an increased volume of personal data needs to be processed as part of the transport system, increasing the possibilities of privacy breach and falling foul of regulations such as GDPR (General Data Protection Regulation).

The adoption and implementation of autonomy and machine learning adds another level of complexity and uncertainty. This sits uncomfortably with the safety-critical nature of transport systems, which demand a high degree of reliability and predictability.

10.2 Research themes

Notwithstanding this daunting outlook from a security point of view, the industry is thriving with research to address the capacity and efficiency challenges, understand the breadth and nature of cyber vulnerabilities and come up with solutions. The remainder of this chapter explores some of this research along a number of themes.

The first theme relates to research of cyber, data and AI solutions, which can help with physical security challenges, for example, relating to crowd safety in public spaces. The next set of themes investigate securing different areas of the infrastructure providing interconnected and ITS, including (i) securing the inputs, (ii) securing the communications links and (iii) securing the decision-making process within autonomous systems. The final research theme addresses trust in the relationship between the human and computer, specifically how to build confidence in the machine to enable its benefits to be realised.

10.3 Theme 1: cyber and data solutions can help with physical security issues

Artificial intelligence is a hyped set of technologies, but machine learning, knowledge representation and image analytics are well placed to add value in security challenges, particularly those that involve processing large volumes of complex data and assembling conclusions at a rate beyond human capability.

Image analytics are a key instrument in physical security solutions and may be enhanced by machine learning, which can be used

- to improve security and safety in physical spaces
- to plan maintenance with improved efficiency and safety
- to reduce the recurrence of accidents

These in turn necessitate a further consideration of the cybersecurity implications of such solutions, as we increasingly rely on them for safety and control mechanisms in transport.

10.3.1 Crowd analysis and monitoring/crowd resilience

One area of opportunity being explored is the use of video monitoring and image analysis, enhanced by machine learning to model crowd behaviour through a physical space, such as an airport, platform or train station.

An example is the development of agent-based models of humans and their everyday patterns of life, enabling the creation of large datasets of actors and their activities [26]. The actors are imaginary people residing at real postcodes undertaking everyday activities such as driving to work, performing the school run and shopping trips. These activities trigger other events such as electronic point of sale transactions, cash machine interactions, and vehicles passing CCTV and being logged by automated number plate recognition (ANPR) systems at tolls and in car parks. In addition, mobile phone usage metadata has been generated between fictional people to create representative personal communications such as phone calls and SMS messages.

Agent-based modelling is predicated on software-based agents representing individual humans, to which theoretical personal characteristics can be attributed. Agents can be programmed to move and interact with other agents, and then subjected to machine learning to evolve their behaviour as a result of those interactions and the outcomes. With large datasets, this can enable a digital thread to be followed, tracking what an individual or group of individuals would do in different scenarios. This can be used to plan or practice strategic responses, such as emergency services, to better prepare for forthcoming events or security incidents.

These types of models have the potential to be extended to develop agent-based models of crowd behaviour and can be exploited as a useful 'ground-truth' dataset, that is, a baseline for validating future algorithms for analysing video streams. An alternative approach to software–based agents that resemble humans is the creation of datasets which are based on actual people but which are derived from anonymised personal data (to protect the privacy of those people). Such a ground-truth set of human data would be useful for validating algorithms in the physical security domain.

Along a similar vein, video streams of actual crowds can also be analysed, and individuals within the scene 'tagged' within the image based on (depending on the size of the image) the space, colour and background features. There is no need to resolve any information about the individual person, simply to establish that they are indeed a person. The individuals can be tracked according to their tags, and each individual can be monitored as they move across the scene captured by the image. Tracking individuals in comparison with the individuals surrounding them, using Deep Learning through Deep Neural Networks (DNNs), allows the establishment of a normal baseline of crowd movement, which in turn can enable detection of unusual behaviour *of a crowd*, and unusual behaviour *of an individual within a crowd.*

Queen's University Belfast (QUB) and BAE Systems Applied Intelligence Labs (AI Labs) implemented such a proof of concept using entity recognition and movement tracking in an ice hockey stadium. The application was aimed at the provision of a real-time situation awareness picture of people and crowd movements in an arena or sporting event. The derived situation awareness picture could be used to alert the security operations centre team with early warning of crowd disruption or to anomalous behaviour of individuals or groups of individuals in the crowd.

In a similar vein, Hacktrain, a technical innovation convention focused on the transport sector, initiated a feasibility study in 2016 called VivaCity [27]. VivaCity demonstrated how deep learning of features associated with individuals in crowds can be used in other venues with crowd management requirements – such as underground metro stations. In 2020 Greater Manchester had also chosen to deploy Vivacity Labs, so they can use AI-controlled road junctions to solve traffic issues [28].

Crowd motion should generally be predictable, and so learning what the flow through a station is like on a normal day allows the possibility of detecting unusual movements by crowds or within them. Scenarios that this capability could be applied to range from detection of health incidents, for example, indicated by an unusual length of time without moving, to preventing public safety incidents such as detecting crowds moving towards a choke point, and auto-signage re-directing them to other exits and even predicting incidents based on historic data. Indeed, AI Labs ran a demo showing how people flowing through a station can be modelled using Communications Packet Flow Network technology. In this demo, passengers are modelled as if they were IP packets in a network, such that their flow can be optimised as if it were network traffic (Box 10.3). The application of optimisation rules or algorithms to the 'packet' flow (or in this case to the flow of people) provided input into automated information displays in the station to assist people in avoiding congested routes and getting caught at choke points.

Additionally, researchers at Cardiff University have proposed a method [29] for assisting human operators who are responsible for monitoring video feeds. The method was developed to cope with the poor quality CCTV video footage often provided by legacy hardware: poor illumination, crowded scenes, occlusion, slow resolution and camera shake. Monitoring multiple feeds makes it harder for the operator to notice abnormalities in one feed and, in this case, the proposed algorithm used computer vision techniques to classify abnormal behaviour in crowds that would signify violence in real time. This ability to detect and classify abnormal crowd flows (when the correct representation of data has been included) would accurately indicate potential violence, and the guidance this gives to the human operator presents the opportunity to detect such events and respond to them much faster.

One key consideration in assuring this technology as it is implemented in different countries is to customise the training datasets used for deep learning systems to provide an accurate representation of social, cultural, ethnic and religious populations. Research has shown that cultural and socio-economic differences have an impact on crowd-mapping behaviours, and many existing spatial crowd-sourcing datasets are therefore biased [29–32]. Ensuring a dataset is representative of a population is, however, non-trivial, and there are long standing issues with regard to accurate ethnic and cultural representation in deep learning datasets, which has led to algorithmic biases, in turn resulting in the misidentification of images and behaviours of ethnic and cultural communities. This ultimately leads to negative systematic biases causing inaccurate action and consequences for underrepresented individuals [33]. Given that such inferences arise from algorithms that often cannot

Box 10.3 Network traffic optimisation

Network traffic optimisation refers to the monitoring and analysis of traffic across a network, with a view to re-directing packets to achieve the greatest overall efficiency of traffic. This generally prefers the shortest path between two points as being the quickest route but will take into account blockages and congestion on the network, as well as traffic backlogs, to adjust the optimum route for individual packets. As a traffic management tool, it is intended to achieve an overall efficiency in traffic routing around a network.

be easily understood by humans (that is, algorithms appear as black boxes), trust can be quickly eroded. These algorithmic biases can be combated by number of methods such as [34]:

1. Interpretable or explainable systems that measure learning models
2. Identifying data sources and including accurate representative data
3. Sharing transparently how data are selected and cleansed

Algorithmic bias correction needs more than data collection issues being solved [30], so further solutions such as careful management of synthetic counterfactuals [31] are worthy of consideration.

10.3.2 Prediction for preventative security

Another opportunity for improving security and safety is in prediction. This involves allowing algorithms to train on historical datasets to infer rules about causation, which can be applied to predict future cases of recurrence. This has been applied in a proof of concept by Avon and Somerset Constabulary's Office for Data Analytics, in which the algorithm underwent supervised learning on historical data about road traffic accidents (RTAs) and generated indicators about the likely causes of incidents. It was then used to apply these indicators to predict where else the same sort of accidents might occur in the future. This offers the ability of preventative maintenance and planning emergency services support to limit the probability and impact of RTAs [32]. A similar approach has been taken in Nevada [33].

Network Rail implemented another proof of concept for proactive track maintenance [34]. A set of algorithms were trained on the historical pattern of trees falling on the track that would have required reactive maintenance. Following training on the historical dataset, the algorithms were then used to predict future tree-falls, enabling more proactive maintenance that minimised the period during which the section of track was unavailable. Shortening maintenance windows not only reduces the downtime of the track and impact on travellers but also reduces the amount of time a human has to be on the track – a major safety concern for track operators. The machine learning model showed a significant reduction in minutes spent on maintenance.

Predictive machine learning has further applications in the cybersecurity space, in the detection of malware, where salient features that have been learnt as indicators can be reported along with the classification of malicious or benign software, and trust scores can be generated to indicate a potential (rather than a confirmed) breach. Research into the ability to predict cyber threats [35] using historical log data that preceded digital intrusions or breaches has already been implemented in digital security products, such as email protection services – a growing market indicating an emerging trend for predictive security. Further examples of machine learning application to cybersecurity are covered in Chapter 7.

Threat prediction is different from anomaly based detection, in which a baseline of normal activity is established, including whether two components normally communicate with each other. The ruleset that is learned from establishing the normal baseline is used to identify potentially anomalous traffic that might indicate an intrusion. Threat prediction uses historical data to indicate early warning signs, which might include indicators of reconnaissance, vulnerabilities and configuration changes, and to calculate a likelihood of a subsequent attack. The observations are distributed through the system components, analogous to a neighbourhood watch system where each component is looking out for suspicious actions on the street from their own field of view. Such indicators might amount to nothing, but they serve as early warning indicators that can be acted on, either to check, deter or respond.

A question researchers are asking is whether and how predictive analysis, supported by machine learning, can be applied in the transport sector, for example, to predict physical safety or security threats within crowded spaces to assist in their detection and response. Given the concepts demonstrated above, a clear opportunity here is the potential to improve both the effectiveness and efficiency of predicting when and where crime is going to happen, based on unusual behaviour of individuals within a crowd. The obvious benefits to public security and safety are clear, as both the occurrence and impact of crime could potentially be reduced – and there are significant applications in public spaces such as stations, ranging from petty crime to violent assaults and even terrorism.

Integration of crowd analysis data with other sources of data about the individuals or the crowd can help address crowd safety issues in transport environments by enriching the data to better understand human flow through. This has been applied by Transport for London (TfL) in a proof of concept in the measurement of human flow through tube stations in London. Previously, it was hard to map the flow of people as it was only possible to see where people entered and exited the transport system. By mapping user WiFi connections in tube stations and tracking devices via de-personalised data, TfL were better able to understand human flow through major connection stations (e.g. King's Cross Underground Station) during rush hour. This could be used to improve the crowd management and self-service information provided to travellers to help them avoid crowds and anticipate travel time, so there is both a crowd safety and efficiency benefit.

There is further interest in integrating social media data to better understand situational awareness and human flows. Rashidi *et al.* [36] review the opportunity for collecting social media data from travellers such as trip purpose, mode of transport, activity duration and destination choice, as well as land use variables such as home, job and school location and socio-demographic attributes including gender, age and income. The attraction of this in capturing richer data is to understand traveller behaviour. Underlining the perceived importance of behavioural analytics to public safety challenges is a recent area of interest in defence and security [37].

10.3.3 Theme 2: securing the decision–making process in autonomous systems

Machines make decisions based on rules that either have been encoded into the algorithm or are learned from processing training sets and receiving feedback on whether its classification was correct. Autonomous cars, for example, need to sense, detect, classify and identify at speed the objects around them, and react to them in the appropriate way; for example, correctly distinguishing a child in the road from a cat or a plastic bag enables it to react differently to each. Machine learning introduces a new risk to the integrity of the decision, reflecting that if it was possible to tamper with the decision-making rules, this could lead to wrong decisions and reactions being made on a systematic basis. If they knew that a vehicle will stop in a certain situation, an attacker could engineer a situation where one or more vehicles react in that specific way at a planned time or particular situation.

The tampering that is being discussed and mentioned currently is much more basic, for example, overriding systems, breaking into cars and spoofing electronic keys. By tampering with the ruleset more sophisticated tampering could be effected, such as overriding navigation or safety systems in order to steal a car or take control of it.

Two main vectors for interfering with the decision-making process could be via the training sets or via the ruleset itself. Gaps in the training data set could result in blind spots in the resulting intelligence and decision-making process, whereas a sufficient volume of skewed situations in the dataset could teach the computer to behave in a way that is flawed.

Even when there is no deliberate subversion of the training dataset, there are legitimate concerns about the integrity of decisions made on the basis of machine learning. Adversarial networks reveal how a neural network can learn to make correct classifications using irrelevant pixels, for example, determining whether it is a cat or a dog depending on whether it is inside the house or outside the house. An anecdote from University of California, Irvine (UCI) [38] demonstrated that machine-learned algorithms can make accurate classification on non-relevant features – in this case accurately distinguishing a dog or a husky depending on whether there was snow in the background. It's hard to avoid the conclusion that we cannot automatically trust the integrity of black-box decisions.

Currently, continual learning through experience introduces even more uncertainty. Autonomous vehicles do not learn as they go because that introduces uncontrolled influence into the training dataset. Autonomous cars could be influenced by bad decisions made by other drivers on the road and make incorrect inferences about the rationale for those decisions, adopting that rationale into its own rule base.

This points to the need for explainable AI to address this uncertainty; by understanding the reasoning behind the AI system, it allows for better human trust and also ensures that the conclusions that the autonomous system has derived are drawn from the correct inputs. This is explored in the final section.

10.4 Theme 3: securing the inputs

Autonomous vehicles in any mode of transport have a range of sensors through which they are taking information from the physical environment [39]. This is then fed through a neural network to determine the appropriate reaction, and a control layer orchestrates the various actuators that bring about the kinetic effect, such as steering, switching lane, braking and adjusting the engine.

Perhaps a more trivial approach to subverting autonomous vehicles, rather than interfering with the decision-making process itself, is to subvert the inputs so that the vehicle generates an incorrect perception of the world and its reaction to the environment is flawed.

Generally, sensors are the weak points of the whole system, both in terms of the protocols and the components [40–42]. Sensors are often implemented on COTS (Commercial Off The Shelf) products that are low cost and not robust, and they are based on well-defined interfaces and protocols that are relatively easy to hack or write exploits for. There is therefore significant value in researching the vulnerabilities and attack vectors for individual components, particularly when they introduce additional interfaces (see the text box, e.g. Box 10.4).

Box 10.4 Vulnerabilities of software-defined radios

Software-defined radios (SDRs) are becoming increasingly popular and commercially available, and are being deployed more widely in the civil transportation sector.

Commercial transport vehicles (as well as military vehicles) use RF communications to transmit information ranging from current and future positioning data, to permission to land/dock and vital voice messages. The use of SDRs in telecommunications is particularly beneficial as their upgradable and reconfigurable nature is essential for the adoption of new technologies (such as 5G and 6G).

However, SDRs used on products and platforms may introduce vulnerabilities. Due to the inherent nature of an SDR and its requirement to run software applications that are flexible and reconfigurable, a software download channel must be present for general maintenance as well as configuration changes. This channel would entail a physical or wireless connectivity, which could be used for malicious software updates or execution. As the benefits of an SDR are realised through the flexibility of the software, it is more at risk of executing malicious software than traditional radio systems. If an attacker can find a way to get malicious software to the device, it may be possible to execute it.

Commercially available SDRs have also proved useful for attackers and as a result, many cheap devices are being used to mount RF attacks. For these reasons, SDRs are increasingly becoming known for targeting unsecured devices such as remote keyless entry devices in garage doors, alarm systems and cars. Such products have had increased their security measures to negate the effect of replay attacks by implementing upgrades such as key expiration and cycling techniques.

Similar security measures should be applied to the use of SDRs as for other interfaces to complex systems. A secure download mechanism should be implemented in order to guarantee the authenticity of downloaded software. This should include signing of digital certificates and maintaining accurate records of trusted devices, in particular for those vendors who are writing software applications for the SDR. For safety critical systems timing constraints in which downloads are only permitted in prearranged timeframes. There is further research in the industry to explore the vulnerabilities and hinder the subversion of SDRs [43, 44].

It is notable that the increasing availability of radio products make subversion techniques usable by less capable threat actors at low cost. An example might be GPS (Global Positioning System) jammers used by truck drivers to deceive their tachographs.

Clearly, the accuracy and integrity of the sensory inputs to the vehicle are critical to getting the correct response. Professor Nick Colosimo* observes that we need a discriminatory sensor capability that can distinguish between matter and thin air with certainty: the technology needs to distinguish between white pixels of a road's central reservation and white space. For example, in 2016 a fatal Tesla crash was attributed to a failure to recognise the side of a lorry against the sky [45]. Colosimo argues that it is better to put more money into the sensors than the processing. After all, making mistakes in physical response could be down to misinterpreting the inputs or not having sufficient inputs. However clever the algorithm is, if it is basing its decisions on inaccurate input, or an incorrect mix of sensors for the situation, then the decision will often be wrong. Multiple sensors add redundancy and perspective – such as the comparative distance to the object (the lorry compared to the sky). Other sensors correctly integrated into the overall input capability such as lidar or even a temperature sensor might have helped as well.

Furthermore, sensors are exposed to potentially hostile environments and inputs. Some of the protocols are known to be weak, such as GNSS (Global Navigation Satellite System), which is known to be possible to jam or spoof. Cameras can be fed obscured, incomplete or false images, as demonstrated by Washington University researchers who have shown that machine learnt algorithms can be tricked by sensory input [46]. The research team defaced a stop sign with a few simple stickers that reflected the light. Humans would always recognise the defaced sign, but a car always missed it because of the subtle adjustments to the image.

Changes that could be made to sensory input vary in terms of subtlety and impact. What would the impact be of a change to the temperature or time in a ship, and what impact would result from a petrol gauge reporting empty when it was full, or reporting full when it was empty? If a lorry reported its weight as being significantly less than it was, so the driver felt able to traverse a weight limited bridge? The potential consequences are comparable with other Industrial Control System (ICS) dependant sectors, as illustrated by a hacker's attempt in early 2021 to alter operational parameters at a water treatment plant in Florida [47]. In exploring the new risks being introduced by sensory products, multiple redundancy and fusion between sensors should be built into the overall security case for the integrated system.

There are two avenues for development: (i) to drive both the production and integration of more robust (and a richer variety of) sensors and (ii) to require manufacturers to only adopt certified products into the vehicle. BAE Systems† is working on standards to drive this adoption in the United Kingdom and European Union, to encourage manufacturers to seek secure and certified products to integrate into their vehicles.

*BAE Systems Lead Engineer, Principal Technologist and Global Engineering Fellow.

†Activity led by Emyr Thomas, Operational Technology and Industrial Security Lead, BAE Systems.

The number of sensors that are being incorporated into autonomous vehicles is increasing and is expected to reach 200 sensors per car. This will exacerbate the risk associated with untrustworthy sensors, particularly if they are all from the same manufacturer and could suffer from common vulnerabilities, and therefore be manipulated or taken offline as a group.

If we tip the balance from poor quality sensors to intelligent and robust sensing systems, we would not only collect better data but also be less dependent on the intelligence portion of processing to normalise inputs and make the right decision.

At the same time, redundancy should be built into the sensory inputs by widening the number and types of inputs – such as lidar, ultrasound, magnetometers and so on. These sensory inputs should be used as checks and balances against each other, prior to basing a safety-critical decision on them. The approach of including more sensors and fusing the information together, known as 'sensor fusion', aims to create more coherent and reliable information than if the sources were used separately [48]. Lessons can be learned from map building in the military and space exploration contexts; these use a range of measurements: fingerprinting (small fluctuations in gravity), inertia sensing, barometric pressure, elevation, light levels, RF levels (civil as well as military instruments), magnetometers and GNSS.

This redundancy approach can be illustrated with the example of GNSS. Given the concern about GNSS jamming and spoofing, there are some commercial industry efforts to provide location–based services [49] and to explore other ways of using information within a range of open transmissions to resolve location (fingerprinting, etc.) [50]. There are a number of different information points in these transmissions which, while not intended for the purpose, could be used to help resolve location.

In either partial or no-knowledge cases, the algorithms have to make hypotheses of where we are to start with (e.g. a map) and where the unknown transmitters might be (using, e.g. direction-finding techniques and analysis of signal strengths). The techniques referred to here are additional algorithms required to resolve the location, when location of some or all of the broadcast transmitters are unknown. The theory is that by using a range of transmission, not just GNSS, vehicles will have a better chance of resolving their location even in a GNSS-denied environment. Furthermore, being able to utilise signals from a wide range of sources and radio frequencies makes it less feasible for a third party to spoof or jam all of those sources at the same time and makes it more resilient than GPS, which uses a known set of frequencies.

The ability to take multiple inputs, cross-check them and provide alerts if they appear inconsistent is likely to be better than reacting to a single sensor. This second check can be provided by human operators in some use cases, for example, to check that the weight of the lorry does not exceed a bridge's load before crossing it, despite the load changing daily. However, it is not always reasonable to expect a human to make such decisions on a consistent basis, in which case technology must provide the checks and balances. Humans and computers alike will rely on the inputs being accurate to base navigation decisions on, and casting doubt on the accuracy of the sensors will render the human cautious and potentially lead to grounding vehicles. A pilot who knows that the aircraft has been fuelled but the fuel gauge reads empty, will have to base decisions on a lack of certain information, in which the only safe course of action will be to land the aircraft.

The investigation of the AF447 air crash in 2009 concluded that inconsistent readings of the three airspeed sensors and a part disconnection of the autopilot confused the pilots, who reacted wrongly to a stall condition [51].

10.5 Theme 4: securing the communications

A key characteristic of future transports systems is interconnectedness. The concept of smart cities includes road vehicles and trains that are digitally integrated with the transport infrastructure. Focussing on the connected autonomous vehicles (CAVs) and its communication infrastructure illustrates the challenges and developments in communications security. Vehicles are broadcasting information about themselves continuously, and exchanging data with other vehicles (V2V) and the road infrastructure (R2V) for the purposes of safety, navigation, traffic management and user services. Safety control applications in the vehicle rely on efficient communications with the road and with other vehicles, for example, lane switching/keeping, emergency response and assisted braking – hence the availability and integrity requirements are high. Messages are exchanged via vehicle *ad hoc* networks (VANETs), which are network architectures designed for this purpose and used for data exchange in ITS, CAV and IoV/V2V, supporting the multiple sensory environment within a vehicle that collects and exchanges data within and outside it. As cars are moving all the time with respect to each other and the road infrastructure, VANETs have to cope with dynamically changing topology, mobility and infrequent connectivity that is inherent with vehicle to road communications.

A key requirement for these communications is for speed and efficiency, with a low latency for supplying safety–related messages to other vehicles and users. At the same time, the communicating units (onboard unit [OBU] and roadside unit [RSU]) both have limited computational capacity for unpacking and processing messages, and so the emphasis for communications research is on lightweight protocols. In the first instance, protocols have been adapted for VANETs (V2R) from open standards, such as dedicated short range communications (DSRC), adapted from the 802.11 range and simplified for lightweight efficiency (Box 10.5).

Box 10.5 Simplifications to 802.11 local area network standard

To enable the performance required by VANETs, the simplifications made to the 802.11p standard include the removal of message authentication and confidentiality features. This enables a vehicle to transmit data to the RSU or other vehicles as soon as they connect, without needing to wait for cryptographic operations to take place, increasing the length of time that a fast moving vehicle can transmit or receive useful information. However, this means that security and authentication requirements need to be fulfilled further up the protocol stack.

Vehicles broadcast information about themselves for the purposes of safety and navigation, and this broadcast of information needs to be resolved with the concurrent privacy requirements, which include connectivity and non-traceability. Messages need to be recognisable as coming from the same vehicle for a short period while they are communicating with a RSU but should not enable longer term tracking of that vehicle or accumulation of pattern-of-life data. Close association between the driver and a vehicle is (currently) such that tracing a vehicle is tantamount to tracing a driver. The requirement to build in privacy is based on inferences about driver privacy preferences, that is, that they would not want their locations and usage profile to be tracked. This assumption on preferences, however, has not been informed by a public debate on the balance between privacy and potential safety benefits. For example, Asuquo *et al.* [52] highlight the trade-off between privacy and safety by a potential use case of RSUs 'beaconing'[‡] to alert vehicles of potentially dangerous conditions [53, 54] in the context of the extreme weather conditions of Hurricane Harvey in 2017. Certain privacy-enhancing schemes constrain the beaconing, which means delaying or even preventing these alerts. Similarly, security mechanisms such as authentication and encryption may enhance security but add overhead and delay to incoming safety warnings.

An established approach to the privacy of vehicle identity is the use of *pseudonyms* as vehicle identifiers, which are not linked to the physical vehicle. Pseudonyms need to be switched frequently to avoid the association of the pseudonym with the physical vehicle. However, too frequent a change in pseudonyms could facilitate association with a particular vehicle. This is explored by Bao *et al.* (2017) [60], which proposes a 'joint pseudonymisation changing scheme' to enhance anonymity and non-traceability. The TESLA scheme relies on the RSUs to handle the distribution and switching of pseudonyms, and group vehicles together according to location and velocity to promote anonymisation within the group. TESLA assesses anonymity according to speed groups so that where slower vehicles stay together for longer, anonymity is preserved for longer.

As well as privacy, the primary security requirements for CAV communications are the need for the message to be accurate, and implicitly to trust the sender of that message, which ideally involves both signing the message and authenticating the sender. Public key infrastructure (PKI) is capable of providing the required trust. However, OBUs and RBUs have limited capacity, and the processing requirements and overhead of PKI protocols can lead to packet loss and delay.

Research into protocols such as TESLA offers potential improvements in the delivery of secure requirements, by reducing the time for verification and increasing the amount of anonymisation that can be achieved. However, keys used for authentication between vehicles and RSUs must be renewed at distance-defined intervals, and even with lighter weight protocols the RSU has a capacity limit – there is a

[‡]Bouk *et al.* state that 'Beacon or safety messages are broadcasted in vehicular ad hoc networks (VANETs) to disseminate network state or emergency incident information to other vehicles in the network' [53].

finite number of vehicles it can authenticate in a time period. The faster the traffic, the greater the demand for re-authentication on the RSU. Furthermore, the RSU becomes vulnerable to flooding attacks.

By their definition, VANETs are ad-hoc and the vehicles connected to any given RSU change on a regular basis. This combined with the lack of authentication in the communication protocol itself makes adopting Zero Trust policies a good idea. The UK's National Cyber Security Centre (NCSC) publishes a series of guidelines on Zero Trust architecture, including ensuring there is sufficient monitoring of the network and services, using assessment of user and device behaviour to enable confidence in systems, and authentication and authorising users at every stage, using a variety of signals [55].

Blockchain solutions could be used to reduce the resource requirements for RSUs and the trust infrastructure. PETRAS [56] and the University of Surrey ran a research project to explore the use of blockchain technology to simplify key management in a group broadcast. A group of RSUs share a distributed ledger of authenticated OBUs. This means that an OBU wanting to authenticate can send a shared secret to an RSU, which can be checked on a ledger instead of repeating the authentication. PETRAS found that using lightweight proof-of-work blockchain platforms, such as IOTA, performed better than solutions with centralised key management and reduced load on the RSU. PETRAS also found privacy gains from the use of blockchain – pseudonym management could be achieved more cheaply with better anonymity than traditional centralised PKI-based schemes.

The use of blockchain technology is an example of processing data at the roadside to avoid sending it to a central data pool (e.g. in the Cloud) for further processing and decision–making. This means that the only processed information is sent to the Cloud, reducing the need for a central pool of data in the Cloud, and such that the data sent to the Cloud are more lightweight and hence bulkier encryption methods can be applied.

Finally, potential applications of quantum communication methods in the transport sector are also being investigated for their advantages over conventional encryption such as, for example, in the detection of potential eavesdroppers. These methods are currently being extensively studied to secure communications in multiple industries and are already applied in both long range (optical fibre, satellite to ground) and short range (free space) applications.

Quantum key distribution (QKD) is an example of a quantum communication method developed for highly sensitive communications and with potential applications in transport. In a QKD method, a quantum channel is used for the distribution of secure keys while the encrypted message is transmitted over a conventional channel. This quantum communication channel uses quantum mechanical effects to instantaneously detect if a key has been intercepted. Other examples of future communication methods include quantum secure direct communication (QSDC), in which communication data are directly transmitted over quantum channels, and methods where quantum technologies are applied to detect physical layer attacks.

10.6 Theme 5: building trust between the human and the autonomous machine

We have discussed how cyber and data solutions can be leveraged to improve physical security challenges, and the cybersecurity considerations and solutions emerging into the communications, inputs and decision-making domains of modern transport systems. There is another key aspect to address, which is the relationship between the computer and the human.

The safety improvements offered by autonomous vehicles and aviation are attractive. Human errors are the major cause of road deaths. However, in some ways the technology is the easy part; we still need to overcome the concerns of the public in accepting autonomy in safety critical applications, such as driving and navigation of cars. Individuals have different perceptions and appetites for risk, and many will need to be convinced of the safety (and security) of autonomous vehicles in order to adopt them. Accidents can spook passengers for years, and an early disaster could prevent uptake of new technology.

A holistic study of the human, social and ethical barriers surrounding the uptake of self-driving cars was conducted as part of the Venturer research project [57]. Venturer conducted several trials using BAE System's Wildcat autonomous vehicle platform.[§] The project discovered factors in the human–machine interaction which indicated potential barriers to wide scale adoption. Drivers became impatient with overly risk averse autonomous cars; while false negatives are unacceptable, the false alarm rate also needs to be low. Conversely, in simulations, drivers were uncomfortable with behaviour that would be deemed risky if a human was in control, but can be safely carried out by an autonomous vehicle with arrays of sensors, V2V communications and fast response times, such as driving at speed very close to a second autonomous vehicle (to support reduction in travel times and congestion).

The drive to replace human intensive effort with automation and autonomy reflects two main benefits: (i) reducing the reliance on and cost of staff, particularly where there are staff shortages and (ii) safety benefits through heightened accuracy and concentration, a great example being the reduction of RTAs. Aviation automation in air cargo could be as a first step. Indeed, in recent years companies such as Amazon [58] have experimented with drone delivery services. In a fully implemented drone delivery service, a small drone could pick up a package from a customer at their home, takes it to a central sorting area in the inner city, where it is transferred to a larger intercontinental drone or automated cargo aircraft for transport to the destination sorting area, and transferred again to small drone to be delivered to someone at their home, all without human intervention. This could reduce congestion within cities and quicken deliveries to remote locations. Delivery drones have also been adapted for other sectors such as entertainment (replacing fireworks [59]), healthcare [60] and emergency services [61], changing the way that paramedics carry out their jobs and how people take medical tests or receive medication [62].

[§]Wildcat is an autonomous vehicle built for experimentation in future battlefield concepts.

Unfortunately, very few studies have demonstrated the public perception for the use of transport drones. However, recent studies in Germany [63], Australia [64] and Nigeria [65] have indicated that people are either ambivalent or distrustful.

A recent development in drones is the combination with autonomous vehicles. Highways England is developing a 'Connected and Autonomous Plant' (CAP) roadmap to revolutionise construction by using drones to monitor, guide and lead autonomous excavators through the use of sensors and image feedback. It is thought that this innovation will deliver a 20–25% improvement in productivity, worth £200 million by 2040 [66]. The key elements to the success of these plans are 5G communication, cloud technology, drone surveying, 3D machine control and AI engagement [67]. While this scheme is being used within the realms of construction, the system is applicable to smart cities, mobility and transportation. For example, smart cities can use drones, automated vehicles and Internet of Things (IoT) to manage transportation [68]. However, without public support of drones for delivery systems it will be equally difficult to introduce them as a core part of transport systems.

10.6.1 How do we gain trust in machine intelligence?

Building sufficient trust in machine intelligence to take the human 'out of the loop' depends on the degree of confidence in the underlying algorithms – or indeed whether we can tolerate a lower level of confidence in its accuracy. In many situations, particularly non-safety consumer applications, convenience is paramount and some unreliability tolerated. In other situations, particularly those that are safety or mission critical such as for navigation and speed control, a high level of reliability is required.

Correspondingly, we may also need a secondary decision-making mechanism that 'fails safe'; for example, deterministic fall-back systems to non-AI–based decision-making systems for safety critical aspects such as collision avoidance.

In 2016, Gartner coined the term 'Centaur Intelligence' [69], which subsequently evolved into an industry term 'Augmented Intelligence' [70]. These terms capture the concept of human and machine working closely together, with the human helping the machine with certain skills such as empathy, creativity or innovation, and the machine helping the human with analytics and connecting data. A driver helping the machine might be passing control to the autonomous car on the highway, with a request to notify in situational changes (such as reaching the exit), and the car passing back to the driver because it does not recognise a situation and cannot determine a response with sufficient confidence. This level of automation, described by SAE as 'level 3' [71], is risky, as the human factors/haptic response and latency of processing is likely to lead to safety issues – human error in emergency cases – and many manufacturers are therefore bypassing this level in favour of level 4. The Venturer project (described above) found that the driver easily becomes complacent or loses concentration, with the effect that there is a noticeable delay in taking control of the vehicle followed by an overcompensation in control – a potential safety issue in an emergency.

There are clearly different situations in which we would be comfortable outsourcing entire decisions to computers, particularly in the security and safety context. If there is room for error then there is a risk for a wrong decision, and therefore computers need to be a decision assistant, not the decision–maker. Until we completely trust algorithms to make decisions that have life-changing outcomes, the human will remain in the loop for decisions.

One might imagine similar relationships between a computer algorithm and the human for other life-impacting decisions based on risk assessment; the use of algorithms in legal sentencing or parole, or travel authorisation as in China's social credit system [72]. This issue has already been anticipated by the European Union's GDPR regulation, which states that 'The data subject shall have the right not to be subject to a decision based solely on automated processing, including profiling, which produces legal effects concerning him or her or similarly significantly affects him or her' [73].

A computer algorithm may only be slightly better at predicting more accurately than a human [74], but often the deliberate use of an algorithm as the authoritative decision–maker is predicated on the assertion that the machine will perform at least as well as the human; that is, it would make the same correct decision as a well-informed or experienced human but could improve the accuracy of its decisions by using more information, perhaps with better accuracy and reliability.

RUSI has investigated [75] whether Machine Learning could move beyond its limited predictive uses. The authors quickly come across ethical barriers, which require more regulatory and governance framework to support their adoption. There is clearly a distinction between the appropriate usages of ML and predictive AI, depending on whether it is used as a decision-maker or a decision-making aid. A focus on gaining certainty and assurance will enable us to outsource more complex problems to AI and adopt more intelligence into the decision-making processes.

10.6.2 Moving forward with assurance and accountability

Part of the need for explainable AI is due to accountability to be able to explain in a court of law how a car came to a conclusion that was retrospectively tested (presumably because it had the wrong outcome). In government-sponsored systems such as space exploration, the government effectively under-writes the risk of something going wrong. However, this accountability is much more distributed in private sector travels with a more complex supply chain, whose culpability in a failure or unpredictable response will be difficult to untangle.

Any sector with the need to explain why something happened or went wrong will have a preference for deterministic systems (e.g. to support a safety case). Certification of systems is a process to assure the safety, robustness and security of the systems. There remains a need for governance frameworks to ensure accountability for cybersecurity risks resulting in an accident involving autonomous systems – will this fall to the driver, to the car manufacturer who designed and integrated the car, or the component or control software vendor?

The defence sector uses system accreditation to prove that security vulnerabilities, threats and risks have been identified, and is increasingly using security cases – adapted from the safety cases, which are already well established – as a means of proving the safety to an acceptable level.

A similar need for assurance applies to the integrated systems within commercial transport. If products or code have been developed by multiple parties, each with their own supply chain, this can obscure the attribution of failures to one particular component or supplier. Investigating and attributing the cause of critical safety failures will be a necessary process if something goes wrong in a multi-component integrated transport system, where accountability for failures will need to be identified.

The commercial transport sector can look to the emerging thinking in the defence sector around assuring complex systems. A common approach to assuring complex systems (system of systems or socio-technical systems) during their design phase has been to assure individual components on a modular basis. The modular approach enables some flexibility to re-use the components in different arrangements. The industry initially looked to adapt the safety assurance models to also assess security vulnerabilities. Having reviewed existing assurance approaches for safety and security, Johnson [76] intended to come up with a model for weaving in the security assurance to the safety assurance to benefit from the rigour that is applied to safety. However, she realised that the basic approach to modular assurance had a flaw in that the instantiation of each module for operational use, and the way it was eventually used, introduced nuances that themselves introduce uncertainty and potentially vulnerabilities – within features such as connectivity and messaging. This is more impactful for security assurance cases than for safety, because while safety deals with a spectrum of accidental and failure situations, in cybersecurity we assume an intelligent adversary who will hunt out vulnerabilities to exploit. Johnson is therefore producing a model, which addresses the integration of components in complex systems and assesses the vulnerabilities that are introduced on their instantiation and the operational purpose of the end complex system. While these vulnerabilities may not be addressed immediately, they can be identified, assessed, registered and prioritised for a proportional approach to their remediation, that is, dealt with in a risk managed way.

Achieving an approach to assurance that aligns and integrates safety and cybersecurity is increasingly important as new platforms and technology are developed and introduced into the public travel ecosystem, with evolving cybersecurity risk profiles that reflect higher speeds, capacity and passenger volumes. The NCAP rating system is a framework for testing and assuring the physical safety of a vehicle, and the rating system has enabled consumers to easily understand the safety of vehicles since 1997. The 5StarS project, from Innovate UK and the Centre for Connected and Autonomous Vehicles, is creating a framework for the assurance of connected vehicles, with a similar rating system to allow consumers to understand the security of their vehicle. The framework identifies regulations and standards to be followed throughout the development of a vehicle to ensure that an independent assessment of the vehicle's cybersecurity can be made. In highlighting the cybersecurity of a vehicle in an easily understandable way, vehicles that follow good practice may become more attractive to consumers.

Security assurance of safety critical systems becomes more complex as we introduce autonomy and artificial intelligence. The behaviour of non-deterministic systems is less predictable. Rules which follow the 'If-then-X-else-Y' are easily certifiable because they are predictable. Autonomous vehicles are based on more probabilistic decision– making, such as Bayesian techniques, and as described earlier, the input data are uncertain. There are challenges in certifying components or systems because of the unpredictability of decisions in new situations, where the machine gets 'stuck'.

In safety critical systems, the decision-making process cannot be a black box; we may need to query decisions it makes in order to gain confidence in the decision-making process. Explainable AI will help build trust in the machine by understanding the decision-making rationale; what features were salient to the decision and what rules were applied.

The answer to probabilistic decision–making may appear to be keeping a human in the loop for critical decisions, as discussed earlier in the chapter. However, in our caution about computer intelligence and decision–making, we must not forget the inevitable risk of user error. Even mechanical failures can often be attributed to human failure at some point along the maintenance and repair cycle. To illustrate, Japan Airlines Flight 123 crashed in 1985 when its tail fin broke off mid-air, but the investigation attributed this structural weakness to a repair and maintenance check that had been improperly completed [77, 78].

Ultimately we need to be able to entirely outsource the decision–making to the machine. Joint or contradictory decision– making are likely to end in more disasters than if a machine was entirely responsible for the navigation and control. However, we need to be able to gain absolute assurance through explainable AI to build confidence in the decision–making and through change control to assure any 'learning through experience'.

References

[1] Energy.Gov. *Biden Administration Takes Bold Action to Protect Electricity Operations from Increasing Cyber Threats* [online]. 2021. Available from https://www.energy.gov/articles/biden-administration-takes-bold-action-protect-electricity-operations-increasing-cyber-0.

[2] crossrail.co.uk. Crossrail is currently set to open in the first half of 2022 [online]. 2022. Available from https://www.crossrail.co.uk/news/articles/crossrail-project-update while HS2 is being built currently according to the HS2 website https://www.hs2.org.uk/building-hs2/.

[3] High Speed Two (HS2) homepage. HS2 will use the best of British skills and innovation, creating thousands of UK jobs and opportunities for SME, and will be one of the most environmentally responsible infrastructure projects ever in the UK [online]. 2021. Available from https://www.hs2.org.uk/building-hs2/.

[4] crossrail.co.uk. Crossrail project update [online]. 2022. Available from https://www.crossrail.co.uk/news/articles/crossrail-project-update#.

[5] Intelligent transport. NEWS The future arrives as Virgin tests hyperloop with passengers [online]. 2020. Available from https://www.intelligenttransport.com/transport-news/111212/the-future-arrives-as-virgin-tests-hyperloop-with-passengers/.

[6] Mack E. *Crazy fast new maglev train is now ready to fly in china*. New York, USA: Forbes Media; 2021.

[7] Defense News. Electronic Warfare: What US Army Can Learn From Ukraine [online]. 2015. Available from https://www.defensenews.com/home/2015/08/02/electronic-warfare-what-us-army-can-learn-from-ukraine/.

[8] Defense News. Electronic Warfare: What US Army Can Learn From Ukraine [online]. 2015. Available from http://www.defensenews.com/story/defense/policy-budget/warfare/2015/08/02/us-army-ukraine-russia-electronic-warfare/30913397/.

[9] Helm UK. UK Heavy Goods Vehicle (HGV) Platooning project [online]. 2017. Available from https://www.helmuk.co.uk.

[10] Strohmeier M., Lenders V., Martinovic I. 'On the security of the automatic dependent surveillance-broadcast protocol'. *IEEE Communications Surveys & Tutorials*. 2014;17(2):1066–87.

[11] Leonardi M., Sirbu G., Network A.D.S.-B.C.-S. 'ADS-B Crowd-Sensor network and two-step Kalman filter for GNSS and ADS-B Cyber-Attack detection'. *Sensors*. 2021;21(15):4992.

[12] Kacem T., Wijesekera D., Costa P., Carvalho J., Monteiro M., Barreto A. 'Key distribution mechanism in secure ADS-B networks'. *2015 Integrated Communication, Navigation and Surveillance Conference*; Herndon, Virginia, 2015. pp. 3–1.

[13] Kacem T., Wijesekera D., Costa P. 'ADS-Bsec: a holistic framework to secure ADS-B'. *IEEE Transactions on Intelligent Vehicles*. 2018;3(4):511–21.

[14] Kacem T., Barreto A., Wijesekera D., Costa P. 'ADS-Bsec: a novel framework to secure ADS-B'. *ICT Express*. 2017;3(4):160–3.

[15] Kim Y., Jo J.-Y., Lee S. 'ADS-B vulnerabilities and a security solution with a timestamp'. *IEEE Aerospace and Electronic Systems Magazine*. 2017;32(11):52–61.

[16] DHS. Niac surviving a catastrophic power outage: How to strengthen the capabilities of the nation [online]. 2018. Available from https://www.dhs.gov/publication/niac-catastrophic-power-outage-study.

[17] Dawda S., MacColl J. *US water plant suffers cyber attack through the front door*. The Royal United Services Institute (RUSI); 2021.

[18] Osborne C. Colonial Pipeline attack: Everything you need to know ZDNet [online]. 2021. Available from https://www.zdnet.com/article/colonial-pipeline-ransomware-attack-everything-you-need-to-know/.

[19] Singapore Public. "open data" is being used to trial transport planning that has, according to Thales, reduced bus overcrowding rates by 92%. Secondly Singapore have just launched the public trial of "autonomous bus pilot" [online]. 2020. Available from https://www.thalesgroup.com/en/worldwide-digital-identity-and-security/iot/magazine/singapore-worlds-smartest-city.

[20] UAE. Government paperless strategy to reduce paper usage and eliminate 1 billion pieces of paper from government transactions [online]. 2021. Available from https://u.ae/en/about-the-uae/digital-uae/smart-dubai.

[21] Government Technology Agency of Singapore. 'Nation Annual Report 'Improving Lives'. *Supporting Businesses*:2016–7.

[22] IMD. Smart City Observatory [online]. 2017. Available from https://www.imd.org/smart-city-observatory/smart-city-index/.

[23] Thalesgroup. Singapore: The world's smartest city [online]. 2020. Available from https://www.thalesgroup.com/en/worldwide-digital-identity-and-security/iot/magazine/singapore-worlds-smartest-city.

[24] Yu E. *Half of vulnerabilities Singapore government finds via bounties, disclosures are valid.* ZD NET; 2021.

[25] UAE. Digital Dubai [online]. 2021. Available from https://u.ae/en/about-the-uae/digital-uae/smart-dubai.

[26] 'A deep convolutional network for traffic congestion classification by Dr Chris Willis, et al'. *BAE Systems, GBR: IST-158 Specialists' Meeting on Content Based Real-Time Analytics of Multi-Media Streams*; NATO: Information Systems Technology Panel Office, 2017.

[27] Viva citylabs. TfGM Chooses Vivacity to Deploy AI-Controlled Smart Junctions to Optimise Traffic Networks [online]. 2020. Available from https://vivacitylabs.com/greater-manchester-chooses-vivacity-labs-to-deploy-ai-controlled-road-junctions-to-optimise-traffic-networks/.

[28] Viva city. TfGM Chooses Vivacity to Deploy AI-Controlled Smart Junctions to Optimise Traffic Networks [online]. 2020. Available from https://vivacity-labs.com/greater-manchester-chooses-vivacity-labs-to-deploy-ai-controlled-road-junctions-to-optimise-traffic-networks/.

[29] Lloyd K., Rosin P.L., Marshall D., Moore S.C. 'Detecting violent and abnormal crowd activity using temporal analysis of grey level co-occurrence matrix (GLCM)-based texture measures'. *Machine Vision and Applications*. 2017;28(3-4):361–71.

[30] Hooker S. 'Moving beyond "algorithmic bias is a data problem"'. *Patterns*. 2021;2(4):100241.

[31] Blanzeisky W., Cunningham P. Algorithmic Factors Influencing Bias in Machine Learning [online]. 2021. Available from https://arxiv.org/abs/2104.14014.

[32] Nesta. State of Offices of Data Analytics in the UK [online]. 2018. Available from https://www.nesta.org.uk/report/state-offices-data-analytics-uk/office-data-analytics/#content.

[33] Cities Today. Nevada AI trial sees reduction in crashes and speeding [online]. 2021. Available from https://cities-today.com/nevada-ai-trial-sees-reduction-in-crashes-and-speeding/.

[34] Rail UK. Maestrano awarded contract with Network Rail High Speed [online]. 2021. Available from https://railuk.com/contract-news/maestrano-awarded-contract-with-network-rail-high-speed/.

[35] Patents. System and method for predicting impending cyber security events using multi channel behavioral analysis in a distributed computing environment [online]. 2015. Available from https://patents.google.com/patent/US9602530B2/en.

[36] Rashidi T.H., Abbasi A., Maghrebi M., Hasan S., Waller T.S. 'Exploring the capacity of social media data for modelling travel behaviour: opportunities and challenges'. *Transportation Research Part C: Emerging Technologies*. 2017;75(3):197–211.

[37] GOV.UK. Behavioural Analytics for Defence and Security [online]. 2018. Available from https://www.gov.uk/government/publications/competition-behavioural-analytics-for-defence-and-security/behavioural-analytics-for-defence-and-security.

[38] UCI. Husky or Wolf? Using a Black Box Learning Model to Avoid Adoption Errors [online]. 2014. Available from http://innovation.uci.edu/2017/08/husky-or-wolf-using-a-black-box-learning-model-to-avoid-adoption-errors/.

[39] Thomas E. Driverless vehicles are no longer the stuff of science fiction fantasy. However, that doesn't mean they will be catching green lights any time soon [online]. 2020 BAE Systems Applied Intelligence. Available from https://www.baesystems.com/en/cybersecurity/feature/security-challenges-for-connected-and-autonomous-vehicles.

[40] UCI. Husky or Wolf? Using a Black Box Learning Model to Avoid Adoption Errors [online]. 2017. Available from http://innovation.uci.edu/2017/08/husky-or-wolf-using-a-black-box-learning-model-to-avoid-adoption-errors/.

[41] Sun J., Cao Y., Chen Q.A., Mao Z.M. 'Towards robust lidar-based perception in autonomous driving: general black-box adversarial sensor attack and countermeasures'. *In 29th USENIX Security Symposium*; 2020:. pp. 877–94.

[42] Cao Y., Xiao C., Cyr B, *et al.* 'Adversarial sensor attack on lidar-based perception in autonomous driving'. *Proceedings of the 2019 ACM SIGSAC Conference on Computer and Communications Security*; 2019. pp. 2267–81.

[43] Li K., Yu X., Zhang H, *et al.* 'Security mechanisms to defend against new attacks on software-defined radio'. *In 2018 International Conference on Computing, Networking and Communications (ICNC)*; Maui, Hawaii, 2018. pp. 537–41.

[44] Cruz Moura D.F., Barbosa da Silva F.A., Ferreira Galdino J. 'Case studies of attacks over adaptive modulation based tactical software defined radios'. *Journal of Computer Networks and Communications*. 2012;**2012**:703642.

[45] The guardian. This article is more than 5 years old Tesla driver dies in first fatal crash while using autopilot mode [online]. 2016. Available from https://www.theguardian.com/technology/2016/jun/30/tesla-autopilot-death-self-driving-car-elon-musk.

[46] Evtimov I., Eykholt K., Fernandes E., *et al.* 'Robust physical-world attacks on machine learning models'. *ArXiv*. 2017;2(3):4.

[47] Reuters. Hackers try to contaminate Florida town's water supply through computer breach [online]. 2021. Available from https://www.reuters.com/article/us-usa-cyber-florida-idUSKBN2A82FV.

[48] Kocić J., Drndarević V. 'Sensors and sensor fusion in autonomous vehicles'. *2018 26th Telecommunications Forum*; Belgrade, Serbia, 2018. pp. 420–25.

[49] Sky hook wireless. Pinpoint Location Accuracy with Skyhook Solutions. 2020. Available from http://www.skyhookwireless.com/products/precision-location.

[50] Zhang P. Collaborative WiFi Fingerprinting Using Sensor-Based Navigation on Smartphones [online]. 2015. Available from www.mdpi.com/journal/sensors.

[51] Salmon P.M., Walker G.H., Stanton N.A. 'Pilot error versus sociotechnical systems failure: a distributed situation awareness analysis of air France 447'. *Theoretical Issues in Ergonomics Science*. 2016;17(1):64–79.

[52] Asuquo P., Cruickshank H., Morley J., *et al.* 'Security and privacy in location-based services for vehicular and mobile communications: an overview, challenges, and countermeasures'. *IEEE Internet of Things Journal.* 2018;5(6):4778–802.

[53] Bouk S.H., Kim G., Ahmed S.H., Kim D. 'Hybrid adaptive beaconing in vehicular AD hoc networks: a survey'. *International Journal of Distributed Sensor Networks*. 2015;11(5):390360.

[54] Bao S., Hathal W., Cruickshank H., Sun Z., Asuquo P., Lei A. 'A lightweight authentication and privacy-preserving scheme for VANETs using Tesla and Bloom filters'. *ICT Express*. 2018;4(4):221–7.

[55] NCSC. Zero trust architecture design principles [online]. 2021. Available from https://www.ncsc.gov.uk/collection/zero-trust-architecture.

[56] 'The PETRAS national centre of excellence exists to ensure that technological advances in the Internet of things (IoT) are developed and applied in consumer and business contexts'. *Safely and Securely*. 2022.

[57] Kent T., Pipe A., Richards A., Hutchinson J., Schuster W. 'A connected autonomous vehicle Testbed: capabilities, experimental processes and lessons learned'. *Automation*. 2020;1(1):17–32.

[58] Time. Amazon Drone Delivery Was Supposed to Start By Here's What Happened Instead [online]. 2018. Available from https://time.com/6093371/amazon-drone-delivery-service/.

[59] The guardian. Fireworks could fizzle out as drones rise in popularity for new year [online]. 2022. Available from https://www.theguardian.com/uk-news/2021/dec/30/fireworks-could-fizzle-out-drones-rise-popularity-new-year.

[60] Drones in healthcare. A Role for Drones in Healthcare [online]. 2011. Available from https://www.dronesinhealthcare.com/.

[61] Doctor preneurs. HomeBlog 9 Drones That Will Revolutionise Healthcare [online]. 2015. Available from https://doctorpreneurs.com/9-drones-that-will-revolutionise-healthcare/.

[62] Kunovjanek M., Wankmüller C. 'Containing the COVID-19 pandemic with drones - Feasibility of a drone enabled back-up transport system'. *Transport Policy*. 2021;106:141–52.

[63] Revistas.uva.es. Drones for parcel and passenger transport: A qualitative exploration of public acceptance [online]. 2020. Available from http://revistas.uva.es/index.php/sociotecno/article/view/4555.

[64] Tarr J., Tarr A., Paynter K. 'Transport, drones and regulatory challenges: risk accountability meets COVID-19 fast tracking of a critical industry'. *Australian Business Law Review*. 2020;48(2):202–11.

[65] Abiodun T.F. 'Public support and prospects of drones or unmanned aerial vehicles (UAVs) technologies for effective transport and logistics delivery in Nigeria'. *Researchgate*. 2021;7(1).

[66] National Highways. Case studies Examples of how we stimulate innovation across the highways sector [online]. 2020. Available from https://highwaysengland.co.uk/industry/innovation/case-studies/#cap.

[67] National Highways. Connected and autonomous plant to 2035 [online]. 2020. Available from https://nationalhighways.co.uk/industry/innovation/connected-and-autonomous-plant-to-2035/.

[68] Dung N.D., Rohacs J. Smart city total transport-managing system. *International Conference on Industrial Networks and Intelligent Systems*; Springer, Cham; 2018. pp. 74–85.

[69] Sicular S. Centaur Intelligence: Half-Human, Half Machine [online]. 2016. Available from https://blogs.gartner.com/svetlana-sicular/centaur-intelligence-half-human-half-machine/.

[70] Gartner. Augmented Intelligence [online]. 2022. Available from https://www.gartner.com/en/information-technology/glossary/augmented-intelligence.

[71] SAE. SAE International Releases Updated Visual Chart for Its "Levels of Driving Automation" Standard for Self-Driving Vehicles [online]. 2018. Available from https://www.sae.org/news/press-room/2018/12/sae-international-releases-updated-visual-chart-for-its-%E2%80%9Clevels-of-driving-automation%E2%80%9D-standard-for-self-driving-vehicles.

[72] The guardian. China bans 23m from buying travel tickets as part of 'social credit' system [online]. 2019. Available from https://www.theguardian.com/world/2019/mar/01/china-bans-23m-discredited-citizens-from-buying-travel-tickets-social-credit-system.

[73] Regulations *Regulation (EU) 2016/679 of the European Parliament and of the Council*. The Publications Office of the European Union; 2016.

[74] The Atlantic. A Popular Algorithm Is No Better at Predicting Crimes Than Random People [online]. 2013. Available from https://www.theatlantic.com/technology/archive/2018/01/equivant-compas-algorithm/550646/.

[75] Babuta A., Oswald M. *Machine learning algorithms and police decision-making: legal, ethical and regulatory challenges*. The Royal United Services Institute (RUSI); 2018.

[76] Johnson N., Kelly T. 'An assurance framework for independent co-assurance of safety and security'. *ArXiv*. 2019;01220.

[77] Aviation-safety.net. This map shows the airport of departure and the intended destination of the flight [online]. 2022. Available from https://aviation-safety.net/database/record.php?id=19850812-1.

[78] The Japan Times. U.S. leaked crucial Boeing repair flaw that led to 1985 JAL jet crash: ex-officials [online]. 1985. Available from https://www.japantimes.co.jp/news/2015/08/11/national/history/u-s-leaked-crucial-boeing-repair-flaw-led-1985-jal-jet-crash-ex-officials/#.XPg229FYZ8Q.

Appendix 1

Assuring the cybersecurity of rail systems

Tom Chothia and Richard J Thomas

Introduction

In this appendix to Chapter 1, we present a much more detailed exploration of the challenges of cyber security to operational transport systems, in this case rail systems.

When designing and operating safety-critical systems, it is essential to assure their resilience to cyberattacks. They should provide sufficient protection such that an attacker, who can intercept and manipulate communications between the train and trackside systems, is unable to affect the safe operation of the railway.

One question that arises, especially when presented with these highly interconnected systems is 'where do we start our analyses?' Systems are no longer monolithic where you could previously assess each component in isolation. Today, we need to assure these systems both individually and in the context of the wider system. As an example, we would want to ensure that data that pass from one system to another are valid, accurate and genuine. If the sending system has a vulnerability which could affect this data, it has the potential of misinforming neighbouring systems.

An additional challenge that transport systems and the wider industrial systems community face is that these systems are typically deployed with an operational lifespan measured in the order of decades, compared to commodity systems, which are typically designed with lifespans of at most 5–10 years. As a result, we have to not only assure the system as it stands, but also need to consider what attacks may be possible in the future, within the anticipated lifespan of the system. Attacks which are known to have flaws, but not yet exploitable may be within an attacker's reach given sufficient time.

In this appendix, we will consider some techniques and principles for secure systems design that have been applied to assure the cybersecurity of the European Rail Traffic Management System (ERTMS), a new signalling standard that is currently being deployed throughout Europe and the world. ERTMS is known as an 'in-cab' signalling standard, where the train receives its signalling information from trackside systems over Global System for Mobile Communications – Railway (GSM-R) is a rail-specific version of the GSM mobile communications standard, displayed within the driver's cab, in comparison to 'on sight' systems which use lineside signals. The standards that specify ERTMS functionality are spread over a significant number of documents where some key details may be subtly defined. We will, in particular, focus on the security of the train-to-trackside communications link, in which safety-critical commands are sent to the train and the techniques in which we can use to assure its security. Figure 1 provides an intuition of the structure and format of ERTMS messages, and the layers of protocols in use to deliver signalling information between the train and trackside systems.

At the bottom-most layer of Figure 1 is GSM-R, an international, rail-specific variant of the GSM mobile communications standard, which supports enhanced functionality including emergency group calls, pre-defined short messages (e.g., 'standing at signal' messages which are sent to the signaller) and shortcodes, location-based functions which route calls to the appropriate destination.

The EuroRadio layer above provides additional authentication and integrity messages for messages sent between the train and trackside through the use of a message authentication code (MAC), a cryptographic checksum which is added to the end of messages. This MAC is generated by the train and trackside systems using a cryptographic key shared between the train and signalling system. This has the added benefit of proving that the train and trackside systems are genuine.

Application data (specifically signalling and location information) are received from the EuroRadio layer and handed over to the onboard systems, which include a relative timestamp which is used to verify message ordering and timeliness of messages being received. Message ordering is important in any system, as conditions may change over time, and timeliness ensures the train and signalling system react in an appropriate timeframe and are 'reacting' to current messages.

Whilst the rail sector has a proven background in ensuring safety, cybersecurity assurance remains a very new area of research. In this section, we will review three techniques that have been applied to assure the security of ERTMS: *formal verification of protocols*, *cryptographic analysis* and *threat analysis*.

Formal verification of protocols

Protocols allow systems to communicate with each other in a standard, uniform manner. Within ERTMS, signalling information is constantly exchanged between the signalling centre and train using the EuroRadio protocol, running over GSM-R. In safety-critical systems, we need to be sure that the protocols are robust and secure

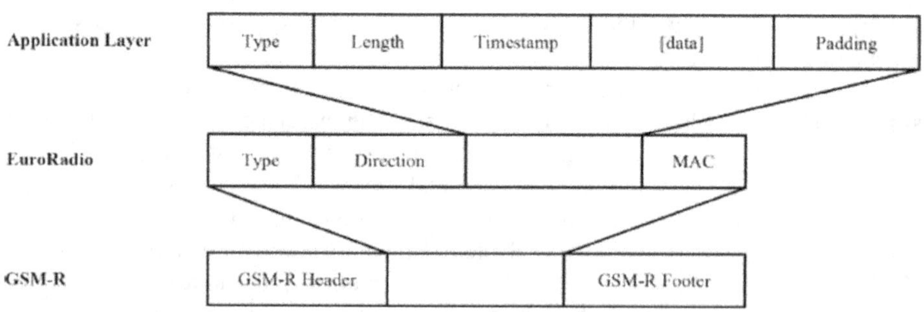

Figure 1 *Overview of the different communication layers in ERTMS. MAC, message authentication code.*

from adversaries. The first principle when designing secure systems is ***understand your attacker***. GSM-R, as previously stated, is based on the GSM wireless communications standard. It uses the A5/1 encryption scheme to provide guarantees of confidentiality and integrity in transit; however, almost 30 years later, it can now be broken [1–3], allowing an attacker to successfully eavesdrop the original messages. Whilst cryptography is one aspect that should be assured, as we will come onto in the next section, we must also be sure that any protocols we design are secure. The next principle in secure systems design is ***verify protocol design***. When we verify a protocol, we are assessing it against a number of security properties which we would expect any good, secure protocol to hold. One way that we can assure the security of protocols is through a method called *formal verification.*

Formal verification is a technique, well established for assuring safety within the rail sector, in particular for signalling systems, to ensure that, for example, two trains cannot occupy the same block of track. A mathematical model of the system is created, where the system is *queried* to assess whether a certain property holds. We can do the same for protocols, where we model the inputs, transformations and outputs for any system participating in the protocol. Formal analysis tools for protocols include ProVerif [4], Tamarin [5] and Scyther [6]. Whilst none of these tools provides assurance about the cryptography used in protocols, they can be used to assure the security of the protocol to ensure that an adversary cannot interfere with it. Moreover, tools like ProVerif are sound, that is that if the protocol has been shown to be secure, then it is secure. False attacks, though, can be returned, where we may need to manually verify the attacks to determine whether they are true attacks.

Technical case study: ProVerif analysis of EuroRadio

In Figure 2, we see the translation of the EuroRadio handshake protocol diagram from Figure 3 into ProVerif syntax from the perspective of the train, where the train process uses in and out statements to describe messages to and from the train. These express how the protocol works in the mathematical model. As previously identified, these tools do not consider cryptography, but we still model its application [e.g., in the encrypt (SECRET, trainKS) statement].

We can assess the security properties of a protocol by monitoring events, which ‘flag’ critical points in the protocol. These allow us to ask questions about the security of a protocol, for example, if the cryptographic keys are leaked to an adversary, or if the adversary is able to replay, delete and reorder messages to name a few properties we would want to assure.

In Reference [7], a formal security analysis of the EuroRadio and Application Layer protocols was carried out, finding the protocol was generally secure, with some issues identified. One flaw, for example, allows an adversary to send unauthenticated emergency stop messages to a train or block the receipt of messages between the train and signalling system. Whilst it would not affect the safe operation of the railway, these flaws have the potential to cause disruption to the network.

```
let Train =
      (* Start of the actual authentication protocol *)
      (* T-CONN.request -- Au1 SaPDU *)
      new trainNonce;
      event trainStartSession(rbc_etcs_id, train_etcs_id, trainNonce, SAF);
      out(c, (TRAIN_ETCS_ID_TYPE, AU1, DF_SEND, train_etcs_id, SAF, trainNonce));
      (* T-CONN.confirmation -- Au2 SaPDU *)
      in(c, (=RBC_ETCS_ID_TYPE, =AU2, =DF_RESP, in_rbc_etcs_id, rbcSaF, rbcNonce, inMAC));
      (* Generate the session key *)
      let trainKS = genSessionKey(trainNonce, rbcNonce, getKey(in_rbc_etcs_id, train_etcs_id)) in
      (* Output encrypted secret to check secrecy of keys *)
      out(c, encrypt(SECRET, trainKS));
      out(c, encrypt(SECRET, getKey(in_rbc_etcs_id, train_etcs_id)));
```

Figure 2 *The ProVerif model of a train carrying out the EuroRadio handshake (the calling party). We use the* ***NEW*** *directive to create fresh values, for example, random numbers used as part of the EuroRadio session key negotiation. We set events, for example,* trainStartSession *to identify the safety feature (a value which determines the cryptography to apply for the MAC) and use the = prefix where we want to directly match against a known constant, for example,* AU2, *a flag sent to determine which stage of the protocol has been reached.*

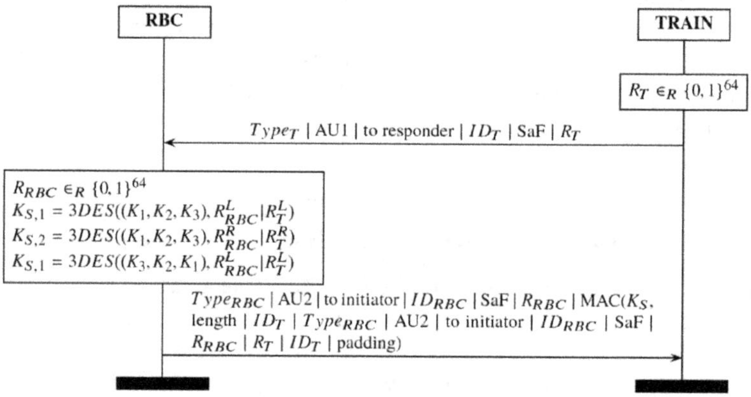

Figure 3 *First two messages sent in a EuroRadio handshake. The train generates a random 64-bit value and sends that to the Radio Block Centre (RBC, part of the signalling system), in addition to its ID and some additional fields. The RBC generates its own 64-bit value, combines the train's value with its own to create a* session key, *and sends the train its random value, some additional fields and an authentication code to provide guarantees of message authenticity and integrity to the train.*

The value that formal methods offer to asset owners, and also system designers, is the reassurance that, given a defined set of security properties, a set of fundamental principles a protocol should hold for it to be considered secure, we can verify whether the protocol complies against those properties. If the protocol does not hold against those properties, we can understand how an adversary might attack the protocol and make appropriate changes to defend against such attacks.

Formal analysis of protocols should be seen as an iterative process, where changes to the protocol should always be assured to ensure that no new attacks have been introduced. Whilst this technique is recommended for standards and implementation, it should be complimented with analysis of the specific cryptography used, in addition to assuring the implementation to ensure that there are no vulnerabilities introduced in the software development process, for example, buffer overflows. As cryptography is deemed 'out of scope' for a number of tools, cryptographic applications must be verified as the protocol might be secure, but is ultimately failed by the cryptography used.

Cryptographic analysis

One limitation of formal protocol verification is that a number of tools cannot model the cryptography in use, that is, they assume the cryptography to be 'perfect', where it has no flaws. Whilst a protocol might be shown to be secure, it might be the cryptography that ultimately fails the security of a system, either due to flaws in its design, or it is no longer suitable for modern use. Another principle of secure systems design therefore is to ***use trusted, established and proven cryptography***.

In ERTMS, all 'normal' messages sent between the train and trackside systems are accompanied with a MAC. These are cryptographic checksums which may prove the integrity of the message sent (to account for errors in the transmission process), but when used with a *key*, can be used to provide guarantees about the authenticity. In ERTMS, the MAC is used to provide a number of guarantees which the GSM-R communications channel cannot provide, in particular, *mutual authentication* of the train and trackside systems to each other through the handshake protocol in Figure 3 and proving that an adversary has not tampered with the message during transmission.

MACs can use different cryptographic ciphers depending on their application, for example, AES or SHA3*. Smart notes that there are three key properties that a MAC should hold [8]:

- Given a MAC, the adversary should only be able to recover the original input with *negligible probability*.
- Given a MAC, the adversary should only be able to recover an input message that results in MAC with *negligible probability*.
- It should be sufficiently *hard* to find two messages which result in the same MAC, a property known as *collision resistance*.

These properties underpin the cryptographic security for any message authentication scheme. In the first property, if the attacker has obtained the MAC successfully during transmission, it should be difficult for them to recover the original

*SHA1 is now deprecated in favour of SHA3 due to its collision resistance.

message that produced that MAC. If they are able to do this, they have recovered the cryptographic key and can forge their own messages. In the second property, the attacker can try and find *a* message that produces the same MAC, perhaps by chance. This brings us to the final property, specifically that if you have two messages which are *not identical*, they should not produce the same MAC.

Collision resistance is a desired property of MACs, but collisions do not necessarily make the cryptography weak. That said, if the cryptography has not been carefully considered and designed, collisions can give an adversary additional detail to break the scheme, as we will see in the case study below.

Technical case study: Cryptographic analysis of EuroRadio MAC

We say that a cryptographic scheme is secure if the adversary has negligible probability of success. For example, if you have a scheme with a 128-bit key, the probability of success for the scheme to be secure should be $<2^{-128}$. This is important when designing a system because it defines the effort required by the adversary to interfere with the scheme.

In ERTMS, the EuroRadio MAC uses a modified variant of the ISO-9797 MAC algorithm 3 [9], also known as ANSI Retail MAC, and is shown in Figure 4.

Given a message, it is split into 64-bit blocks (if the last block is not 64 bits in length, it is padded with zeros to make it 64 bits), and for all but the last block is encrypted with Data Encryption Standard (DES), using one of the session keys generated through the EuroRadio handshake protocol, and XORed (exclusive or) with the next block. When the last block is reached, instead of a single DES encryption with one key, a triple DES encryption is performed using all the keys negotiated during the handshake.

As a first impression, the EuroRadio MAC scheme appears to be difficult for an adversary to forge their own valid MAC because of the final triple DES transformation. They would need to correctly guess the three keys negotiated by the train and trackside systems for the MAC to be correctly generated. This however is not quite secure. One of the three properties that a MAC scheme should hold is collision resistance, where the scheme fails. As all of the blocks, with the exception of the last block are only encrypted with one key, this has the effect that when we reach the last block, two different messages could have the same intermediate block. What this means for the attacker is that they can now guess with relative ease $K1$, used for all but the last block encryption.

We need to also be aware of functionality that we are developing that could be used for malicious purposes. ERTMS messages include a plain text message field, allowing text messages to be displayed on the driver's screen. When the adversary has forged their message, for example, letting a train to go further than it was allowed, they can use this plain text message to effectively null out the last block of the captured message and reuse the MAC captured [10].

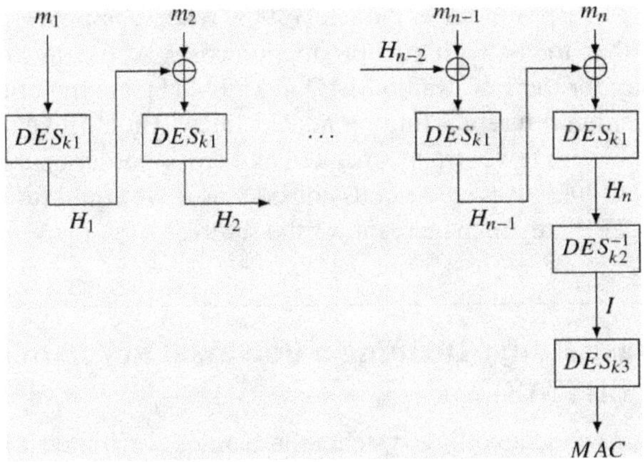

Figure 4 The EuroRadio MAC algorithm. This cipher is made up of n1 DES rounds, followed by a final 3DES round

When developing new systems, we need to appreciate the existing security of ciphers, and when making modifications, understand what the possible implications of the changes could be. Another critical consideration is that we should not limit ourselves to specific schemes, allowing the ability to change schemes such that, if a flaw was found, we can change schemes with relative ease. The final point from this work is that we should use trusted, established and proven schemes – if we modify a scheme in a manner that produced the EuroRadio MAC scheme, we should prove it has not affected the security or introduced any flaws.

Considering the future

Safety verification is typically static in that when a system is proven to be safe, unless changes are made to it, the proof *should* continue to hold. Cybersecurity, however, is a dynamically changing environment within any sector. When we design systems, especially with those with long operational lifespans, we should consider the cybersecurity of that system not only for today but also for the future. In particular, the cryptography used is something which we consider, where attacks not possible today due to limitations in computational power may soon be possible. An example of this is public key cryptography, commonly used in commodity systems to encrypt website traffic. The schemes used typically rely on the fact that it is difficult for existing computers to factorise very large numbers. In the so-called *post-quantum* world, this is seen as a trivial task for a quantum computer. We therefore need to ensure that an adversary cannot break our system today but also in 10, 20 or even 30 years time. One critical issue associated with post-quantum resilience is that an adversary can capture traffic today, and when the capabilities are available, they can 'break' the cryptography and recover the keys. The final principle in this section is ***prepare for the future***. System designers should anticipate changes and advances in technology, allowing for 'plug and play' situations that we can leverage and improve over time. Not doing so increases the risk of exploitation.

ERTMS was first proposed in the late 1990s, when computing power was significantly lower than today, with high production costs. At the time of the standards being ratified, say for the EuroRadio MAC in 1997 [11], computational limitations have limited potential futureproofing for attacks today. ERTMS key management is another example where fewer trains operated and cross-border operation was not as prevalent as it was 20+ years ago. Let us consider how we could futureproof, secure *and* simplify ERTMS key management for the future.

Technical case study: Defining a universal key management scheme for ERTMS

In ERTMS, trains and signalling systems share unique, symmetric cryptographic keys linking that train to a specific signalling system. Whilst a good proposition, it relies not only on engineers visiting all signalling systems and trains in the 'home' country of the train, but also of peer nations where the train crosses borders. It therefore is not scalable. In Figure 5, the lines connecting two trains to a six distinct signalling systems (RBCs) under four area of control (known as NID_Cs) show the unique keying relationship between the trains and signalling systems and give an idea of how many keys are required just for two trains to communicate with the signalling system.

An online variant of the key management system was introduced where trains and the signalling systems 'dial home' and download keys; however, this relies on RSA public-key encryption [12]. As previously noted, RSA-based schemes are not post-quantum secure. For ERTMS, this means that a developing system relies on components which do not consider the potential adversary capabilities of the future. If, during the design of such systems, we allow changes to be made over the course of a system's lifespan, we allow the system to adapt to the changing nature of security.

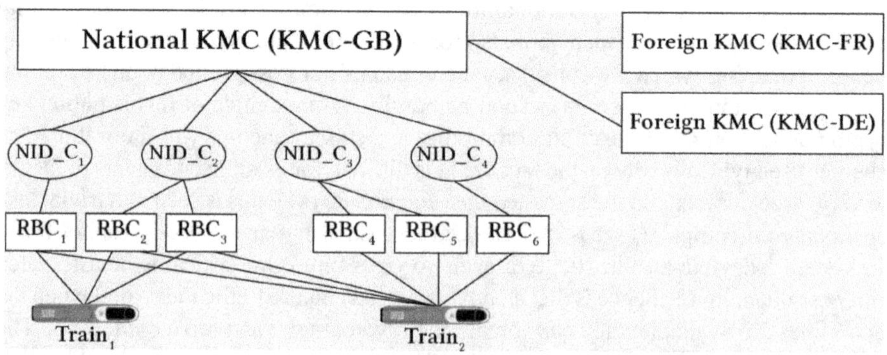

Figure 5 A high-level example of current ERTMS key management. Trains and RBCs share unique cryptographic keys between each other and also a pair of unique keys between themselves and the key management centre (KMC). KMC-XX are 2-letter country codes, GB, FR and DE.

Another factor to consider when designing such systems is to consider the effects arising from the interconnectivity of components. As seen in the industrial space in recent years, attacks are taking place by traversing networks from the traditional corporate IT network to the operational systems, including Stuxnet, BlackEnergy and Triton. By understanding how systems are connected, we can assess their resilience to attacks, how they interact with each other and identify potential pathways to critical systems.

References

[1] Ghafari A.V., Mohajeri J. An improved attack on A5/1. Information Security and Cryptology (ISCISC). 2011 8th International ISC Conference on; 2011. pp. 41–4.

[2] Jain A., Chaudhari N.S., CoRR. Two Trivial Attacks on A5/1:A GSM Stream Cipher. 2013. Available from http://arxiv.org/abs/1305.6817.

[3] Lu J., Li Z., Henricksen M., ACNS. 'Time-memory trade-off attack on the GSM A5/1 stream cipher using commodity GPGPU'. 13th International Conference on Applied Cryptography and Network Security; 2015.

[4] Blanchet B., Smyth B., Cheval V. *ProVerif 1.90: automatic cryptographic protocol verifier, user manual and tutorial*. 2015. Available from https://bensmyth.com/publications/2010-ProVerif-manual-version-1.90/

[5] Meier S., Schmidt B., Cremers C., Basin D. 'Computer aided verification' in Sharygina N., Veith H. (eds.). *The TAMARIN Prover for the Symbolic Analysis of Security Protocols*. Berlin, Heidelberg: Springer Berlin Heidelberg; 2013. pp. 696–701.

[6] Cremers C.J. 'The Scyther tool: verification, falsification, and analysis of security protocols'. International Conference on Computer Aided Verification; 2008. pp. 414–8.

[7] de Ruiter J., Thomas R.J., Chothia T. 'A formal security analysis of ERTMS train to trackside protocols' in Lecomte T., Pinger R., Romanovsky A. (eds.). *Reliability, Safety and Security of Railway Systems: Modelling, Analysis, Verification and Certification*; 2016.

[8] Smart N.P. *Cryptography made simple*. Cham: Springer International Publishing; 2016.

[9] BSIBS ISO/IEC 9797-1:2011 – Information technology. Security techniques. Message authentication codes (MACs). Mechanisms using a block cipher. 2011. Available from https://bsol.bsigroup.com/Bibliographic/BibliographicInfoData/000000000030105371.

[10] Chothia T., Ordean M., de Ruiter J., Thomas R.J. 'An attack against message authentication in the ERTMS train to trackside communication protocols'. Proceedings of the 2017 ACM on Asia Conference on Computer and Communications Security; 2017. pp. 743–56.

[11] Braband J. 'Safety and security requirements for an advanced train control system'. *Safe Comp 97*. Springer; 1997. pp. 111–22.

[12] Thomas R.J., Ordean M., Chothia T., de Ruiter J. 'TRAKS: A Universal Key Management Scheme for ERTMS'. Proceedings of the 33rd Annual Conference on Computer Security Applications. ACSAC '17. New York; NY, USA: ACM; 2017.

Index

www.ingramcontent.com/pod-product-compliance
Lightning Source LLC
LaVergne TN
LVHW011902200926
848484LV00002B/507
9781785616686